COLLECTED PAPERS

IN

PHYSICS AND ENGINEERING

Yours truly,
James Thomson

COLLECTED PAPERS

IN

PHYSICS AND ENGINEERING

BY

JAMES THOMSON, D.Sc., LL.D., F.R.S.

PROFESSOR OF ENGINEERING IN QUEEN'S COLLEGE, BELFAST
AND AFTERWARDS IN THE UNIVERSITY OF GLASGOW

SELECTED AND ARRANGED WITH UNPUBLISHED MATERIAL
AND BRIEF ANNOTATIONS BY

SIR JOSEPH LARMOR, D.Sc., LL.D., Sec. R.S., M.P.

LUCASIAN PROFESSOR OF MATHEMATICS IN THE UNIVERSITY OF CAMBRIDGE

AND

JAMES THOMSON, M.A.

Cambridge :
at the University Press
1912

CAMBRIDGE
UNIVERSITY PRESS

University Printing House, Cambridge CB2 8BS, United Kingdom

Cambridge University Press is part of the University of Cambridge.

It furthers the University's mission by disseminating knowledge in the pursuit of education, learning and research at the highest international levels of excellence.

www.cambridge.org
Information on this title: www.cambridge.org/9781316611876

© Cambridge University Press 1912

First published 1912
First paperback edition 2016

A catalogue record for this publication is available from the British Library

ISBN 978-1-316-61187-6 Paperback

PREFACE

IT is well known—chiefly through references to his work by other physicists such as his brother Lord Kelvin, Thomas Andrews, James Clerk Maxwell, Osborne Reynolds, J. Willard Gibbs—that the scientific activity of James Thomson has left permanent marks on the history of several branches of physical science. The collection into readily accessible form, in one volume, of his published contributions to knowledge, has thus been desirable, especially in view of the scattered local journals, often remote from modern facilities for reference, in which much of the work originally appeared. The project of a posthumous collection of his brother's work was constantly encouraged and looked forward to by Lord Kelvin during his later years, but his numerous pre-occupations prevented active assistance.

On the recent completion of the Collected Edition of Lord Kelvin's own scientific work, the idea has impressed itself on others that there should be available a full record of the work—very different in general type from his own—of his brother who was his lifelong companion and scientific associate. The materials for the present collection have been brought together mainly by the care of his son, James Thomson, of the Elswick Engineering Works, Newcastle-on-Tyne, assisted zealously by his wife who has not survived to see the completion of her work; while the Biographical Memoir, containing passages of high interest as regards the personal aspect of the progress of Physical and Engineering Science in this country in the middle of the last century, has been constructed largely out of narratives and recollections obtained by his daughter Mary Hancock Thomson.

Letters of scientific importance which passed between Lord Kelvin and his brother, on subjects such as the Theory of the Dissipation of Energy and the characteristics of natural flow in liquids, have been included, as well as scientific correspondence with Faraday, Clerk Maxwell, Andrews, H. C. Sorby and other eminent men.

The main characteristic of James Thomson's mind was the marked originality of his way of looking at scientific problems, arising in part perhaps from a combination, then more common than now, of the abstract physical with the practical engineering interest, but also fostered by the independent and unconventional character of his scientific education. This quality was united with singular persistence in brooding over a train of thought, and following it out into all relevant details and ramifications. So constantly was he pre-occupied with the various aspects of his scientific problems, as to sometimes produce an impression that his whole life was concentrated in one absorbing interest. But in the published expression of his results, at any rate, he was usually concise. The assemblage in 500 pages of so much pioneering insight in so many subjects recalls in some respects the work of even the most original of physicists. In many of these subjects discovery and invention have of course progressed far beyond what was feasible when these notes and papers were concerned often with breaking almost new ground; the recent great development of water-turbines for the electric utilization of hydraulic power in mountainous regions is a case in point. But in the main these papers are in no sense obsolete; their acute direct scrutiny of phenomena is of interest to the mathematical physicist by way of contrast and supplement to his own procedure; and where appositeness to practical application may have failed through lapse of time, their relation to the history of science takes its place.

Other examples of the topics treated in these papers may be mentioned. In the first rank there is his own cardinal discovery of the lowering of the freezing point by pressure, which ramifies on one side into the subject of the influence of stress on solution and crystallisation, and on the other into acute observations on the formation of ice in nature, the ground ice of the St Lawrence, and the glacial origin of the Parallel Roads of Glenroy. The investigations on the natural flow of water and its measurement are fundamental in the practical science of hydraulics. He also devoted lifelong attention to the problems of the general atmospheric circulation, summing up in the Bakerian Lecture compiled in the last year of his life. Again, his close association as a colleague with Thomas Andrews, at the time when the latter was occupied by his classical researches on the continuity of the gaseous and liquid states of matter, led them to the well-known developments concerning the unstable regions in the Andrews diagram, on which much new matter has now been added from unpublished manuscripts. The remarkable early papers on elasticity, on the influence of intrinsic strain on the strength of materials, on the dynamics of spiral springs, have remained prominent through Lord Kelvin's expositions. In his later years the foundations of abstract dynamics, in connexion with the fundamental characteristic of relativity, led to fresh and original papers, which have regained interest from the wider physical discussions now prevalent on that subject.

The Editors desire to record their thanks for much valuable assistance received, as regards the printing of the volume, from the staff of the Cambridge University Press.

J. L.

CAMBRIDGE,
July 1912.

CONTENTS

CONGELATION AND LIQUEFACTION

CONTINUITY OF STATES IN MATTER

DYNAMICS AND ELASTICITY

GEOLOGICAL

MISCELLANEOUS PAPERS

APPENDIX

BIOGRAPHICAL SKETCH

THE late Professor James Thomson, during his long and active life, wrote numerous scientific papers which were read before the Royal Societies of London and Edinburgh, the British Association, the Philosophical Societies of Belfast and of Glasgow, and other learned bodies; but, scattered in a fragmentary manner through the Proceedings of many different Societies, they have not hitherto been easy of access. It has been thought desirable that the fruit of his activity should be collected and reprinted consecutively in a single volume.

His life is a record of long days spent in the search for truth, and of the ample reward in new knowledge gained. The fundamental character and value, to science and to engineering, of the discoveries and ideas which he has given to the world, is the feature of his work.

James Thomson and his brother William (afterwards Lord Kelvin) are striking examples of hereditary talent; and it would have given their father much happiness could he have foreseen how fully his own best qualities, as well as those of his beloved wife, had been transmitted to their children. In fact one cannot completely understand either of the two sons without knowing something of their father.

Descended from a long line of farmers who had originally migrated from Scotland to the North of Ireland early in the 17th century (about the time of the Plantation of Ulster), he inherited those sterling Scottish qualities which have made Ulster men so successful in turning the resources of their province to good account. At the same time he acquired the softer qualities characteristic of his Irish neighbours.

A letter from one of the family states that "they nearly all bore the character of being religious, moral, patriotic, honest, large, athletic, handsome men." They had occupied for 200 years a farm called Annaghmore, near Ballynahinch in County Down, 14 miles from Belfast, and within sight of the noble range of the Mourne Mountains. Here, in 1786, James Thomson, senior, was born.

The country in the immediate neighbourhood, like that of County Down generally, is fertile and undulating; the hillocks are so peculiarly rounded that their appearance has been aptly likened to a basket of eggs. Between these hillocks one finds little valleys with glancing streams and green copses, and sometimes a clear lakelet lies open to the sky. Low-lying bogs are few; and drainage has converted into rich meadowland many swamps that were formerly tenanted only by the coot and snipe. A Spa attests the fact that health-giving springs abound; and when the South wind blows it brings with it the ozone from the sea.

James Thomson, senior, the first of the family to leave agriculture for scientific pursuits, was a man of remarkable ability. When quite a young boy, having received only the rudiments of education by the teaching of his father, he made out for himself the art of dialling without the aid of teachers or good books. He thus succeeded not merely in making a sundial, but also a dial to tell the time at night by one of the stars of the Great Bear. He used to tell his children how he had puzzled over an old book on navigation which contained a chapter on dialling, and felt disheartened because he could not understand it; and how he subsequently found that his difficulties had arisen from the fact that the book was all wrong. One night sitting up with his father and elder brother, Robert, to guard the orchard, while he watched the stars slowly revolving he thought out for himself the true principle of dialling, and made a dial for his father's farm which told the time correctly. Robert's son, Hugh, used to give the following account of his youthful discovery :—
" While the ploughing of a field was going on Robert and his little brother, James, then about eleven years of age, rested at times. The boy was then observed to be working with a slate and a bit of stone for a pencil. In the evening after they came home, again he began to work, having brought in a handful of shavings from the workshop to make a blaze till the candles should be lighted. After a little the boy exclaimed to his brother, ' Robert, I have made a discovery, I have found out how to make dials for any latitude.' ' Can you show me ? ' said Robert, and he showed it so clearly that his brother understood it quite well."

Two of these rough sundials and a star dial are now in the possession of his grandson, James Thomson. His early progress in Mathematics was remarkable. His father allowed him to go to Dr Edgar's school at Ballynahinch, where he very soon made himself useful as a teacher; and then for four years, from 1809 till 1814, he studied at the University of Glasgow, taking his degree of M.A. in 1812. He then attended the classes of Theology and became a licentiate in the church of Scotland; and he also commenced the study of Medicine; but on the foundation of the Royal Belfast Academical Institution, he was elected to the chair of Mathematics, and returned to Ireland. He taught in Belfast from 1814 till 1832, when he was appointed Professor of Mathematics in the University of Glasgow. He wrote a number of important and successful school books on Arithmetic, Geography, Trigonometry, and higher mathematical subjects. The first of these was brought out in 1819, and new editions have been in wide circulation up to the present day. His books reveal originality, and constructive as well as analytical power, qualities which we find transmitted to most of his children, but particularly to his two eldest sons, James and William. The gentleness, ready wit, charm of manner, and warm-heartedness characteristic of all three, stamped them as partly Irish in spite of the Scottish blood in their veins.

In 1816, Margaret Gardner, a young, bright and lively girl, came from Glasgow to Belfast, to visit her relations, Dr and Mrs Cairns, who were friends of James Thomson; and the young professor, then 30 years of age, was much attracted to her. She was a daughter of William Gardner, a merchant in Glasgow, who had fought on the British side in the American War of Independence. His watch, which is preserved in the keeping of one of his great-grandchildren (Agnes Gardner King), shows the dint of a bullet which the strong case of the watch saved from piercing his heart. An extract from one of Miss Gardner's letters to her sister, Agnes, afterwards Mrs Gall, gives interesting impressions of Belfast and the people who lived there in 1816.

"*July* 1816.

On Tuesday last we spent a most agreeable day at a Dr Drennan's*. This has been our longest walk. It is more than three miles in a most beautiful part of the country, but

* Cabin Hill, near Belmont.

indeed every way you can turn the country here is fine and always something new; the party was ourselves, Miss Reid, and Professor Thomson. I never saw a more agreeable family. The Dr is a gentleman retired upon a small fortune; he and his wife devote their time almost entirely to the education of their family. She has been a beautiful woman and is said to be very accomplished, and the children from the little we saw of them seem to repay the pains taken in their education. It is just such a place as one would wish for, in the cottage style, white-washed and thatched roof, but containing a good house within; not fine, but neat and comfortable, upon the side of a hill well sheltered with trees; commanding to the front an extensive view of the surrounding country, to the back the Loch with its cultivated shores, and the Cave Hill which is a rugged mountain that forms the termination to a range of hills on the other side of Belfast. Belfast was hid in the valley between, but its smoke showed where it stood. A very fine day and the season of the year gave to these beauties their full charm; agreeable company and good spirits made us enjoy them. But they are far above description; so I leave you to fancy the Loch smooth as a lake, vessels sailing in different directions, mountains sloping into green hills and fertile plains, etc.

Everything about the grounds showed the 'touch of taste.' We walked about them all the afternoon and came home by moonlight, all agreeing we had spent a most agreeable day. I had the honour of the Mathematician as my walking companion. His first appearance is about as awkward as can be; he looks as if he were thinking of a problem and so modest he can scarcely speak, but when tête-à-tête he improves amazingly in the way of speaking. On our forenoon walk we had a most edifying and feeling discussion on sea-sickness and the best mode of preventing it. But in the evening we were much more sublime. I suppose the moon rising in great beauty, and Jupiter shining with un-common lustre, called forth the Professor's energies, and I got a very instructive and amusing lecture upon astronomy."

In the following year Margaret Gardner married the "Mathe-matician," and she became the mother of a family of seven children*, of whom James, the subject of this sketch, was the third child and eldest son, and William, afterwards Lord Kelvin, the next in age.

* Elizabeth, Mrs David King, born 1818, died 1896.
    Anna, Mrs William Bottomley ,, 1820, ,, 1857.
    James   ,, 1822, ,, 1892.
    William, Lord Kelvin   ,, 1824, ,, 1907.
    John   ,, 1826, ,, 1847.
    Margaret   ,, 1827, ,, 1831.
    Robert   ,, 1829, ,, 1905.

She was gentle and refined, pious and truthful, and with a strong sense of humour. She directed her household with the courage, energy, perseverance and practical commonsense so characteristic of the Lowland Scots; and it was a dire calamity to her husband and to the young family, when death carried her off a year after her youngest son, Robert, was born. All this and much more about the influences which moulded the minds and characters of Professor James Thomson's young family has been gracefully told by Elizabeth King in her book on *Lord Kelvin's Early Home**.

James was born in Belfast on February 16th, 1822, and his brother William two years later. They both developed early a zeal for knowledge, and were encouraged in this by their father who educated them himself with tenderest care. In fact during these early years he was both father and mother to them, and they never forgot it.

In October 1832, their father removed with his family to Scotland on his appointment as Professor of Mathematics in the University of Glasgow. Their home was in the Professors' Court of the old College in the High Street, a fine historic group of buildings, but situated in what had even then become a very undesirable part of the city.

The College session lasted for six months, from the 1st of November till the 1st of May. The long summer holidays enabled the professors to live with their families during a considerable part of each year in more healthful and congenial surroundings. Professor Thomson used to rent a house in the Island of Arran, at some place on the Firth of Clyde, or on the coast of Ayrshire. There the boys had sea-bathing, learned to swim and to row, and sailed toy boats which they made and rigged themselves. Thus they lived a healthy, active life in summer, spending some hours each day at lessons with their father. James and William were, then and always, devotedly attached to each other. From their earliest childhood they were closely associated; all through their youthful days, in their studies and their amusements, they were constantly together. Certain characteristics of the two brothers were noticed even when they were boys, which remained through life. James was the more careful and exact, and William the more quick and ready. For

* Macmillan and Co. 1908.

instance, when making their toy boats, James was never satisfied
till his boat was finished with the utmost perfection in every
detail even to the neat painting of it; William was content as
soon as he had his boat finished enough to be able to sail, and
would not spend more time and trouble over it. In after years,
James, before publishing a scientific paper or bringing out one of
his inventions, had the whole thing long and carefully thought
over. His theories thus stood the test of investigation, and his
machine or engineering structure could be relied on to do exactly
what it was intended to do. On the other hand, his brother
William's less patient scientific work was often thrown off at white
heat, and thus subject to frequent correction and adaptation by its
author.

At the ages of twelve and ten respectively they entered
Glasgow University together, and they very soon distinguished
themselves in every subject they took up. The younger brother,
who was even then adored by all the family as a genius, generally
took the first place and James the second. No cloud of jealousy
ever marred the friendship of the pair. The elder sympathised
with his father in the affectionate pride he took in the ability
of the younger; and the younger was ready at all times to check
his own impetuosity by laying his ideas before the elder brother
for criticism or advice. They were unusually young, even for
those days, to be at a University; but there was a reason for it.
Owing to a misunderstanding as to the amount of the emoluments
of his Chair, their father found himself in a very bad position
financially during the first years of his residence in Glasgow.
He could not afford to send his sons to school. Instead of this
being a misfortune, the two sons looked upon it as a very happy
circumstance for them; and they always rated very highly their
obligations to the early grounding their father had given them.

At this time Professor James Thomson, senior, was engaged
in writing his Mathematical books; and he also found time to
conduct a class on Astronomy, to which ladies were admitted
—a novel departure at that time. This strenuous life prevented
him from taking much part in the social life of Glasgow; but
it gave his wife's relations (her cousins Mr and Mrs Walter Crum
of Thornliebank, in particular) an opportunity of showing much
kindness to the motherless family. The intimacy thus begun
with the cousins was to be cemented later by the marriage of

their daughter, Margaret, to William Thomson in 1852; while James owed much of the encouragement he received for his early inventions to her father's and brother's kindly interest.

In 1840 James Thomson took his degree of M.A. with honours in Mathematics and Natural Philosophy. In the holidays of 1839 and 1840, their father had taken the young family to the Continent, spending some time in London *en route*. They thus gained some acquaintance with the French and German languages, and with the general culture, manners and customs of foreign nations. The diaries kept by James show the keen interest they took in all they saw. Mechanical inventions and appliances especially interested them.

They also much enjoyed going to the theatres with their father's friends, and visiting museums, galleries and exhibitions. Macready in *Henry V* made a lasting impression on them; Robert Rintoul (the son of their father's friend, Mr Rintoul, Editor of the *Spectator*) took the young people to this performance. Another friend of their father, Mr Knowles, of the same family as Sir James Knowles, the late Editor of the *Nineteenth Century Magazine*, is also mentioned in the diary.

The greater part of the first year's visit was spent by the boys in Paris, while their father went with his two daughters to Switzerland. The next year they all went to Germany, sailing up the Rhine through Holland. While they were staying in Bonn in June 1840, Professor Nichol* and his family joined them. He took James and William for a three days' geological tour among the hills. At Königswinter they were joined by all the rest of the party. The Drachenfels, and especially the sail up the Rhine to Nonnenwerth by moonlight, made a strong impression on this company of happy young people. This is how one of them, John Nichol†, wrote about it in after years:

"'It was upon a trancéd summer night' that we sailed round the corner of the Rhine which reveals the Siebengebirge, and came gliding into the island of Nonnenwerth. Clear and calm and fair the memory of that night comes back to me from over all the years. One by one the peaks appeared, and stood grandly above the quiet stream, in the grey light which soon faded away

* John Pringle Nichol, Professor of Astronomy in the University of Glasgow, 1836—59.

† John Nichol, Professor of English Language and Literature at the University of Glasgow, 1862—89.

beyond their purpling crests. The moon stood out, a glorious crescent on the ridge of Rolandseck, and a bright star led the host of heaven over the brow of Drachenfels. The lights of the little convent were twinkling through the trees, and the boatmen were chanting their evensong as they came and brought us to the shore, where we stepped hand in hand together to live what seems to me like a dream of the gates of heaven. If my summer on the Rhine is an oasis in my life, Nonnenwerth is the oasis within the oasis, the greenest and most beautiful spot amid the whole of this enchanted ground *."

More than sixty years later, when Lord Kelvin came to visit his nephew in Newcastle-on-Tyne, and his father's diary of this little tour was shown to him, this was the incident that caught his attention; and he enlarged upon it in such a way that those who listened felt that he loved to dwell on it.

Shortly after this, James unfortunately injured his health when walking with his brothers William and John in the Black Forest. He hurt his knee in some way, but did not complain at the time. He had already decided to adopt Civil Engineering as his profession, and, after his return from Germany, he entered Mr (afterwards Sir John) McNeil's office in Dublin, but in three weeks ill health obliged him to give up work and he returned home quite lame. He had a good friend in John R. McClean, who afterwards attained a great position in the early days of railway construction. They had been fellow students in his father's class, and McClean had been one of the Professor's favourite pupils.

A characteristic letter of McClean's refers to one of James's earliest inventions, a boat which, by means of paddles actuating jointed legs reaching to the bottom, is able to propel itself up a river—walking against the stream.

"WALSALL, 15th Nov. 1841.

MY DEAR JAMES,

I was much gratified at receiving your note of the 6th Inst. and if it had been longer would have relished it still more. Whenever you have leisure I will feel obliged by your writing me a few lines. I regretted much not seeing your father when he was in England and also your postponing your visit, but I will hope to see you in Spring for some months when we may be mutually serviceable. I am glad to find that you will be able to attend the Civil Engineering Class this session—no doubt you

---

* See *Memoir of John Nichol*, p. 26 (MacLehose and Sons).

will derive much benefit from it. If Mr Gordon* has published
the heads he proposes lecturing on, I will be much obliged by
your sending me a copy by post.

Your scheme for propelling vessels against the stream is very
ingenious. Altho' you have not been aware of it and have equal
merit as an inventor, it was tried in France many years ago.
If you refer to a Book entitled *Machines approuvées par l'Académie
des Sciences* you will find four schemes proposed.

. . . .

I have given you these in detail that you may be able to
judge whether your scheme is similar. If so, as I said before,
I do not consider that it lessens your merit as a discoverer in
the slightest degree. There is one thing it may however prove,
the great difficulty of hitting upon any useful scheme that has
not been previously attempted, and the necessity of examining
well before one expends time on such occupations.

Indeed were I you, and I am sure you will excuse me as a
senior apprentice, I would, for the present, avoid all attempts at
inventing machinery of any description.

After a few years have gone round and you have become
well acquainted with the principles of mechanism and with the
discoveries and *failures* of others, you will be able to bring a
mind well stored to the task, and from knowing the wants of
practical Science be able also to fill them up.

It seems to me to be nearly as great a waste of time, making
attempts at *useful* discovery without this previous knowledge, as
for a person to labour at working out the highest problems in
Astronomy without having first gone through the Calculus.

I cannot avoid speaking feelingly, my dear James, upon this
subject, as I lost a great deal of time and spent money on making
a rotatory steam engine, which, when finished, I found had been
patented ten years before.

. . . .

Believe me my dear James Yours sincerely

J. R. McCLEAN."

In 1842, having considerably improved in health, Thomson
went for some time to Walsall (not as a pupil nor as an *employé*,
but simply as a friend) to learn what he could by practical experience under Mr McClean's guidance.

Early in 1843 he went to the Horsely Iron Works, Tipton,
Staffordshire, where he remained for a few months as a pupil in

* The first Chair of Engineering in the United Kingdom, probably the first in
the world, was that founded by Queen Victoria in 1840 in the University of Glasgow
Lewis D. B. Gordon held it from 1840 to 1855, and James Thomson was a student
in his class of 1841—2.

the drawing office. In August, to his great satisfaction, he was apprenticed to Mr (afterwards Sir William) Fairbairn, his father paying £100 as his apprenticeship fee, and he began working in Fairbairn's Engine Works at Millwall, London. Messrs Wm. Fairbairn & Co. were well known as the first millwrights and among the first engine makers in Britain.

McClean in writing to congratulate him says "I have always considered your health as the only obstacle to your being a first rate practical engineer. I feel confident that you have the talent for it."

Unfortunately his health was not equal to the strain of working in the damp unhealthy surroundings at Millwall. He suffered from continual colds as shown by the following extracts from his letters to his father.

"MILLWALL, *Oct.* 8, 1843.

The Isle of Dogs I find is a very unhealthy place; and the workmen are often obliged to leave it soon after they come, on that account. The ground is low and marshy, and a kind of rank smelling fog rises for several feet from it in the evenings. I have had a cold now for some weeks, and have been coughing a little in the mornings, but not nearly so much as when I went down to Southannan*; and I think that when I get to my new lodgings, by wearing plenty of clothes and staying in the house almost entirely after work hours I shall be able to get on quite well."

"MILLWALL, *Oct.* 10, 1843.

Since last Friday I have been taking holidays from the works on account of my cold, and taking quick walks (half walking and half running) with a great deal of muffling. In this way my cold is now beginning to get better, and I think that in a few days more I shall be able to go back to the works. In the house I am reading the Hot Blast Trial, which is I think the most useful book I could have on Engineering.

.   .   .   .

I am very glad to hear that the Algebra has been getting on so well. You must let the Trigonometry wait for a while, as it would keep you too busy when the Session begins. One of the workmen with whom I have had some conversation, as he worked beside me for some time, asked me to recommend him some simple book on mensuration; so I lent him your Arithmetic on account of the article in it on mensuration. He has been greatly taken with it, and has been sometimes sitting up late at night and

---

* Southannan was an interesting old house on the Clyde which Dr Thomson had rented for summer quarters that year.

sometimes rising at 4 or 5 in the morning to read about fractions, common multiple, proportion, square root, cube root, &c. and to do the examples and exercises."

"MILLWALL, *Nov.* 24, 1843.

I cannot say that it [my cold] is getting better yet, although I have now been away from the works a fortnight and four days, and have been taking all the care I could."

While he was working at Millwall he was interested in a method for preventing smoke in furnaces, which he had thought of. The gases were to be taken downwards through the furnace instead of upwards; and the fire-bars were to be tubes with water circulating through them.

In October 1844 James Thomson was transferred to Messrs Fairbairn's Manchester Works, a welcome change, seeing that it afforded more opportunities for personal intercourse with Mr William Fairbairn who lived there. He moved into lodgings from which he wrote to his father "One might as well live in a chimney as live near the Works." He remained in the fitting shop until ill-health interrupted his progress in his profession. There was some derangement of the heart's action; his pulse beat too quickly—as much as 120 a minute—and he felt weak and ill. He wrote to his father, "I confess I feel very uneasy about it. I wonder if there is any likelihood of my getting stronger as long as it [the pulse] continues to go so fast as it has been lately or if it is likely to begin to go slower."

At the end of the year he returned home to Glasgow; his medical adviser there, a stern old Calvinist, told him he had heart-disease and might die at any moment, and advised him to put away from his mind all thoughts of this life and prepare himself for the other world. Happily this diagnosis turned out to be wrong. The distressing symptoms were the result of temporary disturbance. After some years he completely recovered his health, and was able during the remainder of his long life to enjoy active exertion both of mind and body. But at the time the verdict seemed a death sentence, and the effect on a man of twenty-three, at the very commencement of his career, might easily have been to render him a complete invalid. Happily his natural energy saved him from this, and when bodily exertion was prohibited, he turned his mind with undiminished activity to such pursuits as remained possible. His correspondence and note books show the

varied interests with which he occupied his mind. The following
letter to Mr Robert Murray, one of Messrs Fairbairn's staff at
Millwall, on the subject of obtaining fresh water from salt by
double distillation, is interesting from its very modern attitude to
the problems of economy of power, and from the partial anticipation
of subsequent ideas on available energy, regenerative action, etc.,
which is tacitly involved in it.

"October, 1845.

I received your two letters regarding the apparatus for
distilling sea water, and I am much obliged to you for the full
account you have given me of it. Two methods had occurred to
me, different from one another, but yet involving the same principle
of making the heat which is given out in the condensation of the
steam serve to generate new steam. The one is exactly the same
as you describe, and in it you will see that the fresh water is
separated from the salt by the expenditure of a certain quantity
of heat. In the other method, which so far as I could judge
appeared to be the preferable one, the effect is produced by the
expenditure of labour. The steam is generated in one vessel
(in the first instance by the heat of a fire) and it is pumped from
that into another which is placed within the first. The pressure
applied to the steam by the pump raises its sensible temperature
slightly, and therefore allows the heat to pass back through the
metal to the first vessel, there to generate exactly the same
quantity of steam as is condensed in the second. Hence if the
salt water be supplied at the boiling temperature, no supply of
heat will be required to produce the chemical effect of separating
the fresh water, but only a very small quantity to make up for
the cooling of the apparatus by the external air. But it is easy
to arrange the apparatus so that the salt water entering the steam
generator will be raised *almost* to the boiling point by the heat
contained in the condensed water and in the brine.

In relation to the quantity of fresh water which could be
produced by the labour of one man applied to a machine of
moderate size, I have made such calculations as convince me that
the plan is not at all impracticable. The only *data* however to
found the calculations on which I have as yet been able to obtain,
are too vague to enable me to come to any decided opinion as to
whether this or the first plan would be preferable.

The apparatus for the second would no doubt be more expensive than that for the first plan, but the supply of coals
required for the second would be but a very small fraction of that
required for the first. The labour of the sailors I suppose is not
of very much consequence, as they have but little to do except in
stormy weather, and when going into and out of ports.

We have now come up to Glasgow from our summer quarters.

I don't know whether you are acquainted with Glasgow, but if you are, you will know that the College is in a very bad part of the town. On this account I am very glad to say that the College grounds are wanted for a general terminus for a number of railways, for which purpose they are admirably suited. There is a great probability that the matter will be arranged to the satisfaction of the different parties concerned, and that the College will be moved to an excellent situation in the West end of the town*."

The plans expounded in this letter for the recovery of waste heat, by raising it to a higher temperature by the expenditure of mechanical work, show how his thoughts were ripening in the direction of his forthcoming thermodynamic discoveries.

The spirit in which he studied Nature is indicated by extracts written on the title-page of his note-book on Natural History, dated March, 1846.

My heart is awed within me, when I think
Of the great miracle which still goes on,
In silence, round me—the perpetual work
Of Thy Creation, finished, yet renewed
For ever.   [*Forest Hymn*, BRYANT.]

Thou art, O God! the life and light
Of all the wondrous world we see;
Its glow by day, its smile by night,
Are but reflections caught from Thee.   [MOORE.]

In the note-book itself we find squares very neatly marked out with an alphabetical index for different subjects: for example, "Temperature and Pressure of Atmosphere" under A; "Design in Creation" under D; "Expansion of Ice" under I; "Glacial Period and Glacial Markings in Glen Spean," G; "Lakes, how formed," L; and "Parallel Roads" under P. The illustrations, pen and ink sketches, of Rock Strata and River Basins are excellent in execution; and so are the sketches of molluscs in the notes on the Classification of Fishes by Agassiz.

Broadly speaking, his most important discoveries may be said to follow in sequence along three main channels or streams of thought; and we can trace each one back to its rise in these years when the note-book was in progress and the waterwheel being worked out.

* "This proposal fell through for a time, but some 20 years later the old College was bought by one of the Railway Companies and the present University building erected on Gilmorehill."

T.   c

His study of the laws of fluid motion led to his discovery of the Whirlpool of Free Mobility, to his Patent Vortex Turbine, and to his Jet Pump. Later on, his professional work in designing turbines, and his duties as engineer to the Belfast Water Commissioners, led him to study the measurement of water in rivers and streams, which resulted in the introduction of the V-notch gauge. Similar considerations led on to a series of important investigations into various aspects of the motion of water in rivers.

At the end of his life he returned to a subject which had first attracted him when he wrote a paragraph on Trade Winds for a new edition of his father's Geography, brought out in 1846, namely that of the general Atmospheric Circulation. Finding no satisfactory explanation of this problem he formulated his own theory and developed it at various periods of his life (see papers Nos. 26, 27, pp. 144, 148). He summed up all his ideas and gave a very complete and detailed explanation of his theory in his Bakerian lecture finished just two months before his death and read at the Royal Society by Lord Kelvin in March, 1892 (see No. 28, p. 153).

The second train of thought, that of Thermodynamics, led him on from the "Lowering of the Freezing Point by Pressure," a discovery first published in 1849, to the principles underlying the Plasticity of Ice; and finally to the papers on the transitions between the Gaseous, Liquid and Solid States of Matter.

The third group, starting with papers on Elasticity, passes on to the Strength of Materials and the Safety of Structures.

Another important series of papers relates to the question of absolute and relative motion, which has recently assumed fresh prominence in the philosophy of physical science. It is interesting to find that in the paper on the Law of Inertia in 1884 (p. 379), he has developed an idea sketched out in a letter to his brother as early as May 7th, 1848.

"Did you ever observe that we might a priori have just the same difficulty in fixing on equal additions of time at different periods as we have in fixing on equal changes of temperature at different temperatures? There is nothing in the human mind to determine what time three months hence shall be equal to a certain time to-day. What one man thinks a long time seems short to another. We might adopt the time from the Sun's coming to the meridian till his reaching it the next day as a unit of time. Then it would be the case that a body not subjected to force would not move over equal spaces in equal times, &c., &c. There would thus be many very complicated and puzzling laws of motion, which would at once be all simplified when any one said, Let us change the scale of time, and take as equal times those occupied in describing equal spaces, or those in which equal velocities are generated by a constant force."

The various methods of utilizing the natural sources of power in our rivers and waterfalls had always a strong fascination for him. His early diary shows that he took particular notice of the waterwheels in Holland. He would naturally be introduced to the French turbine by Professor Lewis Gordon in his classes at Glasgow. With thoughts thus running on these lines, he gradually developed his Vortex turbine, which afterwards became so successful. Extracts from two letters to John McClean show the progress of this invention.

"*Christmas Day*, 1846.

Since our return from the country in the end of October, I have been making a model of a horizontal waterwheel, on what I consider to be an improved plan. The idea of this occurred to me long ago, in fact, before my visit to you at Walsall; but, following your advice, I laid it on the shelf along with some others, trusting that any good ones among them would keep, and that the useless ones would moulder away without occupying time which ought to be devoted to more regular business. Now, however, after having been so long nearly inactive, I enjoy this light occupation very much; and even if the thing should in the end not turn out to be of any pecuniary advantage to me I shall not consider the time to have been thrown away."

"*April* 12, 1847.

About the beginning of the year I mentioned to you that I had made a small model of a new waterwheel. I have since been staying for some time in the country with a friend of ours, Mr Walter Crum of Thornliebank and at his works, which are very extensive and varied in their character, a large model of one of my wheels was constructed and its performance carefully tested by the friction dynamometer of Prony (*frein dynamometrique*, as you may have seen it called by French authors). The model worked to a tenth of a horse-power, and produced between 69 and 70 per cent. of the total work due to the water expended. As the best overshot waterwheels are only estimated to produce 75 per cent., and as there are some defects in the workmanship of the model, the result was very satisfactory. I should have mentioned that mine resembles the French turbine, and Whitelaw and Stirrat's wheel, in being of very diminutive size and of small cost in proportion to its power; and I consider that besides producing more work than they with a given quantity of water, it will be free from some important practical objections which apply to them. I have matured two other forms of wheels depending on the same general principles as those I have already constructed but which I believe will be considerably superior in efficiency and I am just commencing to get large models of them made in Glasgow."

When his invention was completed the question of a patent was carefully gone into by his father and friends, an important factor in the consideration being the fear that his delicate health would prevent his working the invention effectively. Finally the patent was taken in July, 1850, and various engineers were licensed to manufacture the turbine for him. The principal manufacturers were Messrs Williamson Bros., of Kendal.

The following letters show how ready William Thomson was to help his brother with his patent work; while the reference to Joule's early papers gives them a very special interest in the early history of Thermodynamics*.

"St Peter's College.
July 12 (12ʰ 30 a.m.), 1847.

My dear James,

        I am going from Cambridge to-morrow afternoon (or more properly to-day) to London, where I shall remain as few days as possible, before starting for Paris. I enclose Joule's papers which will astonish you. I have only had time to glance through them as yet. I think at present that some great flaws must be found. Look especially to the rarefaction and condensation of air, where something is decidedly neglected, in estimating the total change effected, in some of the cases. Keep all the papers carefully together and give them to me when I return. I am glad your patent is getting on. Tell me immediately what you want me to do in Paris, writing to the *Poste Restante*. I shall arrive there about Saturday.......

Your affcte brother,
WILLIAM THOMSON."

"Paris, *July* 22, 1847.

My dear James,

        ......I was introduced to Poncelet (having accidentally met him at a party to which Le Verrier took me) and I had a great deal of conversation with him. I was quite taken with him, as he seems to be a most excellent old man. I mentioned your

---

* Cf. "Obituary Notice of Lord Kelvin," by J. Larmor, in the *Proceedings of the Royal Society*, Series A, Vol. LXXXI. June 30, 1908, pp. xxv. xxix.—xxxvii.

wheel and asked him about patents etc. He told me, before
I described yours at all, that he has invented a '*Turbine à
injection extérieure*' and that although he has not published any
account of it, yet there are several of them working, in the South
of France. He made an appointment for me to call on him at his
own house, so that he might explain it to me. When I went
there he showed me some sketches and explained it a little, but
still I am not quite sure that I understand it. There are passages
in the wheel, curved I suppose on the same
principle as the common turbine. The sides
of the passages are '*en fonte*,' and are double,
leaving a vacant hollow space (the black-
ened parts) which he considers useful in
modifying the form of the aperture.

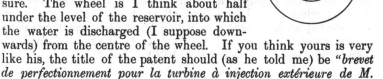

The water is introduced by a number
of, apparently tangential, passages, and the
centrifugal force is overcome by fluid pressure. The wheel is I think about half
under the level of the reservoir, into which
the water is discharged (I suppose down-
wards) from the centre of the wheel. If you think yours is very
like his, the title of the patent should (as he told me) be "*brevet
de perfectionnement pour la turbine à injection extérieure de M.
Poncelet.*"

I do not know however whether any of your modifications
(unless the 1st. kind, with curved passages) could be viewed in
the light of an improvement of Poncelet's. With reference to the
advantages of Poncelet's over ordinary turbines, one which he laid
some stress on is that, when the supply of water is very slight the
wheel works to best advantage, the reverse being the case for
turbines. He gave me the following particulars (dictated partly,
and partly he wrote in my pocket book) relative to one which has
been constructed "Turbine à injection extérieure établie à Toulouse
par M. Abadie, ingr.

  Shute 1·7 met. depense maxm. 650 kil. d'eau
  650 k. × 1·70 m. = 1105 kil. met.

14·73 chevaux dynam. rend au maxm. pour le faible depense
de 140 k. (per minute, I suppose) 0·68 (of the power) pour les
grandes de 170 kil. (there is some mistake of course) rend seule-
ment 0·50."

With the expenditure 160, intermediate between the two
former "0·60 du travail absolu." The velocities of the wheel in
the three cases were 41, 43, 41 turns per minute respectively.

  Exterior diamr. 1 m. 20.   Inter. 0·80.

Poncelet considers his turbine better than Fourneyron's (which
is now falling out of use, or at least of being made) I believe in

every case in which a horizontal wheel ought to be employed, sometimes it would be advantageous, sometimes an overshot wheel, and sometimes a breast wheel (he has been making improvements in the last mentioned "Les Brist-weeles ne vont pas assez vite ") according to the fall or the kind of work wanted. For cotton mills he agreed with me that turbines would probably be preferable. About the end of this year he will have a memoir on waterwheels published.

Since Monday I have been as busy as possible (before that I could find nobody and get nothing done), commencing by calling on Le Verrier to whom I had been introduced at Cambridge by Mrs Hopkins. He engaged me to go to Mr Milne Edwards in the evening, with whom he and Struve were going to dine.

I breakfasted with him next morning and he took me and Struve and a little son of his own to Versailles, where we were for the day; but before I breakfasted with him I saw part of the *Cabinet de Physique* of the Polytechnic, to which he got me admitted, made engagements with Regnault the *preparateur* at the Polytechnic &c. I have been all day to-day and a good part of yesterday at Marloye's shop, where I ordered a great quantity of acoustical apparatus. Mme. Dubuat is away for 2 months. I start for Geneva to-morrow at 10.30.

W. T."

"EDENBARNET BY DUNTOCHER.
July 24, 1847.

MY DEAR WILLIAM,

......Three papers have come for the Journal* from Newman†. None of them is very long but they would cost several shillings if they were sent to you. The first is regarding Fourier's theorem: the second regarding $\int_0^\infty e^{-x}x^{a-1}dx$. The last is a development of $(\cos x)^a$.

I suppose you got two letters at the *Poste Restante* in Paris from me; and I hope you have been able to do something with the one I enclosed to you from Mr MacGregor. I wrote yesterday to Mr Fairbairn on the subject of my wheels, just asking generally whether he could give me any useful information as to the way in which I ought to proceed.

.   .   .   .

I have read a good deal of Joule's papers and I certainly think he has fallen into blunders. There is one blunder certainly. He encloses some compressed air in one vessel, connects that with another which is vacuous, and allows the air of the former to rush

---

* [*The Cambridge Mathematical Journal*, then edited by W. Thomson.]

† [The late Professor F. W. Newman, brother of the Cardinal.]

into the latter till the pressure is the same in both. Both vessels were immersed in water, and after the operation the temperature of the water remains the same as before. Joule says that no mechanical effect has been developed outside of the vessels during the operation, and that therefore the heat remains unchanged. But in reality mechanical effect *was* developed outside *, as the two vessels became of different temperatures. Some of his views have a slight tendency to unsettle one's mind as to the accuracy of Clapeyron's principles. If some of the heat can absolutely be turned into mechanical effect, Clapeyron may be wrong. I think, however, that before coming to a conclusion, we would need to define more accurately what we mean by *a certain quantity of heat* as applied to two bodies at different temperatures. Perhaps Joule would say that if a hot pound of water lose a degree of heat to a cold one, the cold one may receive a greater absolute amount of heat than that lost by the hot one; the increase being due to the mechanical effect which might have been produced during the fall of the heat from the high temperature to the low one.

.   .   .   .

I am your very affectionate brother,

JAMES THOMSON."

"EDENBARNET, BY DUNTOCHER, DUMBARTONSHIRE.
July 29, 1847.

MY DEAR WILLIAM,

I got your letter to-day and I am glad to find that you have seen the gentlemen to whom I sent you introductions. The only thing in regard to Poncelet's wheel which gave me any uneasiness was the expression you used that 'The water is introduced by a number of apparently tangential passages and the centrifugal force is overcome by fluid pressure.' Now are you *sure* the cent. force is overcome by fluid pressure transmitted directly from the stationary part of the apparatus? In other words, does the water in passing through the orifices of the tangential passages attain the velocity due to the *whole fall*; or only to *half the fall* as in mine? From the form of the vanes you drew, as well as from its being thought desirable to make the vanes double with an empty space between the two parts, I think that the water must attain the velocity due to the *whole* fall, or nearly so, as is the case in Poncelet's wheel of which I have seen published accounts. I have often thought myself, that in that wheel, and in the turbine of Fourneyron, it would be advantageous

* [Joule meant of course mechanical work. This passage illustrates the early difficulty in distinguishing the varying mechanical availability of heat from the steady conservation of its amount.]

to have the vanes made double, but in mine there would not be the least use in such a construction.

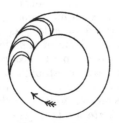

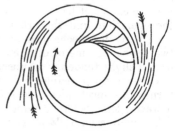

Poncelet's as drawn by you.   My Wheel.

In mine you will see the outer ends of the vanes are in the direction of radii or nearly so; but in Poncelet's they are nearly tangential. This tangential direction makes me think he intends that the water should spout in with a much greater velocity than that of the wheel, in which case it will be impelled towards the centre by its own impetus as it glides along the curved buckets. The action of mine would not be materially altered by even a considerable deviation of the outer ends of the vanes from the direction of the radii: but when the direction comes to be nearly tangential the action is entirely altered, and I think it assumes the character of that of Poncelet's. When the ends are nearly tangential the space between the vanes at the outside is greatly contracted. Then this renders it desirable to make the vanes double, so that the width of each passage for the water may not be greatly increased towards the middle. In mine the water has no rapid motion in relation to the wheel at the parts near the circumference, and hence a moderate altera-tion in the form of the buckets would be of no consequence: but I believe this is not the case in Poncelet's. In Poncelet's old wheel the water was admitted by only one orifice, and the circumference of the

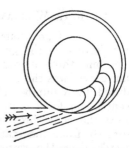

Poncelet's Wheel, of which I have already seen ac-counts.

case was not spiral but circular and concentric with the wheel. Hence the water acted on only one side of the wheel, the rest of the wheel being filled with air, and some of the buckets being partly filled with air and partly with water. His new wheel, which you say has several tangential passages for bringing on the water, must be better and more like mine. I think, however, from what I have said it will appear to you to be different in principle from mine; and if the water in his be discharged *only* below, mine must have a great practical advantage in discharging both above

and below. Did Poncelet say anything to you regarding the way
in which he regulates the size of the orifices so as to make them
suitable for different quantities of water? I have a very good
regulator for mine.......

I am, Your affte. brother,

JAMES THOMSON."

Many turbines were made under James Thomson's patent, and
were erected in various parts of England, Scotland, and Ireland,
while some were sent abroad. They were very successful, and at
the International Exhibition of 1862 a medal was awarded for
"originality of design, practical success, and good work." The
Encyclopaedia Britannica * says:

"Professor James Thomson's inward flow or vortex turbine has
been selected as the type of reaction turbines. It is one of the
best even in normal conditions of working, and the mode of
regulation introduced is decidedly superior to that in most reaction
turbines; it might almost be said to be the only mode of regulation
which satisfies the conditions of efficient working, and it has been
adopted in a modified form in the Leffel turbine, which is now
largely used in America."

At an early age he became deeply interested in questions
dealing with glacier motion. Writing on 13th February 1847 to
his sister Anna (Mrs Wm. Bottomley), he says, "On the Friday
before last Professor Forbes came here [Glasgow College] on a visit
to us, by William's invitation....I had a good deal of conversation
with him; especially in regard to glaciers and their bearing on
the Parallel Roads of Lochaber, about which perhaps Robert has
told you I have been writing." On April 5th of the same year
there appears a memorandum in his handwriting: "This morning
I found the explanation of the slow motion of semi-fluid masses
such as glaciers."

He had read all that had been published on the Parallel Roads
of Lochaber though he had never been in the district himself; but,
studying the subject from books and maps, he supplied the missing
link in the chain of reasoning that was necessary to substantiate
the explanation begun by Agassiz. He showed that certain loca-
tions of glacier barriers which were perfectly possible would explain
the positions of all these terraces. His paper (No. 60, p. 407), read

* *Ency. Brit.* 9th Edition, Vol. xii. 1881, page 527, article "Hydro-mechanics,"
by Sir A. G. Greenhill, F.R.S., and Prof. W. Cawthorne Unwin, F.R.S.

before the Royal Society of Edinburgh on March 6th, 1848, enforces
that view, now the accepted explanation. On the same night, the
Secretary, Dr William Gregory, wrote a long letter to the young
author, telling him of the enthusiastic reception of this, his first
paper, and inviting him to send others. Thus encouraged, he
turned his mind more and more to pure science rather than to
practical engineering. His next two papers appeared in the
Cambridge and Dublin Mathematical Journal of November 1848,
in which year his brother took over the editorship of that
journal. The first (No. 51, p. 334), "On the Strength of Materials
as influenced by the existence or non-existence of certain mutual
strains among the particles composing them" is reproduced *in
extenso* (with a few changes made in it with the author's concurrence) in the article on "Elasticity" written by Sir William
Thomson for the *Encyclopaedia Britannica*, 9th edition, Vol. VII.,
p. 798. The second was "On the Elasticity and Strength of
Spiral Springs and of Bars subjected to Torsion."

On January 2nd, 1849, he communicated to the Royal Society
of Edinburgh his famous paper on the "Lowering of the Freezing
Point of Water by Pressure" (No. 29, p. 196), which showed how
thermodynamics, by passing beyond the mere discussion of heat
engines, could become the ground subject of the physical sciences.
In this paper he not only showed by the application of thermodynamic principles, that the freezing point of water is not a fixed
temperature; but by calculations founded on the known values of
the latent heat of fusion of ice and the pressure of aqueous vapour
near the freezing point, and the known expansion of water in
freezing, he predicted, in advance of all experiment, the exact
amount of the lowering of the freezing point of water which
would be produced by the application of a given amount of
additional pressure.

The importance of this new development was at once grasped
by his brother, William, who soon confirmed it by accurate experiments. The classical papers dealing with this matter have already
been reprinted many times, for example in Sir William Thomson's
Mathematical and Physical Papers, Vol. I. pp. 156, 165. But
James Thomson's paper, as reprinted in *Phil. Mag.* in November
1850, has another title to interest, which is not so generally known.
In it, by means of an alteration of one sentence*, for the first time

----
* See *infra*, p. 201 footnote.

Carnot's principle was stated, and Carnot's cycle described, in words carefully chosen so as not to involve the assumption of the material nature of heat, or rather as Thomson himself puts it, the supposition of the perfect conservation of heat. The principle of liquefaction by pressure, thus established, was afterwards (in 1857) used as the foundation of his well-known explanation of the Plasticity of Ice *, a phenomenon which had been impressed upon Forbes by his observations on the motion of glaciers. Later, from 1857 onwards, for several years the whole subject afforded James Thomson much food for thought; extensions and developments followed on the influences of physical environment on crystallisation and on other changes of state. These pregnant indications of general laws obtained due recognition, when the subject assumed in 1876 its ultimate quantitative form in the profound work of Willard Gibbs.

Many questions of secondary importance but of considerable interest, arising from the foregoing investigations, occupied much of his time and attention. A paper read before the Belfast Natural History and Philosophical Society on the 7th May 1862 (No. 40, p. 258), on Ground or Anchor Ice, offered an explanation of the phenomenon which is so great an impediment to the navigation of the St Lawrence at Montreal, and which has been recently minutely examined by Professor H. T. Barnes and other Canadian physicists under the auspices of the Canadian Government with a view to the discovery of a remedy. The actions which accompany the solidification of crystalline substances out of liquid solutions are used to explain the atmospheric disintegration of stones in a paper read before the British Association in 1862 (No. 39, p. 257). On the 20th April 1864 he brought the same subject before the Belfast Natural History and Philosophical Society, and, at the same time, he made a communication on the "Tubular Pores in Ice Frozen on Still Water"; two years previously he had given in a letter to his brother, printed *infra* (No. 33, p. 220), the substance of this explanation. On the same evening he made some remarks on "Conditions affording Freedom for Solidification to Liquids which tend to Solidification, but Experience a Difficulty of making a Beginning of their Change

* "On the Plasticity of Ice as manifested in Glaciers," *Proc. Royal Society* 1st April 1857 (No. 31, p. 208) and "On the Plasticity of Ice," *Brit. Assoc.* 1857 (p. 211).

of State "; he spoke apparently without notes, but a manuscript in his own hand written on the following day has been printed here (No. 41, p. 268).

A paper probably to the same effect as the paper of 1862 on "Disintegration of Stones" was read before the Philosophical Society of Glasgow on the 11th April 1877, "On some of the Chief Modes of Decay of Stones in Buildings." There is no copy preserved.

In the early months of 1864 he made a minute study of a familiar phenomenon, the pushing up of the level of the ground in times of frost. A sketch—reproduced below, p. 269—explains what he observed of the action of the frost in producing separating spicules of ice in the soil; he made it the subject of a paper to the British Association in 1871 (No. 42, p. 270).

Besides the main lines of thought there are a number of miscellaneous papers. No one who reads these works can fail to be struck by the fact that the author possessed in a higher degree than most men the gift of penetrating to the heart of his subject, and so to say of feeling the cause of the natural phenomena he observed. He had the faculty of nursing theories in his mind, and working them out in every detail. Instances of this completeness will be found in the papers on the "Lowering of the Freezing Point by Pressure"; "On the Flow of Water in River Bends"; "On the Parallel Roads of Lochaber"; on "Columnar Structure"; and "On Atmospheric Circulation."

Another characteristic trait, to quote the words of his favourite pupil Dr Archibald Barr, " was his endeavour at all times to reach the greatest possible clearness in his own thoughts, and his scrupulous conscientiousness in endeavouring to impart to others in the clearest and least ambiguous language attainable the exact ideas he desired to convey." This led him to be unusually careful in his choice of language, and as a result of this trait in his character he often could not find a word that exactly suited his purpose. Thus many new words were introduced, some of which are now well established in scientific language.

This trait made him very chary of expressing an opinion until he felt quite sure of it. He thus produced a smaller amount of scientific work than might have been expected from his constant occupation with it. The same mental attitude debarred him from teaching his classes any new or hypothetical doctrines until he

President in the year 1821—2. There he would meet many friends of his father, as well as younger men who had joined after the Thomsons moved to Glasgow; among others the Rev. John Scott Porter, whose writings and preaching attracted much attention, and whose son, Andrew Marshall Porter (now Bart.), was afterwards the very distinguished Master of the Rolls in Ireland; the two brothers Murphy, Isaac and Joseph John; Dr Neilson Hancock of whose sister we shall hear more anon; and Professor Thomas Andrews with whom James Thomson afterwards did some of his most important scientific work. Elected a member in 1853, James Thomson became President in the session 1864—5, and honorary member on leaving Belfast in 1873.

He read to the Society the following papers:

March 5th, 1855.—On the Parallel Roads of Glen Roy;
May 14th, 1855.—On various plans for warming rooms and buildings;
Feb. 8th, 1858.—On Work and Power; their measures and measurement;
Jan. 10th, 1859.—Ventilation of Apartments;
April 2nd, 1860.—The Theory of Perspective;
Feb. 1st, 1864.—On Bridges and Tunnels;
Jan. 7th, 1867.—On Strength, Safety and Danger of Structures;
March 6th, 1871.—Explanations and Illustrations of Hydraulics.

Some of these papers have never been published, and their titles are given here to show the kind of work that occupied him in Belfast.

In 1851 a new society, called the Belfast Social Enquiry Society, had been formed, of which Dr Neilson Hancock, who held the Chair of Political Economy in Queen's College, was Secretary, and William Bottomley was Treasurer. It was before this Society that James Thomson read a remarkable paper on Public Parks in connection with large towns, on March 2nd, 1852. (See No. 66, p. 464). In it he explained a scheme by which a growing town can provide public parks for the benefit of the community, the money for the purpose being obtained by taxing the land surrounding the town while it is increasing in value. This paper led directly to the purchase, for the town of Belfast, of the large Ormeau Park. Many years later, when he was

T. d

President of the Belfast Engineering and Architectural Association, he put forward in his presidential address a strong plea for the State Purchase of Railways in Ireland. (See No. 67, p. 472.)

To the Belfast Natural History Society he communicated much on the newest scientific thought of the day in addition to reading papers of his own. One of these read on April 1, 1857, on "Capillary Motions Observable on the Mixing of Liquids" contained the views he had expressed two years before at the British Association in a paper entitled "On Certain Curious Motions Observable at the Surfaces of Wine and other Alcoholic Liquors" (No. 21, p. 125).

On the 13th January 1869, he read another paper before this Society on "Capillary Phenomena as influenced by Temperature," and on the 6th April 1870, "Illustrations of the Diffusion of Liquids"; but no reports of these papers are available.

In 1863 he became one of the original members of the Belfast Naturalists' Field Club of which he was elected Chairman in 1868. His most important contribution to its proceedings was the paper on the "Jointed Prismatic Structure of the Giant's Causeway" read in 1869. (Cf. No. 62, p. 422.)

He was a member of the Institution of Civil Engineers in Ireland, and in 1871 wrote for it a paper on the Jet Pump with Intermittent Reservoirs for the drainage of flooded lands.

He was a well-known figure at the meetings of the British Association, which he attended as regularly as possible from 1850 onwards, and he was President of the Mechanical Section at the Belfast Meeting in 1874.

Some account of his original work up to this time may now be inserted. When studying the utilization of water power by turbines, he investigated the actions in a Free Vortex or, as he named it, "Vortex of Free Mobility" and he read an account of this new departure in exact hydrodynamic observation and theory at the meeting of the British Association in 1852 of which no full report is preserved. The short abstract is given below (No. 1, p. 1). It will be remembered that Helmholtz's great Mathematical Memoir on the general properties of Vortex Motion appeared in 1858.

A full description of the Vortex Turbine was read at the same meeting of the British Association and was printed *in extenso* with the Reports (No. 2, p. 2). In 1854 he read another paper on Centrifugal Pumps, of which no report is preserved; and in 1855 a paper on a windmill and centrifugal pump (No. 3, p. 16). The best description, however, of his improved centrifugal pump with exterior whirlpool is to be found in the paper read before the Institution of Engineers in Scotland on 27th October 1858 (No. 4, p. 16).

Two other important lines of thought that interested him in later years come directly from his work on turbines and centrifugal pumps, viz. improvements in the friction brake dynamometer, and improved methods of measuring the flow of water. He contrived the jet pump in order, as he says in a note, to drain the pits of his turbines when this was required for purposes of cleaning: see the paper at the British Association of 1852 (No. 6, p. 26).

Experimental results showing the efficient action of the jet pump were given at the British Association in 1853 (No. 7, p. 27). A subsequent paper, read before the Royal Dublin Society on 1st June 1855 (No. 8, p. 30) gives a further development of this invention, viz. the provision of a series of jet pumps for use in draining swamps, so arranged that more or fewer could be used according to necessity. His own words, in a paper read before the Chemico-Agricultural Society of Ulster in 1859, may be quoted:

"Now, it is to be observed that the jet pump cannot work at all with less than its full supply of water. A simple Jet-Pump is therefore not suitable for working on a varying stream of water. For the drainage of lands, however, it is proposed that a series of Jet-Pumps should be so combined as that one, two, three, or more of them will come into action according to the amount of water supply flowing at any particular time. This is effected by placing any number of them, such as four or five, in a row underneath their supply cistern, and arranging the inlets for the water to them in such a way that when one has got its full supply, whatever surplus water there may be shall flow on to the next of them, and so forth with the others of the series."

In this paper he further stated:

"That a jet pump had recently been erected by William Forster, Esquire of Ballynure, near Clones, a member of this Society, for draining a portion of the wet lands adjacent to his house; and that the performance of the pump in raising the water had been found to be fully successful."

He exhibited to the meeting various drawings explanatory of the principles and mode of application of such pumps; also drawings of the pump on Mr Forster's estate: and his explanations were illustrated by a working model.

A later paper entitled " On a new application of the jet pump by steam " read at a meeting of the Belfast Natural History and Philosophical Society on 21st December 1864 (No. 9, p. 34), illustrates his capacity for observing and noting facts, and his power of turning known facts to use in new directions. The last paper on this subject is the one referred to above (communicated to the Institution of Civil Engineers in Ireland in 1871), on the Jet Pump with Intermittent Reservoir for the drainage of flooded lands or shallow lakes. After a description similar to that given in previous papers he continued:—

" Now since the jet pump, when wanted to be applied for drainage, must be made large enough to do the work of flood times, it would be quite too large to work continuously in dry weather; and, therefore, it is made to work intermittently, by the arrangement in conjunction with, it of an intermittently flowing reservoir. The reservoir is made to receive the continuous but variable supply of water for power coming from the higher ground, and to give it out intermittently to the jet pump. The intermittent action is brought about in a very simple way by means of two syphons, the smaller being used to start the larger."

His professional work in designing turbines, and his duties as engineer to the Belfast Water Commissioners, led him to study the measurement of water in rivers and streams. The first communication on this subject, read before the British Association in 1855 (No. 10, p. 35), merely relates to the experiments made by Poncelet and Lebros at Metz in 1827 and 1828, and gives some practical details that should be attended to in order to obtain accurate results.

Having been commissioned by the British Association to make further investigations, he presented a preliminary report in 1856 which is not here reproduced. In the report which he presented in 1858 (No. 11, p. 36) he made a great advance by proposing the use of triangular or V notches in the weir board instead of rectangular ones. Notches of this form are amenable to exact calculation, because the issuing jet is of the same shape whatever be the extent of the notch that is occupied by the stream. In fact he deduced the law that the quantities of water flowing through similar notches are proportional to the 5/2 power of their linear dimensions.

Another report to the British Association in 1861 (No. 12, p. 42) gives the result of further experiments for determining the single numerical coefficient that is involved in the formula for the V notch. The work on this subject is summed up in the paper which he gave to the British Association in 1876 (No. 14, p. 56). These experiments were all made at trifling cost with simple apparatus on a stream in the open air. It is interesting to note that an important set of experiments made with great care and accuracy, in 1909, in the new James Watt Laboratory, in Glasgow University, by Mr James Barr (under the supervision of Prof. Archibald Barr), while revealing effects due to the bottom and sides of the stream, and to the smoothness or roughness of the weir board, has yet confirmed in a remarkable degree this early work of 1860. Mr Barr says*:—

Dr Thomson gave ·305 as the average value of the coefficient c for heads ranging from 2 to 7 inches. Figs. 21 and 24 show how remarkably near this may have been to the true value for the gauge notch which he used. It is a proof of Dr Thomson's genius as an experimenter that, with the crude apparatus at his disposal, he was able to arrive at such an accurate result.

The necessity for accurate measurement of power led Thomson to study the friction brake dynamometer, and he read a paper before the British Association in 1855 of which an abstract is given below (No. 5, p. 24). The dynamometer submitted to that meeting was put aside for many years on the ground mainly that he " could not see how to work out the principle in an apparatus cheap enough and simple enough for the occasional uses for which such dynamometers are ordinarily applied." Later, he devised a new form of dynamometer or ergometer, and constructed a small working model which he showed from year to year to his classes in Glasgow University. It comprised fast and loose pulleys on a shaft, and a rope, attached to the loose pulley, lapping partially round the fast pulley, with weights at each end. The loose pulley adjusted itself to give the requisite arc of contact, winding or unwinding the cord on the running pulley as required for adjusting the resistance. He did not formally publish this invention, but in a letter to *Engineering*, 29 October

* *Engineering*, Vol. LXXXIX. April 8th, 15th, 1910. pp. 435, 473.

1880 (Vol. xxx, p. 379) he claimed to have had it in public use in November 1879, and also to have devised and tried a regulator for preventing oscillations in the loose pulley and the two weights. The principle of this regulator "consists in introducing what is equivalent to adding inertia without adding weight to the lighter of the two weights of the ergometer. The inertia is applied by connecting a small and finely pivoted fly wheel to the lighter weight, so that the fly wheel revolves forward and backward with a varying velocity always proportional to the downward or upward velocity of the lighter weight. Inertia accompanying the motion of the heavier weight I judge by theory, confirmed by experiment so far as experiments have gone, is injurious as to attainment of steady action in the ergometer; and unfortunately the loose drum introduces inertia injuriously in this way. But, on the other hand, inertia accompanying the motion of the lighter weight I judge by theory to be very beneficial; and my practical trials, so far as they have gone, seem to confirm this theoretical view, and to show inertia introduced in that way to be very remarkably beneficial in producing steady action without entailing any vitiatory conditions*."

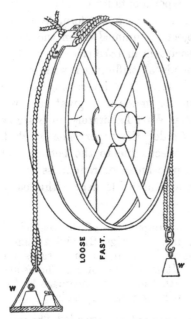

In the discussion of the paper on Engineering Laboratories read before the Institution of Civil Engineers on 21st December 1886 by Prof. Kennedy†, Prof. Barr described Thomson's Dynamometer and gave the illustration of it here reproduced. Prof. Barr said, "It was shown by Prof. Thomson to his class some years before it was published by Carpentier, and it was used at the Glasgow Gas Exhibition (1880) for testing on the very day on

---

* The inertia wheel is regularly used, with most beneficial effects, on friction brake dynamometers in the James Watt Laboratories in Glasgow University.

† *Proc. Inst. Civ. Eng.* Vol. lxxxviii. pp. 109, 110.

which a description of it was published in this country. He
believed it was the only brake which could apply an absolutely
uniform resistance." On a later occasion Professor Ayrton said * :—
"Another plan was to vary the arc of contact. It was curious
that every person who had arrived at that idea seemed to regard
it as his own. It had been invented over and over again....He
believed it was due initially to Professor James Thomson."

On the 20th November 1888, in a discussion at the Institution
of Engineers and Shipbuilders in Scotland on the Horse Power of
Marine Engines, Professor Thomson said :—

"That he placed no confidence in the numerical expressions
put forward under the name of nominal horse power. He had
always felt that such an expression was rather misleading than
helpful. He was in favour of improving the means of ascertaining
the actual power of the engine delivered at its own shaft, and
dealing with that power rather than with the power given to the
driving piston or pistons by the steam. He thought there was a
possibility of measuring the power by the elastic yielding of the
shaft under the torsional forcive action or torque to which it is
subjected, a method that had been proposed by Hirn, and which
was to be found mentioned in one of Rankine's books†. He had
himself, conjointly with his son, about eight years ago, proposed
essentially the same mode of measurement of power, without then
knowing of Hirn's invention or publication in this matter, though
they found it soon after as described in Rankine's treatise. Their
own contrivances differed importantly in some of the mechanical
arrangements from those of Hirn, and their intentions included,
further, the application of the general principle to the measurement of power in screw propelled steamships, and more especially
in their speed trials, with a view to obtaining improved data for
ship design. Their intentions related also to the obtaining of
trustworthy records of the varying violences of torsion applied
during long voyages by the waves and engine jointly to the
shafting, as such information would be valuable with a view to
the better determination of strengths for propeller shafting, and
so to the abatement of breakages at sea."

William John Hancock, the father of Dr Neilson Hancock,
who has been mentioned above, had been land agent on Lord
Lurgan's estates for many years, and was subsequently appointed
Poor Law Commissioner at the time when the new Poor Law was
introduced into Ireland. He literally laid down his life for the

* *Proc. Inst. Civ. Eng.* Vol. xcv. p. 54.
† Rankine, *Machinery and Millwork* § 344, p. 387, "On the Torsion Dynamometer of M. G. Hirn."

poor and suffering, in the year of the terrible epidemic of typhus fever which followed the famine of 1847. In exerting himself to try to minimize the hardships of the poor patients in the over-crowded workhouse hospitals he himself had contracted the fatal disease. His eldest son, John Hancock, had attended Professor James Thomson's classes at the Academical Institution at the same time as William Bottomley, and when the younger brother Neilson was appointed to the Chair of Political Economy at Queen's College, Mr and Mrs Bottomley visited his mother with whom he lived. His only daughter, Elizabeth, a bright intelligent girl who had been his companion during his lifetime, and was then her mother's chief helper, soon became an intimate friend of Mrs William Bottomley. James Thomson, who was constantly at his sister's house at Fort Breda, thus had many opportunities of meeting her friend. Acquaintance ripened into friendship. Elizabeth Hancock's mother was a woman of unusually broad views for that time, and she had brought up her children in such a way that her daughter had no difficulty in reconciling James Thomson's creed with her own. When Dr Neilson Hancock was appointed to Archbishop Whately's Chair of Political Economy at the University of Dublin, Mrs Hancock decided to leave Belfast and join her son. James Thomson found it very pleasant to help Miss Hancock to pack up her brother's books, and before she actually left he discovered that her presence was necessary to his happiness. They became engaged in the autumn and were married in Lurgan Church on December 28th, 1853. Their honeymoon, spent at Bryansford at the foot of the Mourne Mountains, in a little inn where they were nearly snowed up, was cut short by a request from Queen's College, Belfast, that the young engineer would fill the office of Professor of Civil Engineering during Professor Godwin's temporary absence, an offer that was gladly accepted in the hope that it might lead to something permanent. This hope was fulfilled in 1857 when he was appointed to the Chair.

A happier marriage could hardly be imagined. Mrs Thomson devoted herself to her husband and to their three children. He relied on her help and sympathy in his own work for the various Societies to which he belonged, while she enlisted his sympathy for the many public movements, such as the higher education of women and the Married Women's Property Act, to which she gave her mind in later years. The young couple found themselves immediately in the centre of a large social circle, as they were both well known and much liked in Belfast before their marriage. One of their friends was George Fuller (the younger brother of Frederick Fuller of Cambridge, afterwards Professor at Aberdeen), who came to Belfast as assistant to James Thomson soon after the latter opened his own office as an Engineer. This was the commencement of a friendship which lasted through life. George Fuller, who was best man at his wedding, later on went to India. He and Thomson carried on a very interesting correspondence, and when James Thomson left Belfast in 1873 to occupy the Chair of Engineering in the University of Glasgow, Fuller succeeded him as Professor in Belfast.

About the time of his marriage James Thomson was appointed Engineer to the Belfast Water Commissioners, and he was from time to time consulted by various public boards in the North of Ireland with regard to the water supply of their towns. Meanwhile William Thomson had married Margaret Crum of Thornliebank, nearly a year before James married Elizabeth Hancock. William's last letter to his brother before his engagement may well be given here :—

"32 DUKE STREET, ST JAMES'.
Monday, June 21, 1852.

I arrived in London I believe this day fortnight, and ever since I came I have been very much engaged with seeing people, looking after instruments etc. I have not yet had time to do almost anything in the way of sightseeing and other London strangers' occupation. I have seen Faraday, Col. Sabine, and others of the 'savans' a good deal, and I was admitted as Fellow of the Royal Society at one of the meetings, which are now over for the season.

On Saturday last I had 3 hours talk with Faraday, and immediately after had a visit from Mr Tyndall (who has experimented and written on magnetism etc.) which lasted for about two hours. I found it very curious to compare the sort of arguments which had effect on the two, each considering as utterly unintelligible or inadmissible arguments on a certain point with which I believe the other was satisfied; and I think I succeeded in convincing them both of one conclusion which was contradictory to a good many assumptions that they and others had been making. I lived with Sylvester for four or five days after my arrival in town, during which time we had perpetual conversations, either on his algebraic investigations, or on the subjects of some of my papers on electricity and magnetism which he has been reading.

Stokes lodged along with me for a week, and we used to go together very frequently to Darker's (the optician) in Lambeth, always passing through Barton Street going and coming. I called at No. 9 one day and heard that Mrs Turnbull had been having an addition to her family. She was to leave that locality almost immediately.

I suppose you have heard of Stokes' great discovery in optics, but whether or not, I must delay telling you anything of it till I see you.

I was surprised last Friday morning to receive a great packet by post from Paris, which I found contained a proof sheet of my two papers on the Dynamical Theory of Heat, translated into French for Liouville's Journal. I have had rather a heavy job correcting it, and have not nearly done yet. I have a great many projects of papers to write and a few actually in progress, so you may conceive that my hands are pretty full....

Talking of babies, Mrs Joule has had a second which is a girl, notwithstanding which I am asked to be its godfather; so I have an addition to my family of godchildren which is now tolerably numerous."

Another letter from William Thomson may be quoted here:

"PHIL. SOC., SATURDAY, *Jan.* 13, 1854.

Did you see that I have applied for a patent, with Rankine and John Thomson, for an improvement on telegraphic conductors! I accidentally got on the theory of the propagation of electricity by submarine wires, one day in October before leaving Largs, which showed me at once what would be necessary to ensure efficiency for great distances (300 miles or more) which led to this, Rankine having suggested the plan of taking a patent, which I had no idea of at first. In a few days I expect it will be secured to us: in the meantime don't say even as much as I have said to you, on the subject. I am not very hopeful of making anything of it, but it is possible that it may be profitable."

During the twelve years that elapsed till the splendid enterprise came to a successful issue in 1866 and the Atlantic Cable was an accomplished fact, and during the repeated trials and supposed failures described so graphically by Prof. Silvanus Thompson in his *Life of Lord Kelvin*, James Thomson watched and waited full of hope and anxiety, ready at the slightest signal to give his brother any aid in his power; and great indeed was his joy when the news came from mid-ocean through the *Great Eastern's* cable that all was well.

These years were among the most strenuous in the lives of both the brothers; and whenever a new idea or principle or invention occurred to either of them he would send off a note to the other, reserving a deeper discussion of the subject till they had an opportunity of meeting. Some of the letters from William to his brother are written on fly leaves of his manuscript "Green Book"; one is written from the ss. *Elk* off Carrickfergus on the backs of two hotel advertisements borrowed from the steward on board. One of his most brilliant Belfast pupils, Professor John Perry, F.R.S., has put on record* Professor Thomson's method in dealing with his class :—

"As one of Professor James Thomson's pupils, I may say that he was one of the kindest of men, with a very strong sense of duty. He was always nervously anxious that his pupils should have correct notions of the scientific principles of engineering methods. He was, in fact, a pure-minded man, whose good influence is now widely felt, as I have met his pupils in all parts of the world. Every one of us could laugh at his well-known foibles, but, however great or small we might be, there was not one of us who had not a loving memory of him as distinguished from a mere liking and from mere respect. We all respected him.

What is especially interesting in his career, is that he did not publish much—but everything he published was of sterling importance to the physicist and engineer, and no one has ever had to correct his results. Take one branch of engineering, for example, hydraulics. In any treatise on practical hydrodynamics there is a great parade of mathematical calculation; but beyond the simple equations relating to pure fluids, what is there that can really be relied upon by the engineer except the few propositions so thoroughly proved by Professor James Thomson? What a mass of mathematics there is to be found upon the gauging of water—but is there anything *demonstrated*, anything that the engineer can rest upon as a principle in this subject, except Professor James Thomson's proposition concerning the flow from similar and similarly placed orifices? Two deductions from this are the rational formulæ concerning triangular and rectangular gauge notches. Again, he, as early as 1847, applied Carnot's cycle and first principles to the study of the pressure and temperature of melting ice, and he may be said to have really shown to Clausius, Rankine, and his brother (Lord Kelvin) the method of attack which has led to the development of thermodynamics.

I will relate one incident to illustrate Professor Thomson's sense of duty. In 1870 I was a student at Queen's College, Belfast, and about to go up for my engineering degree examination

* *Industries*, 13 May 1892.

to Dublin. Now, I had not particularly cared to obtain a degree, and I had neglected to prepare at the college the requisite show of drawings. But at a time when he much needed a holiday, he took the trouble of going to my house and afterwards to the Lagan Foundry, having heard that during my apprenticeship there I had made many drawings. He spent a long time examining them, and made such a report to the University authorities as caused them to put aside their rule in my particular case. Now this is not all. It was found subsequently that my attendances at his classes were not up to the standard, and he utterly refused to give me the certificate which would enable me to go in to the examination. It was, in his opinion, wrong to do this; and, in spite of a battery of persuasive argument brought to bear upon him by Dr Andrews and others of his colleagues, his refusal was absolute. The difficulty was got over through some newly discovered technical right of the President of the college to exercise grace. Taking the two incidents together, they may be regarded as unprecedented in the history of the professorial sense of duty."

As further illustrating his methods, it may be remarked here that he constructed a sliding balance, by the use of which the examiners could mechanically place the students according to the marks gained in class and during examination, thus eliminating any possibility of favouritism or prejudice in deciding the results.

In the autumn of 1861 James Thomson and his wife and six year old daughter spent ten days very happily with Prof. William Thomson and his wife on the Island of Arran, where the William Thomsons had rented for summer quarters a small country place called Kilmichael, the only part of Arran not belonging to the Duke of Hamilton. There were two other guests staying at Kilmichael, namely Dr Joule and Prof. Forbes, and endless discussions arose among the four at table—where so unimportant a matter as food was liable to be entirely forgotten—or out boating on the lovely September evenings. Sometimes the conversations on deep questions of science, such as the effect of stresses in causing the melting of solids, became so absorbing that one or more of them forgot to row, while the others, not perceiving the omission, continued rowing steadily, so that the boat went round in graceful *gyrations* or loops and circles, in spite of the steerswoman's efforts to keep a straight course. One evening when the sea was quite smooth except for small ripples, they noticed long sinuous streaks of perfectly calm water among the ripples, and, rowing out to the spot to investigate the cause, James Thomson pointed out to the others that these calm lines were caused by small floating objects, bits of seaweed, withered leaves, etc., which collected at the place where two currents met and perforce the water had to descend, leaving the small floating particles on the surface at the line of meeting, where they had the effect of deadening the tiny wavelets and thus caused the calm line. His explanation is developed in a paper read before the Belfast Nat. Hist. and Phil. Society on the 7th May 1862 and published in the *Philosophical Magazine*, (No. 25, p. 142). This paper was reprinted long afterwards as an appendix to one read in 1882 before the Phil. Society of Glasgow on "Changing Tesselated Structure in Certain Liquids" (No. 24, p. 136), which he traced in a most interesting and suggestive manner to a vertical circulation establishing a sort of cellular or rather columnar pattern of convection currents.

The subject of the melting of ice under pressure led to important developments with regard to geological questions, besides its application to glacier motion already referred to. It brought him into friendly relations with Faraday and with H. C. Sorby, and the scientific correspondence which ensued is printed *infra*, p. 212 and p. 252, in its natural sequence. His brother William, writing to J. D. Forbes on August 20, 1860, sums up as follows:—

"His (Faraday's) 'experiments' are beautiful. But I think the results are in very strong accordance with my brother's view. The first contact between the convex blocks takes place at a mere point, and with a slight impact. It must therefore give rise to considerable internal stress in a very small part of each block. This, according to my brother's theory, is enough to make the two freeze together on the ceasing of the instantaneous pressure. After they are thus united, the tension which Faraday applies and keeps applied tending to separate them, will raise the temperature at which the solid neck or isthmus joining them could melt; it will also instantaneously raise the actual temperature of the ice-isthmus, and tend to thin it viscously (according to my brother's explanation of the viscosity of ice). But unless it is strong enough to crack or quickly thin down to nothing the isthmus, and so separate the pieces again, the temperature of the isthmus will very quickly sink to that of the surrounding water—viz. 32° Fahr.—then there will be an isthmus of solid *at a temperature below its proper freezing point*, constantly washed, for hours and days, by liquid of the same composition which therefore cannot but freeze upon it and render it thicker and stronger as in Faraday's experiment."

The following letter refers to a controversy with Prof. Tyndall on the same subject.

"2 DONEGALL SQUARE WEST, BELFAST.
1st Oct. 1861.

MY DEAR WILLIAM,

...In a letter which I wrote to you on the 1st of Aug. 1861, I said that I was not sure whether in a vortex of free mobility, extending outwards to an infinite distance radially, but which I meant to be of a finite thickness as measured parallel to the axis of rotation, the mechanical work-required to generate the motion would be finite or infinite; but that I thought it would be easy to calculate this. I have now made the investigation for that purpose, which comes out quite easy, and which I think is of interest, and I enclose to you a copy.

I have also written out, a little more distinctly than when I was with you, the statement of the general principle which I think holds good—and of which I think my view of the effect of stresses in causing crystals in their melted fluid, or crystals in their saturated solutions, to melt or dissolve on the application to them of stresses tending to change their form, is a particular case. If you can readily let me know your opinion of this axiom or general principle, I shall be glad to receive it. If you think the principle is correct and the statement of it good, I would probably introduce it into my intended paper for the R. S., on the effect of stresses tending to change of form on crystals in their molten fluids or in their saturated solutions*. I met Dr Stevelly yesterday and he had seen Tyndall at the British Association.... He told Tyndall that I had succeeded in squeezing crystals of wet salt into a hard mass, or something to that effect.

Joseph John Murphy† is just home from Switzerland. He tells me that he saw Tyndall on the Eggischorn, and that they spoke together as to my views of ice, and that J. J. Murphy said he thought Tyndall had quite misunderstood my view, and that Tyndall replied: 'Oh, well, if Mr Thomson thinks so, let him explain.' This makes a little more reason for publishing my reply to Tyndall....

Your affte brother,

JAMES THOMSON."

During this autumn he suffered from sleeplessness and was advised to take complete rest from College duties. He went to Dublin to stay with his brother-in-law, Dr Neilson Hancock, and meantime Dr Andrews kindly undertook to conduct his classes.

A most interesting and remarkable correspondence took place

* See infra, p. 236.

† Author of a notable book *Habit and Intelligence*, and an early writer on evolutionary subjects.

on the limitations of the law of dissipation of energy, and the influence of vitality on matter, which is here given in full.

"2 DONEGALL SQUARE WEST, BELFAST.

April 7th, 1862.

MY DEAR WILLIAM,

According to the Theory of Thermodynamics which you and others have lately worked out and adopted, it appears that 'although mechanical energy is *indestructible*, there is a universal tendency to its dissipation, which produces gradual augmentation and diffusion of heat, cessation of motion, and exhaustion of potential energy through the material universe' (quoted from your writings, see 'On Age of Sun's Heat*'). I understand that you consider that when mechanical energy disappears, it is replaced by an equivalent amount of *vis viva* in the molecular or other motions, which are supposed to constitute heat; and that while man can apply *mechanical vis-viva*, or *mechanical energy of motion* (by which terms I mean to express the energy of the motion of bodies of finite size or of size decidedly larger than what can be regarded as molecular; and to distinguish it from the energy of such motions as may be supposed to constitute heat), so as to produce potential energy, he cannot possibly turn to such account, by any actions of matter on matter induced by him nor in any way, the *vis viva* of heat, if the heat he has at command be all at one temperature: also that he can obtain potential energy from heat only through allowing other portions of heat at different temperatures to equalise themselves in temperature, and that so the continual tendency of man's obtaining potential energy from differences of temperature and expending it, must be towards bringing about a state of uniform temperature from which it would be impossible for him to obtain potential energy again.

With the foregoing I am quite disposed to agree, except that I think it may perhaps be necessary to qualify the theory, or to qualify what might be taken as implied in it, by excepting from its decided statements, and leaving as still unknown and uncertain, the mysterious influence of spirit, life, or the vital principle in animals and plants over the matter composing their living bodies. I think we must look on this influence as being contrary to, or predominant over, the ordinary laws of matter. Thus for instance at any particular moment the molecules composing a living body are in a certain position and connexion with respect to one another. They are possessed of much potential energy, and if left to themselves at that instant without the control of what we call spirit or life, they would go through certain actions, continually passing to states of lower potential energy until arrival at some minimum of potential energy though not necessarily the lowest

* *Popular Lectures and Addresses*, Vol. I. p. 349.

minimum. (What I mean by this may be understood on consideration that in certain circumstances decomposition or putre-
faction may be arrested indefinitely.) Now when spirit or life,
connected with these material particles, can hinder the actions
which the material particles would of themselves go through
in virtue of their potential energy, and can cause them to
go through quite different sets of actions, and through sets of
actions which, if we believe in free will, as I do, we must consider
as not following one another in a natural* sequence of cause and
effect according to the physical† laws of matter; as not being
fixed for the future by the present condition in which the matter
exists; and as being absolutely at the command or under the
predominant influence of something which is not matter; it
cannot be disputed by anyone who believes the mind or life to be
distinct from matter and to be no mere result of arrangement of
particles of matter and of their energies, that the vital principle
has the power of regulating and applying the potential energies of
the matter composing the living body with which it is connected,
so as to keep them from applying themselves as they would do if
left to themselves. This influence of mind or life over matter is
altogether mysterious. It implies I think a perfect suspension or
contravention of the physical† laws of matter; and may be
regarded as a part of a great miracle which constantly goes on
around us (although most people will not call anything miraculous,
unless it be extremely rare, or unless it be unique, so that all the
human race except a very few, must either believe it or reject it
by judging from the testimony of others, or from geological records
of events which have occurred even before the human race was on
the earth).

Now if we allow that mind or life really has the power of
counteracting the physical laws of matter—that is the laws of
inanimate matter—it seems to me to be a perfectly admissible
supposition that mind or vitality may have the power, in the
living body, of collecting, and applying as potential energy, the
energy of the heat motions, or of bringing out in the living body
from the *vis viva* stored indefinitely everywhere as heat, the
potential energy which you regard as a *true equivalent* for the
vis viva of the heat, and for which you consider the *vis viva* on
the other hand to be also a true equivalent. Without asserting
that we have a right to assume that vitality really has that power
of bringing out the equivalent of mechanical energy from an
equivalent of heat (that is of effecting a reversal which by merely
the physical laws of matter is impossible to be brought about)

* I here use 'Natural' and 'Physical' not exactly as you are proposing to use
such terms. I mean them both to apply to the phenomena of the energies of
inanimate matter.

† I use the term 'physical' here in its ordinary sense at present in common
use; namely as distinguished from vital or psychological.

I want to guard against what I consider is equally a doubtful supposition but is yet generally assumed: namely that *all* the animal and vegetable energies are derived *solely* from previous potential energies, or other energies capable by ordinary physical means (such as air engines, chemical actions, etc.) of being converted into potential energies. I want to say that perhaps the vitality of animals and plants may be able to reverse processes which would go on if not influenced by the vitality. We know, as I have said before, that the vitality *can and does divert the course of action of energies*: does it not then follow that it at least stops some of their courses of action or even reverses some? And farther is it at all incredible that vitality should be able to cause the passage of heat into what you regard as neither more nor less than a true equivalent for it, namely, into the Joule's "equivalent" of potential energy or mechanical energy? I do not think this supposition would be at all more wonderful than the one already generally admitted, that mind has the power of controlling the courses of energies during their passage to the lower state, or state of dissipation, and of equivalent augmentation of heat.

I suggest the above ideas for consideration, and I think there is much in them in regard to which it will be safest to suspend judgement.

Your affte brother,

JAMES THOMSON.

P.S. I would not have you to suppose that I am *against* the prevailing opinion that animal power is derived (or mainly derived) from potential energies stored in the food eaten and the air breathed: but what I chiefly want to say is that I do not think we are entitled to set down as a *proved impossibility* the conversion, through vital influence, of heat into mechanical energy; and I would say may not the impulse of volition communicated from the mind to the body consist in a reversal of an action or set of actions in the body, perhaps in the brain or nerves, which would go on under the ordinary physical laws; the antecedent and subsequent conditions, whether in the forward or reversed action, being (as I believe you say) truly equivalent for one another. Then when one part of an action is reversed or stopped it is easy to conceive that the whole remaining course of the actions may be thereby turned into new channels.

J. T."

"COLLEGE, *April* 21, 1862.

MY DEAR JAMES,

I have not had time yet to read Mr Murphy's paper which you sent me, but I hope to do so soon, and to write to you accordingly.

T.

e

Your previous paper on thermodynamics of animals I received and sent on to Tait. I agree with it on the whole. You will see that I have excepted plants and animals from any assumption (founded merely on what we know of inorganic matter) of either the First or the Second Law. In fact neither can be held to be an *axiom* when life is concerned, in the same sense as it can for dead matter. Life being not perpetually reproducible in a cycle, the sound objections to assumptions involving the 'perpetual motion' are not applicable to the assumption that energy may be called into existence by free will. But analogy from known facts renders it probable, (almost certain) that energy is *not* so called into existence. This I have stated in an old Article*, *Proc. R. S. E.*, about 1850 [Feb. 1852].

In my 'Dyn. Theo. Heat' I except action involving life from the axiom on which I found the Second Law. Physiology as yet proves nothing as to whether the 2nd Law is really violated or fulfilled, although I think it probably is fulfilled, in the vital processes.

Please answer Poggendorff's questions (see enclosed letter) and send back the letter and your answer to me. I have delayed too long, but I shall forward my own answers along with yours as soon as I receive yours.

Your affecnte Brother,
W. THOMSON.

P.S. Thanks for your paper on Spiral Springs. I shall return it if you wish. I must have a copy or copies somewhere.

P.P.S. You should send the separate copies to Faraday, or others notwithstanding they are F.R.Ss. I shall be glad to have one."

In 1863 the British Association met for the second time at Newcastle-upon-Tyne with Lord Armstrong, then Sir William Armstrong, as President. Professor and Mrs James Thomson and their brother and sister-in-law, Dr and Mrs Neilson Hancock, joined forces and went over to England to be present. This meeting left very pleasant memories in the minds of the little party from Ireland. They were entertained with the other members of the British Association by the President at a dinner given in the new Banquet Hall which Sir William Armstrong had just built at Jesmond Dene, and they saw a great deal that interested them in the neighbourhood, and as usual met a number

---

* "On the Mechanical Action of Radiant Heat or Light, on the Power of Animated Creatures over Matter; on the Sources available to Man for the Production of Mechanical Effect." From the *Proceedings of the Royal Society of Edinburgh*, Feb. 1852. Reprinted in *Mathematical and Physical Papers*, Vol. I. p. 505.

of their friends. The visit terminated with a tour in the Lake District, where they all agreed that the natural scenery in England could hold its own with the beauties of Scotland and Ireland. Soon after their return to Belfast, Prof. Thomson found his pleasure in his work at Queen's College much enhanced by the advent from Dublin of the young, brilliant and distinguished Professor of Mathematics, John Purser. The delight with which the three friends, Andrews, Thomson and Purser worked together in the following years, until James Thomson left Belfast in 1873 on his appointment to the Chair of Engineering in the University of Glasgow, is best told in Prof. Purser's own words*:

"...But of all my colleagues the one with whom I had the most congenial intercourse was my friend James Thomson who held the chair of engineering. Of the scientific men I have come across he most fulfilled the idea of a philosopher, his ever-working brain always thinking of the causes of things, pondering on the why and wherefore of different natural phenomena—why, for instance, the blocks at the Causeway assumed their characteristic prismatic form, or by what process the solid ice of the glacier, to all appearance so unyielding, was able to move along its rocky bed. Most modest and retiring, it was only on closer acquaintance you realised his original power as an investigator and the tenacity of his hold on physical principles. You learned, too, that this same grasp of principles found expression also in the moral sphere, in a character singularly just, anxious to carry out at all costs what he felt to be true and right. My intercourse with him, which continued after he had left us to take charge of the Engineering School in the University of Glasgow—indeed, to the end of his life—was to me most stimulating and instructive. He was far the profounder thinker, but I knew a little more of the technique of mathematical analysis, and used to give him pleasure when I pointed out quicker ways of proving some of the theorems he had discovered. Both James Thomson and I followed with keen zest the magnificent results that Andrews arrived at in his renowned investigations in the early 'seventies,' the investigations on the behaviour of gases under pressure at different temperatures, and the great discovery of the continuity of the liquid and gaseous states of matter. Indeed, it has seldom been the good fortune of an institution of our modest dimensions to include on its staff at one and the same time two such wonderfully original thinkers as Thomas Andrews and James Thomson. And Ulster may well

* Taken from his speech at a dinner given to him at Queen's College, Belfast, on the occasion of his retirement from the professorship: *Belfast News Letter*, Jan. 8th, 1902.

be proud of both. Both were Ulstermen born and bred. If his fellow citizens believed in Andrews, certainly he on his side believed in Belfast, its energy and enterprise, and was wont to speak with pride of the old worthies it had produced. He even, though some of us found a difficulty in following him so far, believed in its climate, declaring that the Belfast winter closely resembled the Italian. Such were the men with whom in those early days my happy lot was cast—men all of them interesting, some of them, as I have tried to show you, deserving the name of 'great'."

In 1869 Dr Andrews gave to the Royal Society his famous Bakerian Lecture on the "Continuity of the Liquid and Gaseous States of Matter" which closes with the following words:

"We have seen that the gaseous and liquid states are only distant stages of the same condition of matter, and are capable of passing into one another by a process of continuous change. A problem of far greater difficulty yet remains to be solved, the possible continuity of the liquid and solid states of matter. The fine discovery made some years ago by James Thomson, of the influence of pressure on the temperature at which liquefaction occurs, and verified experimentally by Sir W. Thomson, points, as it appears to me, to the direction this inquiry must take; and in the case at least of those bodies which expand in liquefying, and whose melting-points are raised by pressure, the transition may possibly be effected. But this must be a subject for future investigation; and for the present I will not venture to go beyond the conclusion I have already drawn from direct experiment, that the gaseous and liquid forms of matter may be transformed into one another by a series of continuous and unbroken changes."

Both James Thomson and his brother took the keenest interest in Andrews' work, and the substantial contributions made by James to the general philosophical development of this subject are recorded in several papers.

In 1866 the house in Donegall Square, where he had lived for many years, was wanted for business sites, and Prof. Thomson bought a house in University Square quite close to the College. Finding serious defects in the drainage there, his attention was directed to the whole question of the drainage, warming, and ventilating of houses. He read several papers on these subjects before various societies and public bodies, but complete copies have not been preserved. An abstract of a paper read on 4th December 1867 is given *infra*, p. 477. His ideas and principles have in the main been taken up, and have formed the basis of many of our modern improvements in these matters. For instance in those days it was the custom to carry down soil pipes inside a house and under the floors, and so to join them with the public sewers, often with very defective connections. Thomson was among the first to urge the necessity of carrying the soil pipes through the wall above ground, and letting them enter the main sewer by a properly constructed trap, clear outside of the house. It is not generally known that he was the first to use smoke as a test for defects in house drains. He devised a very simple method of applying his "smoke test" in a house at Castlerock which he had taken for the summer months in 1872, and the idea was quickly taken up by others and applied successfully to the detection of minute leaks.

In later years he gave renewed attention to the ventilation of public buildings, taking a patent in 1891 for warming and ventilating. The invention consists in the exchanging of old air for new by means of a mechanism—called the "exchange wheel"—resembling a fan with septums or blades radiating out from the axle to the circumference. The wheel revolves within a fixed casing fitted closely to the circumference at two parts diametrically opposite, each of them extending through an angular space as great as that comprised between two of the blades of the wheel. Thus whether the wheel be standing still or revolving, the two halves into which the portal is divided by the axle, are always both closed, each by at least one of the radiating blades; and, in consequence, the action of the wheel is not affected by any wind that may be blowing against the walls of the building. For a large hall, in addition to the ventilation, a circulation of air would be maintained by an ordinary fan with a view of preventing the heavy cold air from lodging in the lower part of the hall, the cold air being drawn away from the floor by apertures there, or as low down as may be convenient, and the circulating air would enter the apartment at the ceiling, or at some conveniently elevated place, so that the circulation is in the main downward through the apartment. His death prevented the working of the patent; but part of the invention has come largely into use in recent years for revolving, draught-free doors of banks, hotels and offices.

In the summer of 1867 he was engaged in building a new weir near Belfast for the Lagan Navigation Company, to which he was at this time appointed engineer.

Two years later Margaret King, the eldest daughter of his sister Elizabeth, was married in Edinburgh to the distinguished physical chemist Dr J. H. Gladstone, F.R.S. The whole of the Thomson family went over for the wedding, and after the event they had a delightful little tour through the Highlands.

During the session 1867—8 Professors Thomson and Purser were much interested in watching the progress of Dr Andrews' far-reaching researches into the different states of matter; the following letter from his brother, then Sir William Thomson, shows that he too watched the development with keen interest.

"LARGS, *Friday, Mar.* 4/70.

MY DEAR JAMES,

...You may tell Dr Andrews that I have four different reasons for knowing that molecules of water, glass, etc. cannot be less in linear dimensions than about one 1000th of a wave length of light: or say about $\frac{1}{20 \times 10^6}$ of a centimetre: or that there cannot be more than about 16×10^{21} molecules of water in a cubic centimetre. I am getting a short sensational article on the Size of Atoms ready for Nature[*], Dr King acting as amanuensis. He and Elizabeth came here about three weeks ago, for his sake, as he had been very poorly. They are in lodgings, but we see a good deal of them, and are doing all we can to help Elizabeth to keep Dr King's mind off wearing thoughts.

He proposed himself taking down this article for me as a mental relief for himself. He came two evenings ago when we made a beginning, and he is coming again to-night.

. . . .

Your affectionate brother,

W. THOMSON.

P.S. If you see him (Dr Andrews) thank him for the copy of his paper which he sent me. I am very glad indeed that it is now out. It will make an era in science.

P.S. The LL.D.[†] is owing originally to Blackburn. Rankine made a very good and appreciative speech in proposing or seconding and Dr Rainy spoke very warmly."

About this time an exhibition of mechanical and useful arts for working men was got up in Belfast. It is worth recording, as it was one of the first movements of the kind started for the benefit of working men in this country. Prof. Thomson exhibited

[*] See Lord Kelvin's *Math. and Phys. Papers*, Vol. v. p. 289; or *Nature*, Vol. I. pp. 551—553.

[†] Degree conferred on James Thomson by the University of Glasgow.

his Jet Pump and Intermittent Reservoir, and was one of the judges. The exhibition was a great success.

Meanwhile Sir William Thomson was passing through a period of great anxiety. His wife had a serious illness soon after her marriage and had suffered much from that time onwards. She became much more ill in the spring of 1870, and on the 17th of June died at Largs, surrounded by her much loved sisters and brothers. Her death left a blank in her husband's life which could not easily be filled. He threw himself into his work with, if possible, greater concentration than ever, and the letters that passed between the brothers for the next few years are full of descriptions of the syphon recorder or designs for the new telegraph ship *Hooper*. But before this interesting vessel had been fully planned, Sir William had found at Cowes and bought his yacht, the *Lalla Rookh*. Her first voyage from Cowes was to Belfast Lough where she arrived on October 16th, 1870, and was much admired. Prof. Thomson, David T. King, and some of the young Bottomleys were taken for a cruise to Scotland; and plans were made for a longer cruise in the following year. The next year Sir William Thomson took a party including Professor Helmholtz for a cruise, and a visit was made to Prof. and Mrs Blackburn at their beautiful home on the West coast of Scotland. The following extracts from letters from James Thomson to his wife give an idea of life on board the *Lalla Rookh* in the holidays. In Mrs Blackburn's cabinet of drawings there is a sketch of the group watching the flight of sea-birds, drawn in her own inimitable style.

"(LOCH AILORT), 8th Sept. 1871, OFF ROSHVEN,
Friday morning. On board L. R.

...We had a delightful sail with a pleasant breeze, and Atlantic waves moderated by shelter from Hebrides and Coll and Tiree. Then coming up into Loch Ailort (on south side of which Roshven Mountain and Roshven House, both belonging to Prof. Blackburn, are) we had shelter against waves from the Islands of Eigg, Muck, and Rum. Then we came farther up the Lough inside of Goat Island, and anchored in shallow water in front of Roshven house. We went ashore and saw Prof. and Mrs B. and their 3 sons and 1 daughter, also Mr Ferguson* Lecturer or Demonstrator of Chemistry in Glasgow College, under Prof. Anderson who is unwell for some years and has been mostly unable for duty. Also John Thomson†.

* [Now Professor of Chemistry, University of Glasgow.]
† [John Millar Thomson, now Professor of Chemistry, King's College, London.]

We all went to Goat Island; and climbed to top. It is very steep. On top there is a great vitrified fort of quite prehistoric times. Its walls go round a great enclosure like an Irish fort, only irregular to fit the rocky hill top. There is a natural grassy hollow like a shallow bowl in interior. The walls are only a few feet high, rough and irregular, and of course partly broken by people and by exposure generally. I have got a good specimen of the vitrified structure to bring home. It is of lumps of stone like the sizes of macadamizing stones and paving stones, but I think mostly under 3 inches thick. I did not break it off, but got it loose where it had been thrown or taken down the hill.

We arranged last night that if to-day would be fine enough the Blackburns and their party would come out to the *L. R.* and we would all sail in the yacht to Eigg where there is the very remarkable lava top called the Scuir of Eigg. The day is wet and very windy: and I suppose we shall stay partly in the yacht and partly on shore with the Blackburns: but this will depend on the appearance of the weather later in the morning or in the day.

The Blackburn party came on board and have kept their boatmen waiting till letters be written to send ashore: so I have written in haste. We bathe every morning. Most of us swim....

We saw some admirable drawings yesterday evening done by Mrs Blackburn. One was a drawing of a baby cuckoo shoving the other young birds older than itself out of the nest. She sketched it while the process was going on, and while she was watching the proceeding with almost breathless interest. That original sketch* she shewed us, and also a nicely coloured finished sketch picture of same, done afterwards on a paper about the size of a page of this letter. These I think are highly interesting; very few people have ever had an opportunity to see that proceeding: and of those few I am very sure no one has been such an artist as Mrs Blackburn, with such wonderful power to sieze and represent the appearance seen in its most characteristic moment...."

"Thursday 14*th Sept.*, 2 P.M., 1871.

In *Lalla Rookh*, off Easdale Island, and Seil Island, Firth of Lorne, in very nearly the latitude of South end of Mull.

...We set sail from Gair Loch early on Tuesday morning, with a good favourable breeze, and after great battling against strong tide flow in the Kyles where we were detained 2 hours, got almost into Tobermory Bay when the wind fell off completely, and we were becalmed, and could not get in, but remained floating

---

* [This sketch is published in *Birds from Moidart and Elsewhere* by Mrs Hugh Blackburn : Edinburgh, David Douglas, 1895. Cf. also *Nature*, 14 March 1872, and *Lancet*, 2 July 1892; also Dr Jenner, *Phil. Trans.* 1788, Vol. LXXVII. pp. 225, 226.]

in the Sound of Mull till next morning, the Captain and some of the sailors taking care to keep us from being carried ashore by tidal currents, and to be ready for taking suitable action if wind should come on. On the next morning (Wednesday, yesterday) we were almost completely becalmed except that we had light favourable breezes in the afternoon. William during the almost perfect calm, noticed the very slight speed of the yacht through the water as being fit to enable him to make experiments on ripples and waves regarding which he had been making out mathematical theories and discussing them with Professor Helmholtz.

He did it by having a nearly vertical fishing line hanging in the water; and observing by himself with aid of W. B. and of me and of J. T. junr. the conditions as to the mode of spreading of the waves, and as to the speed of the yacht that he wanted. He found the results very satisfactory to him: and I think he finds them suitable for publishing in a paper* he will be writing in which he has been making some quite new discoveries.

There was very little wind yesterday, and we only got on about 20 miles when evening came on and we anchored in a glassy calm in Duart Bay in Mull, where the Sound of Mull widens for a vessel going Southward. Shortly after we anchored two herring merchant sloops or wherries were moving slowly about the bay, one on each side of us, the masters of which kept up for half an hour or an hour an eloquent conversation in Gaelic across us. The whole scene was better than could be acted in a theatre. Our mate understood what they were saying and told us some of it. One had his ship loaded with herrings which he had bought in Loch Hourn off Sleat Sound, and was taking them to Glasgow to sell. I think he had bought them for 5s. the barrel, from the fishermen and expected to realize £1. 10s. 0d. ? or £2. 10s. 0d. ? the barrel for them in Glasgow...."

The gradual refraction of rays of light when passing through the atmosphere claims the attention of engineers in connection with levelling operations. James Thomson, dissatisfied with the existing theories then current, had discussed the subject in 1863 with Prof. John Purser in Belfast, when they agreed that an investigation by the usual method of rays, did actually, when pushed to the limit, prove curvature of a ray exactly horizontal in an atmosphere stratified horizontally, though the transition to this limit is somewhat perplexing and, in fact, had led to loose and erroneous statements on the subject. As a result of continued attention to the problem he sent in 1870 to his brother William a mode of investigation by means of wave fronts which placed the

---

* [Cf. Lord Kelvin's *Mathematical and Physical Papers*, Vol. IV. p. 88.]

subject in a new light, and which has become fundamental for all such questions relating to the gradual bending of trains of waves. This investigation was forwarded by William Thomson to Tait with the characteristic note: "This wants for perfect validity only the approval of T'!" Tait's letter to James Thomson in reply is given below (p. 451). The paper containing an account of Purser's solution was communicated by James Thomson to the British Association in 1872, and is here reprinted*: *infra*, p. 441.

In 1872 through the death of Prof. Macquorn Rankine the Chair of Civil Engineering at the University of Glasgow became vacant, and Sir William much desired his brother to apply. Anxious correspondence followed. Prof. James Thomson and his wife and three children had taken root in their home in Belfast. The Professor was doing very good and important work with his colleagues at Queen's College; his youngest daughter, Bessie, had passed through a severe illness in the summer of 1870 which left her permanently delicate; his wife too had been far from strong during the summer they had just spent at Castlerock. Living in Glasgow would be more expensive than in Belfast; so that unless the emoluments of the Chair were much greater than those at present at his command it would be wiser to stay where he was. Sir William however was able to overcome all these objections. Finding that his brilliant younger brother really considered him the best man to fill the vacant chair, the Professor could not resist. Sir William immediately set to work to promote the candidature. He drew up a statement of his brother's qualifications, particularly with regard to practical work, and, as it gives a clear, concise account of James Thomson's achievements as an engineer in Belfast, it is here quoted in full.

"Professor James Thomson was a pupil in the Horseley Iron Works Staffordshire, and afterwards a Pupil of Sir William Fairbairn at his iron shipbuilding works on the Isle of Dogs. He has been Professor of Civil Engineering in Queen's College, Belfast, for 15 years. For the last five years he has been Engineer to the Lagan Navigation in County Down. He was Engineer to the Belfast Water Commissioners, when pumping steam-engines for supplying water to the Town were erected under his charge.

* [This method of wave fronts is further developed by J. D. Everett in a paper "On the Optics of Mirage," *Phil. Mag.* 4, xiv. pp. 161, 248, March and April 1873. For the important application to rays of sound see Lord Rayleigh, *Theory of Sound*, Vol. ii. p. 288.]

A water pressure engine for supplying the town of Kilrea with pure spring water by power derived from a river, was proposed by him and executed under his superintendence. This was quite a novel design; and it has proved in the highest degree satisfactory.

He was inventor and patentee of the vortex turbine, renewed by Privy Council, and has designed and superintended the construction of many of these wheels now working successfully in various parts of the world: designed and superintended the construction of several large centrifugal pumps for drainage of sugar plantations in Jamaica and Demerara: has invented and brought into practical use the Jet Pump and Intermittent Reservoir for Drainage of swampy lands. This arrangement has been in successful use now for many years in County Monaghan near Clones.

He designed and had executed under his personal superintendence, a flood-sluice-weir on the Lagan River near Belfast, for the purposes of the Navigation. This, on account of the natural difficulties and contingencies of weather, was a critical piece of engineering and required good planning and energetic superintendence. It has proved quite successful.

He proposed an improved method of measuring the flow of rivers and streams by V notches. He worked out the dynamical theory of this method; made the requisite experimental investigations, and published practical rules for allowing of their being brought into general use. He has had ample practice and experience in Land Surveying, and Levelling, and has regularly given field instruction and exercise in this department to his pupils."

Of the many testimonials that were received, three have been selected as specimens, those from Professors Helmholtz and Tait and Dr Joule. The final extract is taken from a private letter written to Sir William by Dr Andrews, which accompanied the more formal testimonial.

"BERLIN, 14*th January*, 1873.

DEAR SIR,

As you wish to have a statement of my opinion regarding Mr James Thomson's merits as a scientific investigator and his qualifications for the chair of Engineering at the University of Glasgow, I have the honour to answer that I regard Mr James Thomson as a man of very sound and unusually acute judgement in questions relating to physical, mechanical, and mathematical science, and a very extended amount of knowledge in these same branches. His theoretical prediction of the alteration of the freezing temperature of water by pressure, was an original idea of first-rate importance. Also his hydrodynamical considerations, and their practical applications for the construction of water wheels, for his jet pumps, for the theory of oceanic and atmospheric currents, are of a very original turn of mind, and prove that he is able to follow out his speculations till to the very fundament of the question.

Besides I know from personal acquaintance his scrutinizing way to penetrate into the very heart of scientific questions, and I should think that such a man ought to be the very best teacher for young engineers.

Believe me to be yours most truly,

DR. H. HELMHOLTZ.

To Sir William Thomson, Glasgow."

"UNIVERSITY OF EDINBURGH.
7 January 1873.

MY DEAR THOMSON,

I am very glad to hear that your brother is a Candidate for the Chair of Engineering at Glasgow.

I was for some time his colleague in Belfast, and I can thus state of my own knowledge that he is an excellent teacher, and works in thorough harmony with his brother Professors.

As an original discoverer in Physics, and especially in the Dynamical Theory of Heat, he holds a very high place indeed. There are but few discoveries of the last fifty years which make such a mark in the history of science as that of the lowering of the freezing-point by pressure: for it has led, not only to numerous kindred discoveries in pure science, but to the true physical theory of glacier motion.

He has made many valuable contributions to Engineering, among which I would specially mention his Turbine and his Vortex Pumps, as being admirably based on true physical principles, and thoroughly successful in practice.

I certainly do not believe that there can be found in Britain any one so well qualified as he is to fill the place of our lamented friend Rankine.

Yours ever,

P. G. TAIT."

"343 LOWER BRAUGHTON ROAD,
MANCHESTER.
7/1/1873.

DEAR THOMSON,

I wrote a testimonial for Mr . . . . . two or three days ago but when I did so I had no idea that your brother would be able to leave Belfast. In the event of his seeking the Chair of Glasgow College, it is impossible there can be any serious competition, inasmuch as your brother's European fame as a

discoverer combined with his eminent success as a teacher of
Engineering science, are such as must place him altogether without
a rival.

Yours most truly,

JAMES P. JOULE."

"*Jan.* 7, 1873.

MY DEAR SIR WILLIAM,

      I hope the enclosed will meet your views.   As you
desired me to be very brief, I have avoided all details.   Your
brother will indeed be a sad loss to us all here—and to no one
more than to myself.   I have long been greatly attached to him,
and his departure will be a sad blank to me.   He is indeed a rare
character, so truthful and gentle and wise withal...T. A."

The candidature was successful, and Prof. and Mrs Thomson
spent the Easter holidays in Glasgow on a visit to Sir William.
On the 17th April he was inducted to the Chair of Engineering,
and he commenced his duties in the following November.

The professors at the University at that time were an excep-
tionally interesting and genial set of men, and the social life of
the College was delightful.   There was Principal Caird (an old
classfellow of student days)*, perhaps the greatest preacher in
Scotland of his generation, and certainly the most genial Principal
the University had had; Dr Dickson, Professor of Divinity, with
his marvellous knowledge of apparently all the books in the College
library; Hugh Blackburn, Professor of Mathematics, learned, gentle
and retiring, and his charming wife, vivacious and accomplished,
who could draw, from memory, any scene she had witnessed;
Lushington, Tennyson's brother-in-law, was in the Greek Chair;
George Ramsay, a nephew of Thomson's old professor, held the
Chair of Latin; Duncan Weir, another old classfellow, was Professor
of Hebrew; and the Principal's brother, Edward Caird, afterwards
Master of Balliol College, Oxford, was Professor of Moral Philosophy.

---

* In the Logic Class Session 1837—8 the following is the list of Prizemen in
the junior division:

1.	Robert Douglas	Kilbarchan (see pp. xxxvii, xxxix)
2.	William Thomson	Glasgow College
3.	James Thomson	Glasgow College
4.	William Ker	Glasgow (see pp. xxxix, lxxxvii)
5.	Claud Marshall	Greenock
6.	John Caird	Greenock
7.	Henry D. Taylor	Dumfries
8.	Peter MacFarlane	Paisley

All these lived in the Professors' Court in the dwelling houses attached to their respective chairs. Then there was the genial and enthusiastic Dr Grant, Professor of Astronomy, residing at the Observatory, where he was ever ready and willing to devote time and effort to showing the stars through the great telescope to any of his colleagues, or even to the youngest members of their families; and John Nichol, Professor of English Literature, was eager to renew acquaintance with the friend of his youth, and took pleasure in relating reminiscences of early days to James Thomson's children. There was Sir James Roberton, one of the law professors, a most charming conversationalist with endless information about people and events of Scottish history; also Robert Berry, afterwards Sheriff of Lanarkshire; Sir George McLeod, Professor of Surgery, tall and handsome, a brother of the celebrated preacher Norman McLeod, who had died shortly before; Allen Thomson, Professor of Anatomy, to whose exertions the success of the new University buildings was largely due; and many others.

The old somewhat narrow spirit with regard to affairs of religion had vanished; Principal Caird's sermons were not only eloquent, and full of deep thought and earnest feeling, but they inculcated a broad toleration. At the time of James Thomson's appointment to the Engineering professorship in Glasgow, his brother William was still a widower; but in the summer of 1874 he sailed to Madeira in his yacht the *Lalla Rookh*, and there married Frances Anna, daughter of Charles R. Blandy. Her home-coming brought new brightness to the College Court, and Sir William's lonely house was once more filled with cheerful company.

James Thomson's introductory lecture to the Engineering Class was a paper on Safety and Danger in Structures (No. 53, p. 349), urging the importance of more frequent and more consistent application of force tests. This subject had been before his mind for many years. It was suggested to him by Professor Lewis D. B. Gordon, who wrote to him on the 13th April 1862: "I wish you would write a neat paper on the principles which should decide the *coefficients of safety* to be used in the application of materials in construction. There is great confusion and error abroad on the subject." His attention being thus directed he read a paper on "Strength, Safety and Danger in Structures, with reference to

Amendment of Prevailing Practices" before the Belfast Literary Society in 1867. This paper is not reprinted here. A paper on "Comparison of Similar Structures as to Elasticity, Strength and Stability" was read before the Institution of Engineers and Shipbuilders in Scotland on the 21st December 1875 (No. 54, p. 361). Professor Archibald Barr has since extended and developed this line of thought in his paper on "Similar Structures" read before the same institution on 21st March 1899.

Early in 1874 Prof. James Thomson was asked to give the James Watt lecture to the Philosophical Society at Greenock, and selected for his subject "The Gaseous, Liquid, and Solid States of Matter." The report of this lecture elicited the following letters from Dr Andrews.

"QUEEN'S COLL., BELFAST, 5th Feb. 1874.

MY DEAR THOMSON,

I have received yours of the 28th and also the *Glasgow Herald* containing a notice of your Watt Lecture. I am glad that the experiments were successful, and that Bottomley was also able to show the liquefaction of carbonic acid to the Nat. Phil. Class. There is only one point on which I am afraid I should remonstrate. You seem in your kindness to have given my poor work a little too much prominence. However we are all vain and unprofitable servants (I have been reading lately 'Old Mortality') and bear a good deal of flattery.

I was confined for 10 days to bed with this rheumatic attack, but I am now nearly as brisk as ever. On the whole it was a luxurious illness.

Cumine*, poor fellow, lost his wife about 6 weeks ago, and attended her so faithfully that he brought on so serious an attack of his bronchitic disease as seriously to endanger his life. He has an order from Edinburgh for two of my compression apparatus, and I expect he will soon execute your brother's order as well as it. We have had an uninterrupted succession here of the finest weather since Novr. To-day is magnificent and the noise of the elections (if any) has not reached us. I was glad to see that Mr Gordon has been returned again by Glasgow and Aberdeen. Home Rule has been absolutely repudiated everywhere in Ulster.

Ever yours,

THOMAS ANDREWS.

P.S. Bessy and I are hard at work on Diderik van der Waals!"

* The mechanician who constructed Dr Andrews' apparatus.

The fundamental memoir of Johannes Diderik van der Waals —a Leyden degree dissertation—published under the title *Over de continuiteit van den gasen vloeistoftoestand, Academisch proefschrift* (Leiden: A. W. Sijthoff, 1873)—reduced the Andrews-Thomson diagram to analytical form, by giving for the system of curves an equation which applied over the whole range of the gaseous and liquid states. The thermodynamic content of the diagram thus became matter of quantitative deduction. The earliest important public notice of this work was an elaborate and appreciative, but withal critical, review by Clerk Maxwell in *Nature* (Vol. x., pp. 477—480, Oct. 15, 1874), reprinted in his *Collected Papers*, Vol. ii., pp. 407—415. The subject is also touched by Clerk Maxwell in the lecture to the Chemical Society, Feb. 18, 1875, "On the Dynamical Evidence of the Molecular Constitution of Bodies" (*Nature*, Vol. xi., p. 359), where, after referring to the exceedingly ingenious thesis of van der Waals, since amply verified, he proceeds "his final result is certainly not a complete expression for the interaction of real molecules, but his attack on this difficult question is so able and so brave, that it cannot fail to give a notable impulse to molecular Science. It has certainly directed the attention of more than one inquirer to the Low Dutch language in which it is written." As regards the phenomena of mixed gases, on which so much progress has more recently been made by the Dutch school, cf. also letters from Clerk Maxwell to Andrews in the *Scientific Papers of Thomas Andrews*, p. liv.

The following letter relates to an apparent discrepancy with theory:

"QUEEN'S COLL., BELFAST,
April 15, 1874.

MY DEAR THOMSON,
          ...It is strange that we both overlooked a chapter in Regnault's large work on steam (*Mém. de l'Acad.* 26. p. 751 et seq.) in which he investigated the tension of acetic acid both in the liquid and solid forms; and obtained the same result which I had done with the solid, freed as well as possible by crystallization from the liquid; viz. a higher tension for the solid than the liquid. But suspecting as I had done that a trace of liquid was still present with the solid, he distilled his acetic acid from anhydrous phosphoric acid, and with the new body he attained *your* result, a lower tension for the solid. He found however that a little acetone was formed at the same time, and concludes that the *differences* observed were

due to foreign matters. I think I could now arrange an experiment which would give decisive results; but I must wait for a more convenient season. In a recent No. of the *Annales de Chimie* there is I observe a paper on the subject. I have only seen the title. Let me have the apparatus, if you have an opportunity before the middle of June or beginning of July.
With very kind regards to Mrs T. and all yours

> Believe me to be ever yours,
>
> > T. ANDREWS."

> "QUEEN'S COLLEGE, BELFAST,
> > 16 *April,* 1874.

MY DEAR THOMSON,

...I am deep in the Dutch paper, and also fully occupied with my own results which I believe will turn out more important than I supposed. But I miss you sadly,

> Ever yours,
>
> > T. ANDREWS."

In 1874 the British Association met at Belfast, Prof. James Thomson returning as President of the Mechanical Section. He and Mrs Thomson were on that occasion the guests of their old friend Prof. George Fuller. A few weeks later the Social Science Congress met in Glasgow; and Prof. Thomson, who was living in a pleasant, roomy, old-fashioned residence "Oakfield House" close to the University, had the happiness of welcoming a number of Irish friends and relations to his new home. About this period, Sir William Thomson was engaged on the invention of integrating machines for the purpose of predicting the tides and, as usual, he talked the matter over with his brother James, who remarked "I wonder if my disc, ball and cylinder apparatus would help you." The drawings and descriptions of it were brought out and considered, and a model was made; and the result was the well known series of tide-calculating machines of Sir William Thomson *. James read a paper entitled "On an Integrating Machine having a new Kinematic Principle" before the Royal Society in 1876, No. 64, p. 452. This mechanism which his brother was thus able to turn to useful account in the harmonic analysis of the tides had been designed many years before. In the earliest form it had a wheel revolving edgeways on the disc instead

---

* See *infra* p. 452; also Thomson and Tait's *Natural Philosophy*, Part I. 1879, Appendix B. especially see pp. iii. iv. v.

T. *f*

of the ball; but the edge of the wheel partly slipping and partly rolling did not please him as a mechanical contrivance, and he substituted the ball, which, touching as it does only one spot, is free from any kind of slipping. In an old note-book there is a sketch of the early form dated August 1845.

Sir William Thomson brought the subject of the initial contraction of free liquid jets, as they issue from the orifice, before the Philosophical Society of Glasgow on 23rd February 1876, by communicating a letter containing a theory, now well known, of the *vena contracta*, which he had received from William Froude; and he appended an interesting letter of Sir Isaac Newton on the same subject, then recently published in the Cotes Correspondence. This led James Thomson to go over the ground with some extensions, and to communicate an interesting note on the *vena contracta* to the same Society (No. 15, p. 88).

In June 1876 Prof. and Mrs Thomson went up to London to see a special loan collection of scientific apparatus at South Kensington Museum in which both brothers had numerous exhibits; among others the Disc, Ball and Cylinder Integrator, and the model surface to illustrate Dr Andrews' experiments on carbonic acid.

In the same year he read before the British Association a paper on Metric Units of Force, Energy, and Power, larger than those on the Centimetre-Gram-Second System, and suitable for practical use, *infra* p. 372.

In 1878 he communicated a paper on Dimensional Equations and on related nomenclature in Numerical Science, *infra* p. 375, which aims at simplification and precision in the language of this fundamental subject.

In 1876 he wrote a paper (No. 16, p. 96) on another subject which he had been considering for several years, viz. the behaviour of water in river beds. The explanation there given, why the inner bank near which the velocity is greatest does not wear away more than the outer one, had occurred to him in 1872. The British Association was to meet, under the presidency of his friend Dr Andrews, in September, and a few days before the meeting commenced, it occurred to him to try the thing experimentally, in order to confirm the theory he had published in May. On his brother's lecture room table he constructed a clay model of a river bend, and turned on the water; and to his great satisfaction, he found that the particles at the bottom, sand and rounded seeds of various sizes, were carried across to the inner side of the bend exactly as he had said they would go, while little streamers of thread attached to pins showed the direction of flow at the surface of the water. A wooden model was made for the Exhibition in Kelvingrove Museum which was prepared for the British Association. He had intended to clear away the original clay model after showing it to his brother, in order to leave the lecture room table tidy for the use of Section A. But when Sir William saw the experiment he insisted on keeping the clay model as it was, in order to let members of the Association see the experiment and hear the explanation from the author himself. A short communication describing the experiments was sent afterwards to the Royal Society. Later, on the suggestion of Prof. Barr, the course of the stream lines was indicated by little specks of aniline dye introduced so as to adhere to the bed of the channel at various places. (See paper No. 18, p. 106.)

In 1877 James Thomson was elected a Fellow of the Royal Society. The following year the honorary degree of LL.D. was conferred on him by the University of Dublin; he had enjoyed the same honour in his own University of Glasgow since 1870. He also received the D.Sc. degree from the Queen's University in Ireland.

At the end of 1877 Graham Bell came to Glasgow, and caused much interest by explaining and showing to a large audience his wonderful invention of the telephone, of which Sir William Thomson had shown a rough example in his Presidential address to Section A of the British Association at the meeting in the previous year above referred to. He dined with the Thomsons, and he and the Professor had much interesting talk both about the telephone and about the system of visible speech invented for the use of the deaf by Bell's father, Melville Bell. Prof. Thomson was particularly interested in this, as both he and his father had been strong adherents of phonetic spelling from the time Isaac Pitman had first introduced it into this country.

As has been stated above, James Thomson studied many geological questions. Thus he minutely examined the jointed prismatic structure of basalt as seen at the Giant's Causeway in Ireland, and elsewhere; and he concluded that the older theories for explaining this phenomenon were untenable. As

*f*2

early as the 12th September 1861 he advanced, in a letter to
Dr Neilson Hancock, the germ of his own theory, which, after
much study, he gradually matured. Papers on this subject were
read before the Belfast Natural History and Philosophical Society
on 26th November 1862, before the British Association in 1863,
and before the Belfast Naturalists' Field Club on 17th November
1869. None of these papers are reprinted here, as the whole
theory was summed up in the paper read before the Geological
Society of Glasgow on the 8th of March 1877 (No. 62, p. 422).
It explains the columnar structure as being the result of shrinking
and cracking of the material during solidification; and it holds
that the cross joints are, in reality, circular conchoidal fractures
commencing in the centre of the column and flashing out towards
the circumference, and that the tendency to cross fracture arises
from the expansion of the outer parts of the columns through
chemical action caused by infiltration of water.

During the sixteen years of James Thomson's tenure of the
Glasgow Professorship, the six months' vacation not only left
leisure for scientific work and engineering business, but enabled
him to spend several months annually at one or another of the
beautiful places of summer resort on the Firth of Clyde, the
Kyles of Bute, or Loch Long. At several of these places the
traces of glacial markings are visible. He used to take great
pleasure in studying these, and in particular in observing every
indication which might show the direction of the motion of the
ice. The result is recorded in the paper "On Features in Glacial
Markings noticed on Sandstone Conglomerates at Skelmorlie and
Aberfoil," No. 61, p. 420.

As one might well imagine, Professor Purser's visits, whether
in the country or at Oakfield House, were always a great pleasure
to the Thomsons. The following extracts from some of his letters
speak for themselves of the affectionate regard with which the
younger man approached his older friend and cheered him to the
end.

"*25th June* 1879.
RATHMINES CASTLE, DUBLIN.

Fred was greatly pleased to receive the warm congratulations
you all have sent him. The more so as he knew you were amongst
the few who appreciated the difficulties of his position and approved
of his action when seven years ago he was obliged to forfeit the

fellowship he had won*....I should like very well to go to the
B. Assn. meeting for a part of the time at any rate, and help to
support our friend [Dr Johnstone] Stoney in Sec. A....

I confess to a strong feeling of satisfaction that my last visit
to Oakfield House has not been without its fruit, when it has
so completely revolutionized the mind of Dr Thomson on the
subject of $\sqrt{-1}$, that whereas on that occasion his attitude towards the 'imaginaries' was one of such derisive scepticism as
to express doubt about the well recognised truth of two circles
meeting in two imaginary points at infinity, now on the other
hand he affectionately adopts these imaginaries and they are
henceforth to be no longer ostracised but lovingly included in
that great family who are to rejoice in the name of numerics!
See Note to Thomson and Tait, new edition p. 389. 'Quod
facit per alium facit per se.'

Fred is rejoicing in the change of his 'rôle' from an examinee
to an examiner, and has been busy the last few days examining
in the *omne scibile* as comes natural to a junior fellow."

"May 13, 1886. Professor Thomson devoting himself to the
serious study of Hegelianism is a phenomenon!"

The following extracts from letters from Mrs Thomson to
her daughter, Bessie, describe two cruises taken during the
summer of 1880, the first in the *Lalla Rookh*, the second with
Dr Young† in the *Nyanza*.

"Saturday, 3 *July* 1880. OBAN.

Went ashore after kettledrum‡ to post letters, then had a
beautiful drive to Dunstaffnage and on up Loch Etive, past
Connel Ferry where the Falls of Lora or Connel are seen when
the tide is ebbing. This current (or rapids) is caused by the
narrow exit to which the great extent of Loch Etive is confined,
and the large quantity of fresh water accumulated in it during
the flood tide, and partly due to a reef of rocks. It was blowing
quite a gale from the N.E., so we were glad to be at anchor in
Oban bay. There was a faint red gleam at sunset; it was cold
but we had tea on deck and turned in at 10.45.

* In 1879 Frederick Purser, afterwards Professor of Natural Philosophy in
Trinity College, was elected a Fellow of Trinity College, Dublin. He had won a
fellowship there seven years before, but at that time the religious tests enforced
at Trinity College formed a barrier to his making the necessary declaration
on accepting it, because he was a Moravian. Seeing that the doctrines of the
Moravian community differed but little from those of the Church of Ireland, some
of his friends had tried to persuade him to accept the tests; but he felt that such
a course would not be right and Thomson sympathised in his scruples.

† Dr Young of Kelly, the friend of David Livingstone, introduced the manufacture of paraffin oil from shale in Scotland. Dr Angus Smith of Manchester and
the Rev. Dr Robertson of Irvine were also of the party.

‡ Afternoon tea.

Sunday, a lovely bright windy day, clear sky and more sun-shine than we have seen since the day at Iona on 22 June. Went to church at Oban and had a little walk before coming on board. We hope to have a walk on Kerrera after kettledrum. We start tomorrow for home and will go either outside Islay or down the Sound between Islay and Jura as we find the weather will permit; any way we expect to be back in Largs by Tuesday some time....

E. T."

" KYLE AKIN, SKYE. 5.45 P.M., Saturday 7 Aug.

Left Isle Ornsay a little after one today and sailed up the Kyles with a bright sun and a clear sky but not much wind. We got on nicely through Kyle Rhea and Loch Alsh and had a pretty view of Loch Duach and Loch Long. We then entered the narrows of Kyle Akin when the wind failed and the tide had previously turned against us at a critical part just opposite the lighthouse. We got a boat out to put her head round and had to get out the anchor to hinder us going aground, so here we are in Kyle Akin and as the tide won't turn for some hours I suppose we are here for the night instead of getting on to Portree...."

" KYLE AKIN ; Sunday, 8 August 1880.

This has been a showery day with bright gleams of sunshine. The scenery is very beautiful, mountain, island and sea. We in-tended to go ashore to hear the Gaelic service in a mission church here. (The regular church is at Broadford 8 miles away.) Dr Robertson and Dr Smith were to go for the whole service, while the rest of us intended going to hear the end of the Gaelic service and stay for the English service, but there was no minister here and no service. The Missionary had gone to assist at a Communion service at Balmacara, so Dr Robertson was asked to preach ; and they hailed us from the beach to come ashore which we did and a congregation was got together about 54 individuals. Dr Robertson gave a very interesting sermon or rhapsody, which though not logical carried one on by the eloquence of the man and the poetical declamation. It all lasted about an hour....

Dr Young poured some oil over the side of the yacht and we watched the calm spot for a long distance and saw what we had intended—the gulls taken in by this calm spot and come to it in search of food."

In 1882 another move, the last in Thomson's life, was forced on his family, as the site of Oakfield House was wanted for the building of a church. Mrs Thomson was then suffering from a weak heart, and the removal to 2 Florentine Gardens was con-ducted with great difficulty. An extract from a letter from

Mrs Elizabeth King to Mrs Thomson recalls that period, and is interesting as it gives Ruskin's opinion of her daughter Elizabeth King's skill in portrait painting. The next letter written a year later gives a graphic picture, in her daughter Agnes' words, of the heads of the Thomson household at that time.

"15 BLOMFIELD TERRACE, *May* 2, 1882.

I am heartily glad that the house affair is so pleasantly settled and that you have got the house you liked the best of all you saw. I wish you much happiness in it with all heavenly blessings. You will have a little while to enjoy the garden of Oakfield before you leave, for which I suppose you will not be sorry; and then will come the confusion and fatigue of moving, which certainly is not enjoyable, but when it is over it is very pleasant to feel settled and in order....

I must tell you that Ruskin has been admiring greatly the portrait of James. He went to the Dudley with Mr Murray of the British Museum, a gentleman George knows, and in walking along he paused and remarked that was an excellent picture, and desired Mr M. to turn up the catalogue to see who did it."

"*April* 23, 1883.

...Agnes brings a pleasant report of you. She says that though not quite so strong as usual, you are so bright that you are the life of the house. The house she says is beautiful and so comfortable. Uncle James she says is full of fun and always making merry jokes. Altogether she has had a very happy time —a time that she will look back to with pleasure all her life. It is a possession for ever...."

An extract from a letter dated October 1, 1883, from Professor Thomson to his wife alludes to the "Law of Inertia" on which he was writing a paper, No. 57, p. 379.

"...We have been at the College with William and Fanny. We intended to stay very short and to post letters at Wilson St. Pillar before 10.50: but there arose so much to speak of and to listen to that the time for posting went past....William says he approves of the view I had expressed to him as to Cayley's non-Euclidian, or dreamland, space. He has tonight quickly glanced at my last writings about the 'Laws of Motion,' and so far as brief time permitted consideration, he seems to approve of my teaching as to the laws of motion, or what I designate as *The Law of Inertia.* We spoke a good deal about Typhoid Fever organisms or germs &c.

Your loving husband, J. T."

In 1884 James Thomson was elected President of the Institution of Engineers and Shipbuilders in Scotland. In his presidential address he gave his views on the fundamental principles of the kinetic branch of dynamics, a subject on which he had long felt the need of improvement in our modes of thought. "Lately," he said, "I have been able, I think, to clear up some parts of this subject a little, and I have within the present year submitted papers upon it to the Royal Society of Edinburgh [see Nos. 57, 58, pp. 379, 389] and my sayings to you this evening will include some passages from those papers." At the end of the address he reverted to the question of safety in engineering structures and insisted once more on the importance of more frequent and more consistent application of force tests, on which he seized every opportunity to influence the practice of engineers. He dealt with the same question in more detail the following year when he was again President. These views he repeatedly expressed on public occasions, and insisted on in private conversations; at first they met with considerable opposition on the part of practical engineers, but they gradually moulded public opinion in the direction of more definite criteria of safety.

During the summer of this year (1885) James Thomson went up to London to give evidence, along with his brother William, on the Manchester Ship Canal Bill which was then before Parliament. He stayed with his sister, Mrs David King, at 24, Hamilton Terrace. Extracts from letters received by Mrs Thomson from her husband, and from his sister, show how he passed the time while there.

"24, HAMILTON TERRACE, *July* 16, 1885.

James is off to spend the evening in the House of Commons. Mr Brown* he expects to introduce him. He asked me to write as he had no time. This morning he was at Thames Street seeing about a stove for the studio, and he thinks he has found a good plan for it. He is very bright and happy looking, and he says he thinks his eye is about well. There have been no 'commas or tadpoles' troubling him of late, only sometimes a little dimness. He won't be going home for a few days, as he must have a little play now that his work at the Committee is finished. He is to take our telescope with him, I do not need it at all....

Your affecte. sister, E. K."

* Now Sir Alexander Hargreaves Brown.

"YACHT *Lalla Rookh*, R.Y.S., COWES, WEDNESDAY, 22nd *July*, 1885.

We came here yesterday afternoon from Portsmouth Harbour having a very pleasant sail in the Solent.

.   .   .   .

We were at anchor off Cowes during the night. Along with E. King I have been intending to go back to London to-day; and then to return home to Scotland with as little delay in London as possible. I have been feeling some incipient uneasiness on the score of liability to accusations of wife and family desertion: as to which I got some premonitory suggestions in Mary's letter of Saturday last. Wm. has however this morning persuaded me to stay till tomorrow (or possibly till Friday morning) so as to continue the cruise to..."

"YACHT, *Lalla Rookh*. At anchor, COWES.
Thursday, 23rd *July*, 1885.

There are great festivities going on all round for the wedding of the Princess Beatrice. The actual wedding ceremony is to be today at 1¼ p.m. or therabouts. The Prince of Wales' steam yacht *Osborne*, and the Queen's steam yacht *Victoria & Albert*, are at anchor very near us. The Prince of Wales has just gone ashore in a small steam launch which we saw passing near us. Capt. Fisher* who has been away in the evolutionary squadron round the coasts of Ireland, has just arrived in the English Waters at Spithead, and has come on here without any landing at Portsmouth. Wm. spoke him as he was going ashore a few minutes ago to Cowes to be at the wedding. Mrs Fisher and her three daughters have now come on board the *Lalla Rookh* and are on deck. Miss Fisher is much interested in drawing, which she has been learning at the Gosport School of Art†. She is involved in the mysteries of perspective examples, taught by rules mostly (Qy all?) not understood:—but I suppose knowledge is longed for‡...."

About this time politics were attracting much attention. In the winter of 1885 a *ballon d'essai* was sent up by Mr Herbert Gladstone (now Viscount Gladstone) suggesting the adoption by the Liberal Party of the policy of Home Rule. James Thomson immediately perceived the danger impending and pointed it out very clearly to his friends. He used all his influence to oppose the scheme, both then and later when it was adopted as the official

---

* Now Admiral of the Fleet Lord Fisher.

† Miss Fisher designed one of the supporters to Sir William Thomson's Coat of Arms when he was raised to the peerage, and his niece Bessie Thomson designed the other.

‡ Professor Thomson had always been much interested in perspective; finding the teaching in books on Art in his day very defective, he planned out a book on the subject but unfortunately did not live to completed it.

policy of the Liberal Party.   It may be said that his and his
brother's influence was largely felt in Glasgow and the West of
Scotland, where the most powerful Liberals ranged themselves
with the Liberal Unionist Party then being formed by Lord
Hartington and Mr Chamberlain.

To Mrs Thomson.

"*July* 7, 1886.   Wednesday, 5.45 P.M.

I got to Glasgow all right yesterday.   I posted the Post cards
at the General P.O.   I dined here with Fanny.   Wm. was gone
to a meeting at Johnstone.   Then I went in haste to a meeting
of Craig Sellar's in the Burgh Hall, Partick.   I have been out
with Wm. at a good many places before luncheon.   Have been
canvassing in Athole Gardens since.   Everybody (almost) away
except in evenings and mornings.   I am now hastening to the
Western Club to dine there with Mr Crichton in order to go
with him to a meeting in Borough Hall, Kinning Park, of
Mr Shaw Stewart's candidature for East Renfrewshire.   It is
likely that I shall have to make a short speech, no more than
to second some motion.   Mr Shaw Stewart is a moderate con-
servative, and I may do good as a decided liberal favouring him as
unionist.   In haste.

J. T."

The next letters from Sir Wm. Thomson show how in their
scientific as in their political work the two brothers kept in close
touch to the end.

"*Aug.* 15/87.

NETHERHALL, LARGS, AYRSHIRE.

DEAR JAMES,

I have been under intense stress since you left us, till
yesterday morning when I got off a 3^d & last instalment of a paper
for the Sept^r *Phil. Mag.*[*] on the stability of laminar motion of a
uniform broad river on a plane bottom; and so have been prevented
from thanking you sooner for your two letters with enclosures all
of which came most opportunely and acceptably.   I return the
Ice Abstract[†] which came before Stokes left, and we were both
much interested in it.   It settles absolutely Tyndall's & Huxley's
contention *against* Forbes, and shows conclusively that there is
viscosity, even without your thermodynamic considerations.   So
Forbes is now proved quite unqualifiedly to have been right.

I am now going on to the 'turbulent' motion of a river or of
water between two planes, & I am very glad to have your 'Uniform

[*] *Math. and Phys. Papers,* Vol. IV. p. 308.
[†] "Note on Some Experiments on the Viscosity of Ice" by J. F. Main, M.A.,
D.Sc., *Proc. Royal Society,* April 1887.

Regime *.' I shall carefully keep both the printed articles, & return them to you as soon as I have done with them, which I suppose will be as soon as my paper is read at the B. A. in Manchester, or as I have got the MSS. of it sent off to the *Phil. Mag.* which I hope may be before the meeting. But now I want very much a practical formula for the flow of water in as smooth a channel as may be. Do you remember the coefficient of 'wetted perimeter' which we used, in the investigation before the House of Commons Committee on the Manchester Ship Canal? If you have not it by you I could easily get it by writing to Deacon. I want to compare the actual flow with what it would be on the same slope if the motion were laminar (rectilineal). Did I tell you that I found Froude's resistance to a plane 50 feet long moved at 10 ft. (300 centimetres) per second through water in a wide deep canal is the same as it would be if the fixed plane were 1/30 of a centimetre from it, & the motion laminar? A corresponding comparison as to the flow of a river will be most interesting & important...

Your Aff^{te} brother

WILLIAM THOMSON."

"*Aug.* 18/87.

NETHERHALL, LARGS, AYRSHIRE.

DEAR JAMES,

      I am greatly obliged to you for your 'Notes'! They give me exactly & all what I want. I wanted the circular tube also, and was going to ask you for it next, but you have happily anticipated me.

      Weissbach's formula for the tube agrees beautifully with Osborne Reynolds' conclusion that resistance $\propto V^{1\cdot73}$ for velocities above those for which the flow is rectilineal, in a straight tube of circular cross section. For, taking

$$\eta = \frac{\zeta}{8}\left(AV^2 + BV^{\frac{3}{2}}\right),$$

we find

$$\frac{d(\log \eta)}{d(\log V)} = \frac{2AV^{\frac{1}{2}} + B \times \frac{3}{2}}{AV^{\frac{1}{2}} + B}$$

$$= \frac{2V^{\frac{1}{2}} + \frac{3}{2}B \div A}{V^{\frac{1}{2}} + B \div A} \doteqdot \frac{2V^{\frac{1}{2}} + 1\cdot8}{V^{\frac{1}{2}} + 1\cdot2}.$$

Hence when $V = 1$ foot per sec.,

$$\frac{d(\log \eta)}{d(\log V)} = 1\cdot727;$$

* See infra, p. 106.

and through a wide range of velocities the difference of $\dfrac{d(\log \eta)}{d(\log V)}$ differs little from Osborne Reynolds' value 1·73. It increases from 1·5 when $V = 0$ to 2·0 when $V = \infty$.

I had come to the conclusion that if resistance $\propto V^n$ be taken, as Reynolds (& Froude) took it, as an empirical formula, n must probably increase towards, or to, 2, as V increases. The mathematical theory of Turbulent Flow between two planes is coming out beautifully, and I am almost sure now that it will bring out your, and the observational, result, of max^m velocity at some distance below the surface of a broad open (not ice-covered) river flowing in uniform regime over a uniformly inclined plane bed. I enclose a rough proof which you may burn[*]. Read, if you care, the § 41 & the non-mathematical part of § 48 and § 49.

Your affcte brother, W. T.

I am now at work on the 'Turbulent Flow' for the B. A. meeting & the Oct. number of the *Phil. Mag.*

Fanny joins me in love to all. It is a great comfort and advantage to me to have your 'Notes.'

P.S.—Osborne Reynolds founds entirely on Darcy & Bazin's observations besides his own. He does not use their empirical formula and uses their results very unclearly and unexplicitly. Altogether although his paper is very important it is made quite needlessly difficult. Do Eitelwein and Weissbach depend on Darcy and Bazin?"

Long previously a correspondence on the modulus of viscosity of water, which may here be reproduced, had passed with a view to preparation for Thomson and Tait's *Natural Philosophy.*

"GLASGOW, Feb. 19/62.

MY DEAR JAMES,

...I write now chiefly to ask you about the flow of water or other fluids through pipes. The theoretical expression I have just worked out, and I find it to be

$$Q = \frac{\pi}{8\mu} \cdot \frac{h r^4}{l}.$$

h. the head of liquid.   r. radius of tube.
l. length. μ. coeff^t measuring the viscosity.
Q. the number of units of volume flowing per second,

when the motion is so slow as to be *steady*, i.e. simple slipping in parallel layers, without eddies or inequalities. I can easily send you the proof if you please.

μ denotes the tangential traction per unit area required to keep a plane layer of fluid, of which one side is held fixed in such

---

[*] *Math. and Phys. Papers*, Sir W. Thomson, Vol. IV. pp. 330, 335.

a state of regular internal sliding (i.e. continuous change of shape) that portions of it at unit distance from the fixed plane shall move at unit velocity. I call μ simply 'the viscosity.' Stokes derives its value for water from Coulomb's experiments on the oscillations of discs under water, when influenced by the elastic force of torsion. The rate of decrease of the arc of vibration is the subject of measurement. When this comes to bear a constant proportion to the amount of the arc at any time, the resistance must be proportional to the velocity simply, & the motion must be regular, free from eddies, and consisting of pure, steady, internal slipping or continuous change of shape by motions in circles parallel to the plane of the disc and having their centres in its axis. But I suspect the effect of eddies and irregularities cannot have been perfectly kept clear of, because the value of μ deduced by Stokes seems too great. It is $\cdot82 \times 10^{-5}$ when 1 square inch is the unit area and 1 inch head of water the unit pressure. Substituted in the preceding formula, this gives

$$Q = 3000 \frac{hD^4}{l}$$

where $D = 2r$ (the diameter of the tube). Now it appears that Poiseuille by pure experiment arrived at the theoretical law $\frac{hD^4}{l}$; and, reducing his formula to the units stated above I find it to be

$$Q = 3400 \left(1 + \cdot03368t + \cdot000221t^2\right)\frac{hD^4}{l}.$$

This is 10 per cent. *too great* a discharge for Stokes's μ to allow, even at temperature 0° Cent. At 10° or 15°, the probable temperature of Coulomb's experiments, the discrepance would be considerably greater, as Poiseuille's result indicates a rapidly decreasing viscosity as temperature rises.

G. Hagen arrives at very nearly the same result, so far as the cases in which his tubes were long enough. His value at freezing for Q, is just 1 per cent. less than Poiseuille's. A little above 0° Cent. it will agree perfectly, and still higher it will be somewhat larger than Poiseuille's. This seems to confirm P.'s result and to show that Coulomb's gives too great viscosity.

On the other hand, Stokes finds that Dubuat's observations on the times of vibration of pendulums oscillating under water lead to considerably smaller value of μ than Coulomb's, and I daresay may agree with Poiseuille's. But Bessel's results on pendulums agree very closely in the μ Stokes finds from them, with Coulomb's; and hence I suppose it was that Stokes preferred Coulomb. I want to give good information on this in the Book (Tait and Thomson's elements of *Natural Philosophy*) and I should be much obliged by any information you can give me; including the formulae you use and consider best.

Your aff^{te} brother W. T."

From William Thomson to James Thomson.

"*Feb.* 25/62.

I have looked up Helmholtz's paper since I wrote last; and I have no doubt that he is right that there is finite slipping of water over some solids, for instance, polished gold or silver. H.'s experiments are of a most unobjectionable character—the time of oscillation and the decrease of arc of a hollow globe filled with water, and made to oscillate by torsion or bifilar suspension, about a vertical diameter—the time of vibration was about 24 seconds. The decrease of the arc was in *perfectly constant proportion*, which proves the resistance of viscosity to have been simply as the velocity and no disturbance to have taken place from eddies or other irregular motions."

A Post Card.

"NETHERHALL,
Sunday morning, *Aug.* 28, 1887.

Many thanks for copy of statement regarding 'viscosity of metals *' which is very clear and satisfactory. Remember me kindly to Pursers, and tell them the turbulent flow (rivers) question has led me to a solution for propagation of laminar motion through an infinite turbulently moving inviscid liquid (the magnitude of orbits and other dimensions of the turbulent motions exceedingly small) which gives velocity of propagation of a plane wave (as of light in the luminiferous ether) $= \frac{1}{3}\sqrt{2}$ of mean velocity $\sqrt{}$ mean square of the turbulent motions. This seems almost to establish the vortex theory for the constitution of the interstellar ether. W. T."

In 1888 when Prof. and Mrs Thomson were staying at Coulport on Loch Long, her brother Dr Neilson Hancock and his wife came over from Ireland to spend the summer with them. The two men had been friends before they became related to each other by marriage, and there had always been a close bond of union between them. A few days after their arrival on a beautiful evening in July, Dr Hancock was seized with angina pectoris, and in twenty minutes he was dead. Thus was sudden gloom cast over the home that had been so full of happiness. Their children

* See infra, p. 406.

noticed that after this, both Prof. and Mrs Thomson began to feel
their years. Still, the same patient spirit which had tided them
through earlier sorrows and anxieties did not fail them now, and
their gentle resignation was a lifelong lesson to the younger
members of the family.

Prof. Thomson continued the lectures to his classes during
the succeeding winter. He contributed a paper to the British
Association on "Flux and Reflux" (No. 20, p. 123), his last
published work on the motion of water in rivers.

The following summer was spent at Toward Point, the
widowed sister-in-law being one of the family party. In the
autumn, shortly before the time for returning to Glasgow, a new
calamity befel Prof. Thomson, the failure of his sight. The
retina became detached in the middle, with the result that in
a few days he could no longer see to read and could only write
with difficulty, because the part of the page before him directly in
the middle of the field of view seemed always to disappear, or to
become so distorted, that the words written on the paper could
not be distinguished. Happily total blindness never came on;
even to the end of his life he could see light and colours and could
to a certain extent recognize the faces of friends. When he became more used to the deprivation of clear sight, he learned to
write with a blunt black pencil on large sheets of cartridge paper,
or better still, with chalk on a large slate, for his wife or one of
his daughters to copy. Dictating always seemed to be difficult to
him. By the aid of a magnifying glass and by directing his eyes
a little above or below the thing he wanted to examine, he contrived sometimes to study a diagram or a formula which thus had
its image on an uninjured part of the retina. The immediate
result of his failure of sight was that he felt obliged to resign his
professorship. Under this affliction his wonderful patience again
asserted itself. He never complained nor was the sweetness of
his disposition ruffled in the slightest degree. He still employed
his mind with scientific work, even under all the inevitable dis-
advantages, and took relaxation in listening to the reading aloud
of literature, and continued his interest in politics. The visits of
his brother and of the friend of his youth Mr William Ker, who
came to see him almost daily and walked out with him, were a
great source of pleasure. There was also much happy intercourse
with Prof. William Jack and others among his former colleagues,

who managed even in the midst of the strenuous work of the College session to find time to come frequently to see him.

It was a source of great gratification to him when his former assistant Prof. Archibald Barr was appointed as his successor in the Engineering chair. Prof. Barr had been one of his students, and has taken many opportunities to express appreciation of his teaching; the following is from a letter written to Mrs Thomson just after his appointment as Professor of Engineering at Leeds, in 1884.

"I have taken this opportunity of writing to you, because I cannot allow myself to settle down to my new work without expressing to you my sense of the great kindness shown to me by yourself and family during the past eight years.

I can assure you that your kindness has been much felt by me, and will always be gratefully remembered.

It will be needless for me to say how much I am indebted to Professor Thomson. If I am in any way fit for my present position I owe it—I need hardly say—to him. I took the opportunity offered by my opening evening lecture to acknowledge somewhat publicly in Leeds how much of my knowledge of Engineering Science I owe to Prof. Thomson. I find it difficult to make any statement of a fact or a principle to my classes without echoing what I have learned from him, proving that I am one of the many echoes of which Goethe speaks in the quotation which I made use of in my lecture. I can only hope that I may be a faithful echo, of which there are, I fear, very few*...."

Prof. Thomson continued to take a great interest in the welfare of the Engineering department. He had some correspondence with his old colleague Prof. Purser on the subject of how practical work might be combined with the theoretical teaching of engineering. Prof. Purser has written as follows:

"QUEEN'S COLLEGE, BELFAST,
2 Dec., 1889.

In Medicine the students learn theory and practice simultaneously, walk the hospitals while they are attending lectures. It would appear very desirable that they should follow a similar plan in Engineering. FitzGerald† tells me there would be great difficulties in this, that in places like Harland and Wolff's or Combe and Barbour's young men who did not give them their

* "There are many echoes in the world, but few voices." GOETHE.
† Prof. Maurice FitzGerald, George Fuller's successor in the Engineering Class at Belfast.

whole time would be very much in the way, and would idle the other pupils, and that any plan for practical millwright work in College would be very costly."

On the 11th December, 1889, a dinner was given in honour of James Thomson and two of his colleagues, John Nichol and Richard (afterwards Sir Richard) Jebb, who were all three resigning their chairs in Glasgow University, the last named having been appointed Public Orator in his own University of Cambridge.

Mr James A. Campbell, M.P. for the University, in the course of his speech said:

"I have now the honour to propose the toast of the evening— 'Our Guests.' We are assembled to-night to do honour to three distinguished gentlemen who have lately retired from the position of professors in our University. For many years past they have exercised the function of professors with great credit to themselves and signal advantage and honour to the University. And now that they have demitted office—two of them, I regret to say, because they were no longer feeling equal to the strain of its duties—it becomes us to take the opportunity of acknowledging our gratitude to them, for we know what their services have been —not only our gratitude, but the gratitude of many others who sympathise with us—and also to wish them health and happiness in the future...."

Principal Caird in responding at the close of the dinner said:

"* * * The resignation of a professor is not always an unmitigated misfortune to a University. It is hard indeed for any man to believe that his best days are over—to let the conviction dawn on his mind, for instance, that the same old lectures that have so often done good service in the past are a commodity that by long exposure to the air has lost much of its original pungency. Yet though it is not in nature that a man himself should find consolation for retirement from office in the thought of lessened efficiency, other people may. Need I say that on the present occasion no such consolation can be ours? In the case of Dr James Thomson our regret for his retirement is indeed deepened by the fact that it is due, as its immediate cause, to a serious, but we all hope, only temporary physical ailment. That it is not due to any abatement of intellectual activity, of scientific interest or ingenuity, or of delight in expounding to others the results of his thought and research, any one, who, like myself, has visited and conversed with him since his retirement, can bear confident witness....

But gentlemen, let me say in conclusion it is not in his writings, however excellent, that the best memorial of a great teacher is to be found. We can perhaps recall the names of

honoured teachers who, whether from lack of literary ambition or of
productive activity of mind, have done little to enrich the literature
of their subject. But be their books many or few, it is not in these
that the teachers whom we revere and honour, and on the thought
of whom we love to linger, have left in our minds the deepest impression of their powers. By their mastery of their subject and
their power of awakening something of their own love for it in other
and younger minds, by their marvellous skill in the art of communicating knowledge, their forbearance with the dull, and stimulating
sympathy with the bright and clever, by their patient and un-
grudging expenditure of time and labour on our behalf, and above
all, by the impulse derived from daily contact with the personality
of a great teacher,—by these and similar means they have created
for themselves in the intellectual and moral life of their pupils
a memorial better and more precious than books, graven, while
memory lasts, in the living tablet of human hearts and minds."

For nearly three years after his resignation Dr Thomson lived
happily and quietly. He now found time to complete his work on
a subject which had been long before his mind, i.e. the theory of
the grand currents of the atmospheric circulation. His first paper
on this subject had been read at the British Association in 1857
(No. 26); another one given to the Belfast Natural History and
Philosophical Society on 6th April, 1859, was similar in terms and
has been omitted here. Later on, the same topic formed the
subject of the "James Watt Lecture" delivered before the Philosophical Society of Greenock on 15th January, 1886. The paper
on Whirlwinds and Waterspouts read at the British Association
in 1884 (No. 27, p. 148), treats of a part of the same general
theme. In his opinion the diminished pressure in the centre of a
whirlwind is the cause rather than the effect of the rapid whirling
motion, as against the view of other writers on this subject that
the diminished pressure was due to the gyratory motion of the air.
He urged the encouragement of accurate observation, and the
collection and scrutiny of observed facts and appearances, as a
foundation for careful theoretical consideration. The last paper
he wrote, when his eyesight was nearly gone, was a recapitulation
of all his work on Atmospheric Circulation; this valuable Memoir
(No. 28, p. 153), obtained the distinction of being chosen to be the
Bakerian Lecture of the Royal Society for 1892, and was read on
the 10th March of that year by his brother, then President of
the Society. The following extracts from two letters received
from Prof. John Purser show a sympathetic interest in the work.

"*Feb.* 10, 1891.

I am very glad to learn that you are taking up again your old work on Atmospheric Circulation, and that you are proposing to put together what you have written and what you may now add in a new paper. I think I discern your brother's direction turning your thoughts towards this subject."

"RATHMINES, DUBLIN,
26 *Dec.* 1891.

I am much interested to hear that the paper on Atmosph. Circn. is going forward and that this enquiry is leading to these others about the phenomena connected with the rotation on their axes of the Sun and Jupiter."

Barely two months after the Bakerian Lecture, on May 8th, 1892, he passed away, his active and courageous mind still in pursuit of knowledge. Within a week his youngest daughter Bessie and his beloved and devoted wife followed him.

A most fitting close to this sketch is afforded by the words of Principal Caird in conclusion of his opening sermon to the University of Glasgow in November, 1892. They allude to the three Professors who had died in that year, Sir George McLeod, Dr Grant, and Professor James Thomson.

"I have no intention to-day to lay the individual character and career of each before you, further than to say in a single sentence that they were one and all men of no little grade in scientific attainment and of most enthusiastic devotion each to his special department, whether as teachers or as original enquirers. All of them were men of pure and lofty character, who drew to themselves the admiration and respect of all connected with them, whether as colleagues or pupils. We, who have been for long years associated with them in this place, can never cease to cherish their memory. They were devoted to noble ends, and of untiring, unflagging application, and pure and blameless lives. Can such lives be for ever lost to the world? Are they, and such as they, given to God's Universe only as shadows appearing for a little and then vanishing for ever away? Has all this lofty intelligence been created only for a few short years of a life—short at the longest—and then to be flung recklessly away? Can we hold our belief in God after such a thought? Do you believe it? When I can believe that man is more wise and merciful than God, and human thought more trustworthy than divine, then, and not till then, will I believe that He spake falsely, who declared 'I give unto them eternal life, and they shall never perish.'"

OBITUARY NOTICE BY DR J. T. BOTTOMLEY, F.R.S.

[From the *Proceedings of the Royal Society*, Vol. LIII. 1893.]

JAMES THOMSON, lately Professor of Civil Engineering and Mechanics in the University of Glasgow, was born in Belfast on February 16, 1822.

His father, a mathematician of very high order, was, in the first instance, Mathematical Master and Professor of Mathematics in the Royal Belfast Academical Institution; but in 1832 became Professor of Mathematics in Glasgow University.

James Thomson, and his brother William, Lord Kelvin, entered the classes at Glasgow University at an unusually early age. They were never at school, having received their early education at home from their father. They passed through the University together, both with high distinction, the two lads usually obtaining the first and second prizes in each of the classes they attended.

At a very early age also James Thomson showed evidence of considerable inventive genius. When he was about sixteen or seventeen he invented a mechanism for feathering the floats of the paddles of paddle steamers. Steamboats, even on the Clyde, were comparatively novel in those days; and the invention was looked on with much interest by the Clyde engineers to whom it was shown. Unfortunately, from a commercial point of view, however, it turned out that another method of accomplishing the same object had been invented and patented only a few months earlier.

After passing through the University curriculum, James Thomson took the degree of M.A. with honours in Mathematics and Natural Philosophy at the age of seventeen [eighteen].

As he had decided on adopting civil engineering as his profession, he went, in the autumn of 1840, to the office of Mr Macneill (afterwards Sir John Macneill), in Dublin. But unfortunately his health had, shortly before, to some extent, broken down. He was obliged to leave Mr Macneill's office after about three weeks and return home.

In 1840 a new departure was made in Glasgow University, which proved of great importance, and which has had far-reaching influence on the practical teaching of engineering in this country. This was the foundation, by Queen Victoria, of the first Chair of Civil Engineering and Mechanics in the United Kingdom. The first professor was Lewis Gordon, who was succeeded fifteen years later by Macquorn Rankine. James Thomson, at home and in delicate health, attended Professor Gordon's classes in engineering, and was busy with inventions of various sorts; and particularly with a curious boat, which, by means of paddles and legs reaching to the bottom, was able to propel itself up a river, walking against the stream.

In 1843 he was able to resume work as an engineer, and he went to Millwall, to the works of Messrs Fairbairn and Co., of London and Manchester. He was not, however, able to remain with them for the full time of his apprenticeship. Illness returned; he was obliged to go home; and this illness proved the commencement of a period of delicate health which lasted for years, and, indeed, produced a permanent effect on his whole life.

During the months which he spent at Millwall, he was busy with improvements in furnace construction for the purpose of prevention of smoke. The gases of combustion were to be taken downwards through the furnace instead of upwards; and the fire bars were to be tubes with water circulating through them.

After his return to Glasgow he was obliged to confine himself to work which did not involve bodily fatigue. He occupied himself much with invention; and particularly gave his attention to machines for the utilisation of water power.

He constructed a horizontal water-wheel, which he named a Danaide, being an improvement on the Danaide of Manouri d'Ectot; and somewhat later, after much investigation and research, he invented a wheel which, from the nature of its action, he called the vortex water-wheel. This form of wheel was patented in 1850. It was an important advance on water-wheels of previous construction. The moving wheel was mounted within a chamber of nearly circular form. The water, injected under pressure, was directed, by guide blades, to flow tangentially to the circumference of the wheel; and was led through the wheel to the centre by suitably formed radiating partitions. Thus the water yielded its kinetic energy derived from one half of the fall, and its potential energy from the other half, to the wheel by pressure on the radial partitions, as it passed inwards to the centre, whence it quietly

flowed away in the tail-race. A considerable number of these
wheels were designed by him for various factories and for different
purposes. They were made and supplied by Messrs Williamson
Bros., of Kendal, and gave much satisfaction.

In 1847 his mind was also busy with a question to which at a
later date he gave much thought and labour, and to the solution
of which he made contributions of great importance. On April 5
of this year there appears a memorandum in his handwriting:—
"This morning I found the explanation of the slow motion of
semi-fluid masses such as glaciers."

During 1848 his first three important scientific papers were
published. The first of these was on "Strength of Materials as
influenced by the existence or non-existence of certain mutual
Strains among the Particles composing them." The second was
a remarkable paper on "The Elasticity and Strength of Spiral
Springs and of Bars subjected to Torsion." In this paper the
action of a spiral spring was explained, and important principles
connected with the subject of torsion were brought forward.
These papers were published in the *Cambridge and Dublin Mathematical Journal*, November, 1848.

The third was, perhaps, yet more remarkable. It was con-
tributed to the Royal Society of Edinburgh, and was on "The
Parallel Roads or Terraces of Lochaber (Glenroy)." These remark-
able *terraces* or *shelves* had attracted much attention. Darwin,
Lyell, David Milne, Sir G. Mackenzie, Agassiz, Sir Thomas Dick
Lauder, and others had discussed the causes of their formation.
James Thomson, however, gave in this paper what is now the
accepted explanation.

Curiously, Professor Tyndall seems not even to have known of
the existence of the paper when he gave his admirable exposition
of this wonderful natural formation at the Royal Institution in
1876. He attributes the explanation of the Parallel Roads to
Jamieson, 1863; whereas the whole theory had been given by
Thomson in 1848 in the paper just mentioned, with details as to
necessary climatic circumstances not noticed by Tyndall.

In January, 1849, he communicated to the Royal Society of
Edinburgh a paper of great importance, which was printed in the
Transactions of the Society, and was afterwards republished, with
some slight alterations by the author, in the *Cambridge and
Dublin Mathematical Journal*, November, 1850. The title of this

paper was " Theoretical Considerations on the Effect of Pressure in lowering the Freezing Point of Water." The principles expounded in this paper were afterwards, in 1857, used as the foundation of his wellknown explanation of the plasticity of ice, discovered by Forbes; and later, from 1857 onwards, for several years, the whole subject afforded him much food for thought; and extensions and developments in various directions followed. The paper of 1849 was of great intrinsic importance. In it, by the application of Carnot's principle, an absolutely unsuspected physical phenomenon was discovered and predicted, and the amount of lowering of the freezing point of water was calculated. The phenomenon was shortly after experimentally tested and confirmed by his brother, Lord Kelvin.

But the paper has another title to interest, which is not so generally known. In it [as reprinted in November, 1850, by means of an alteration in one sentence] for the first time Carnot's principle was stated, and Carnot's cycle described, in words carefully chosen, so as not to involve the assumption of the material theory of heat, or rather, as Thomson himself puts it, the supposition of the " perfect conservation of heat."

For the sake of clearness, it may be well to leave here for a moment the chronological order of James Thomson's life, and to explain briefly the subsequent development of the ideas first disclosed in this paper of 1849.

Forbes had discovered, by observations and experiments on the Swiss glaciers, the property of *plasticity* in ice. The fact of plasticity in ice was at first doubted; but it was afterwards admitted, and various explanations were offered of this property, so remarkable in a brittle and, above all, crystalline substance.

In this connection, Faraday called attention to the freezing together of two pieces of ice placed together in water; and from this arose a partial explanation, by Tyndall, under the designation of " Fracture and Regelation." But the theory, and even the not logical juxtaposition of the two words, did not satisfy James Thomson. There was nothing to show why or how reunion (or " regelation ") should take place after fracture. He saw, however, that an extension of his own previous principle of lowering of the freezing point by pressure allowed him to apply it to the effect of distorting stress on solid ice, and would give a perfect explanation of all Faraday's observations and experiments on the union and growth of the connecting link between two pieces of ice under water, pressed together by any force, however small.

By this extended thermodynamic principle he also accounted for the yielding of a mass of ice crystals (dry snow, for instance) at *temperatures lower than the ordinary freezing point.* He demonstrated that the mutual pressures must melt the ice at, and close around, the points of contact; and that, when there is relief from the internal stress by this melting, the low temperature of the main solid mass, and the extra cold due to the latent heat required for liquefaction of the yielding portions, cause the melted matter to re-freeze in the places to which it has escaped in order to relieve itself from strain. Thus a complete explanation, based on a demonstrated physical principle, was offered of the phenomenon.

Thomson's explanation did not, certainly at first, commend itself thoroughly to Faraday. A very interesting correspondence between them ensued; and Faraday made a number of beautiful and interesting experiments, with the object of showing that the placing of two pieces of ice on opposite sides of a film of water (between them) would give rise to the conversion of the film of water into ice, and cause the union of the two pieces of ice, the principle being that of the starting of crystallisation in a supersaturated solution by means of a crystal of the solid. James Thomson, however, showed that, in the experiments adduced by Faraday, pressure between the ice blocks was not absent. For example, in an experiment in which two pieces of ice, with a hole through each, were mounted on a horizontal rod of glass, he pointed out that the capillary film of water between the slabs draws them together with not inconsiderable mutual pressure, and hence the freezing. Thomson further showed that when two pieces of ice are brought to touch each other at a point wholly immersed under water, and thus free from capillary action, the most minute pressure pushing the two together causes the growth of a narrow connecting neck, which may be made to grow by continued application of the pressure; while the application of the smallest force tending to draw the two asunder causes the neck to diminish in thickness, and finally to disappear.

In later years James Thomson further developed the theory of 1849. He showed that stresses, of other kinds than pressure equal in all directions, can relieve themselves by means of local lowering of the freezing point in ice; and he showed, by theory and by experiment, that the application of stresses may assist or hinder the growth of crystals in saturated solutions. Some of these conclusions are of such importance that they deserve to be better known. The title of the paper in which the last-named results were given is, "On Crystallisation and Liquefaction as Influenced by Stresses tending to Change of Form in the Crystals*," 1861. It included the amended and extended theory of the plasticity of ice.

In 1850, James Thomson was engaged in perfecting his design for the Vortex water-wheel. He had soon some orders for the wheel; and in 1851 he took the important step of settling down as a civil engineer in Belfast.

His business grew by degrees. His health improved, and we find him occupied in the next two or three years with scientific investigations as to the "properties of whirling fluids." This led to improvements in the action of blowing fans on the one hand, and, on the other, to the invention of a centrifugal pump and to improvements in turbines which were described to the British Association at Belfast in 1852. At this meeting, also, he described "A Jet Pump, or Apparatus for drawing up Water by the Power of a Jet"; and these investigations led to the designing, on the large scale, of pumps of this kind. Some of these pumps have done important work in the drainage of low lands at places where a small stream, capable of supplying the jet, can be found in the immediate proximity. His investigations on the mechanics of whirling fluids, again, led to the design of great centrifugal pumps, the largest of which are now at work on sugar plantations in Demarara.

It will thus be seen that he was giving much attention to water engineering; and in November, 1853, he became resident engineer to the Belfast Water Commissioners, a post which he occupied till the end of 1857.

In this year he was appointed Professor of Civil Engineering and Surveying in Queen's College, Belfast. He became fully occupied with the duties of his professorship, and gave up his office and business as a civil engineer, except for the connection which he retained with some of his former clients, and for business in consultation.

The professorship in Belfast he held till the death of Macquorn

* *Roy. Soc. Proc.* December 5, 1861.

Rankine in 1872. By this event the Professorship of Civil Engineering in Glasgow became vacant; and James Thomson was in the next year appointed by the Government to succeed him.

In 1888 his sight unfortunately began to fail; and the malady, from which both his eyes suffered, proceeded so far that it became necessary for him to resign his University work. This he did after the end of the session 1888–89. Happily, however, he retained more or less of his eyesight till the end of his life; and as he became more accustomed to the condition of his eyes he was better able to make use of what remained to him, and was able to move about freely with but little assistance, and even to read and write a little, and to make on a large scale the diagrams which he used to illustrate his Bakerian Lecture on "The Grand Currents of Atmospheric Circulation."

His death was almost sudden and was the beginning of a sadly tragic time in his family. In a single week Professor Thomson, his wife, and youngest daughter were all attacked with cold, which was quickly followed by inflammation of the lungs. The next week saw the death of all three; his daughter surviving him only three days, and Mrs Thomson seven days. Professor Thomson's death took place on the 8th of May, 1892.

It is not possible in the limits to which this notice must be confined to refer to all James Thomson's papers, nor to give a complete list of the many subjects which occupied his attention.

Already some of his contributions to thermodynamics have been mentioned; but it must be further remarked that during the portion of his life which was occupied with teaching, he gave great attention to this subject, endeavouring to improve the nomenclature and modes of expression of the various principles and propositions connected with it, and to simplify modes of explanation and of statement.

Another very remarkable contribution to thermal science and thermodynamics was his extension of Andrews' discoveries on the subject of the continuity of the liquid and gaseous states of matter. Thomson's mode of conception of the whole subject, which led to the construction of a model in three dimensions to show the mutual relations between pressure, volume, and temperature of such a substance as carbon dioxide under continuous changes of pressure, and volume, and temperature, was perfectly new and most important. The model itself threw a flood of light on the question; and the imagining of the extension of the three-dimensional surface so as to include an unstable condition of the substance, partially realisable and even well known in the phenomena of a liquid passing its boiling point without forming vapour, and in similar unstable conditions, was an advance in the theory of this important question, the consequences of which are not even now completely realised. The verification of Thomson's theories on this subject has proved a fruitful field of experimental investigation for many workers.

Another subject of great importance to which Professor James Thomson devoted much thought and attention was that of safety and danger in engineering structures, and the principles on which their sufficiency in strength should be estimated and proved. He made more than one weighty communication on this subject to engineering societies; and on his appointment at Glasgow, in 1873, he made it the subject of the Latin address which it is the custom for a newly elected Professor to read to the Senatus of the University of Glasgow. An address in English on the same subject became his inaugural lecture to the students of his class in engineering.

When he took up the question, about 1862, he felt that ordinary engineering practice as to testing of structures, boilers for example, was both illogical and unsafe. He considered that the tests usually applied were quite insufficient to permit of an engineer feeling justified in risking the lives of men and the property of his employers to the dangers of breakdown. It was then a common opinion that severe testing should not be applied lest the structure should be weakened by the test itself; but Thomson denied that the test does weaken the structure if the structure be good; and pointed out that the real reason for not applying a proper test was, frequently, fear lest the structure should be found far inferior in strength to that which it was intended to have. The truth of Professor Thomson's contentions is now admitted by the highest engineers; and the best engineering practice has, happily, undergone a thorough reform in this respect.

Certain geological questions possessed much interest for James Thomson. We have seen how, at an early age, he investigated the parallel roads of Glen Roy; and on many subsequent occasions he examined with great care the places where he chanced to be residing, and found and described glacier markings. He traced out, on more than one occasion, specially interesting features of the ice action, endeavouring to determine, by means of an examination of the markings, details as to the motion of the ice, whether in the form of glacier or in the form of icebergs taking the ground in shallow waters.

His attention was also directed to the jointed prismatic structure seen at the Giant's Causeway in Ireland, and elsewhere. No satisfactory explanation of this remarkable phenomenon had been given. The old theories, involving a supposed spheroidal concretionary tendency in the material during consolidation, seemed quite untenable. He examined with great care the appearances presented in the surfaces of the stones, and concluded that the *columnar* structure is due to the shrinking and cracking during cooling of a very homogeneous mass of material. The *cross joints* he considered to be in reality circular conchoidal fractures commencing at the centre of the column and flashing out to the circumference.

A very interesting subject, and one of very high importance, to which Professor Thomson gave great attention, is the flow of water in rivers. He investigated, with great care, and from a theoretical point of view, the origin of windings of rivers in alluvial plains, and his conclusions were published in the *Proceedings of the Royal Society*, May 4, 1876. Later in the same year he constructed, in clay, on a table, a model with which he investigated the movements of the different parts of the water in passing round the bends in this artificial river; and, finally, he made a large wooden model of a river flowing on a nearly horizontal bed with many bends and various obstacles. By aid of fine threads, small floating and sinking bodies, and coloured streams of fluid coming from particles of solid aniline dye dropped into the channel, he was able to follow from point to point the movements of the fluid, and thus to give not only beautiful and striking ocular evidence of the truth of his early conclusions, but also to extend his theory. Papers on this subject were communicated to the Royal Society in 1876, 1877, 1878. The paper of the last-named date was entitled "On Flow of Water in Uniform Régime in Rivers and in Open Channels generally." It contains a very clear and striking account of what does occur in the motion of a river down its inclined channel; and, in particular, of the fact which seems to be ascertained, that the forward velocity of the water in rivers is, generally, not greatest at the surface with gradual abatement from surface to bottom (as would be required under the conditions supposed in the laminar theory); but that, in reality, the average velocity down stream is greatest at some depth below the surface, from which, up to the surface, there is a considerable decrease, and down to the bottom a much greater decrease. This phenomenon he showed very clearly to be due to the rising of masses of slow-going water, from the bottom, on account of directing action of bottom obstacles. These masses of slow-going water, when they reach the top, spread themselves out, and, mingling with the quicker surface water, give to it, on the whole, a less rapid movement than it should otherwise possess. The paper, as a whole, forms a masterly exposition of this important subject.

Finally, in this brief summary must be mentioned the paper which was made the Bakerian Lecture for the year 1892. In 1857 Professor Thomson read a paper to the British Association, on "The Grand Currents of Atmospheric Circulation." It appears that his attention was first called to this subject when, during the period of his early delicacy, his father asked him to look into the question of the Trade Winds and write a short account of this atmospheric phenomenon for a new edition of Dr Thomson's *Geography*, which was then in preparation. This was done; but young James Thomson found so little satisfaction in the information and theories which he then studied for the purpose that his mind was keenly directed to the question; and in 1857 he himself had formed a theory which he expounded to the British Association.

The subject was before his mind during the rest of his life; and though on account of other pressing work the complete publication of the theory was from time to time deferred, yet it was always his intention to return to the question. When in the last years of his life the affliction of partial blindness came upon him, and when he had somewhat recovered from the first depressing effects of finding himself thus sadly crippled, he set himself in his enforced leisure to complete this work, and, with the assistance of his wife and daughters, to produce the important paper which was read before the Royal Society on the 10th of March, 1892. In this paper a historical sketch is given of the progress of observation and theoretical research into the nature and causes of the trade-winds and other great and persistent currents of atmospheric circulation. Previous theories are discussed and criticised and their merits duly recognized, the theory of Hadley, in particular, being shown to be substantially true. A much more complete theory is then expounded in full detail; and charts and diagrams in illustration show the nature of the aërial motions.

Here this memoir must close. There are many papers of Thomson which have not even been alluded to in it. Nor is it possible or necessary for the present purpose to refer to all the subjects to which his ever active mind directed itself. A character so truly philosophic it is very rare to meet. His was a singularly well ordered and well governed mind. It was, if one may venture to say so, almost too philosophical and too well governed for the business of every-day life. He could scarcely realise a difference between greater and smaller error or untruth. Great or small error and untruth were to be condemned and resisted; and, perhaps, in the matter of public business and in this hurrying nineteenth century pressure, there were those who, thoroughly conscientious themselves, could not yet feel perfect sympathy with his extreme and scrupulous determination to let nothing, however small, pass without thorough examination and complete proof. To temporise was not in his nature; and this extreme conscientiousness gave rise to a want of rapidity of action which was perhaps the only fault in a singularly perfect character.

Purity and honour in word and deed and thought, gentleness of disposition, readiness to spend his labour, his time, his mental energies for others, and for the good of the world in general, all were conspicuous in his life both in public and in private.

Professor Thomson was elected Fellow of the Royal Society in 1877; and he received the honorary degree of D.Sc. from the Queen's University in Ireland, and of LL.D. from his own University of Glasgow, and from the University of Dublin.

In 1853 he married Elizabeth Hancock, daughter of William John Hancock, Esq., J.P., of Lurgan, Co. Armagh, a lady who devoted herself to every minutest interest of her husband's life. They had one son and two daughters, of whom the son and elder daughter survive.

J. T. B.

LIST OF SCIENTIFIC TERMS INTRODUCED BY
PROF. JAMES THOMSON.

1. 1873. *Radian*, used in General Class Examination, Queen's College, Belfast, June 5, 1873. See College Calendar: Thomson's *Arithmetic*, 1880 edition, p. 62. Cf. correspondence with Dr Muir, *Nature*, Vol. 83, pp. 156, 217, 459, 460.

2. 1874. *Interface*, see Sir W. Thomson, "Kinetic Theory of the Dissipation of Energy," 1874, *Math. and Phys. Papers*, Vol. v. p. 12; see also Maxwell, *Theory of Heat*, 1875 edition, p. 95; see also *infra*, pp. 65, 424; also p. 327, 1869.

3. 1875. *Unital, e.g.* unital stress proportional to unital strain, see "Comparisons of Similar Structures," *infra*, p. 364 note.

4. 1875. *Poundal*, see Maxwell, *Theory of Heat*, 1875 edition, p. 83; also "Flow of Water through Orifices," *infra*, p. 66 note.

5. 1875. *Gravity*, instead of weight; see *infra*, pp. 362 note, 454.

6, 7. 1876. *Crinal, Funal*, see "Metric Units," B. A. 1876, *infra*, p. 373, also p. 66 note.

8. 1876. *Ergometer*, instead of dynamometer; see paper "On an Integrating Machine," R. S. 1876, *infra*, p. 452 note.

9. 1876. *Free Level*, see "Flow of Water through Orifices," B. A. 1876, *infra*, p. 64, also p. 97.

10. 1878. *Change ratio*, "Dimensional Equations," B. A. 1878, *infra*, p. 375.

11. 1878. *Numeric*, "Dimensional Equations," B. A. 1878, see *infra*, p. 376; also Thomson's *Arithmetic*, 1880 edition, p. 4.

12. 1879. *Hesion*, for tangential force between two pieces of matter when there is no motion, see letter to Clerk Maxwell, *infra*, p. 458.

13. 1880. *Ratio Equation*; a statement of the equality of two ratios, see Thomson's *Arithmetic*, 1880 edition, p. 95.

14. 1880. *Rate Equivalence*, see Thomson's *Arithmetic*, 1880 edition, p. 108.

15. 1880. *Ply, Plication*, instead of multiply and multiplication, see Thomson's *Arithmetic*, 1880 edition, p. 152.

16. 1880. *Disply*, for the reverse of *Ply*, instead of Divide, see Thomson's *Arithmetic*, 1880 edition, p. 155.

17. 1880. *Round*, used for an angle of 360°, see Thomson's *Arithmetic*, 1880 edition, p. 61.

18. 1882. *Forcive*, for "any system of force," see Lord Kelvin, *Math. and Phys. Papers*, Vol. iv. p. 369 note.

19. 1884. *Apocentric* force, instead of Centrifugal, see "Whirlwinds and Waterspouts," B. A. 1884, *infra*, p. 148.

20.   1884.   *Centreward* acceleration, instead of centripetal, B. A. 1884 ; see also *infra*, p. 69 note, 1876.

21, 22.   1884.   *Clinure, Posure*, see " Law of Inertia," *infra*, p. 380 note.

23.   1884.   *Torque*, for force-pair or couple, used in printed examination papers, April 22, 1884 ; see also *Transactions of the Institution of Engineers and Shipbuilders in Scotland*, 1885—6, Vol. XXIX. p. 91.

24.   1888.   *Fluifaction*, see *infra*, p. 272.

25.   1892.   *Revolutional Momentum* instead of " Moment of Momentum," or " Angular Momentum," see *Bakerian Lecture, infra*, p. 191.

26.   *Expansity*, for converse of Density. "Bulkiness of a body relative to its quantity of matter."

27.   *Ward*, instead of Direction.

28.   *Ways of Freedom*, instead of Degrees of Freedom.

PAPERS RELATING TO FLUID MOTION

1. ON SOME PROPERTIES OF WHIRLING FLUIDS [VERTICAL FREE VORTEX], WITH THEIR APPLICATION IN IMPROVING THE ACTION OF BLOWING FANS, CENTRIFUGAL PUMPS, AND CERTAIN KINDS OF TURBINES.

[From the *British Association Report*, Section G, Belfast, 1852, page 130.]

THE author pointed out several properties possessed by masses of fluids revolving in the circumstances of one of the most ordinary kinds of whirlpools, that, namely, which is formed when water is supplied at the circumference of a widely extended vessel, with a very slight rotatory motion, and is allowed to flow away by a central orifice in the bottom. Of these properties, the following, in which the influence of friction is left out of consideration, may be cited:

The equation of the curve whose revolution would generate the curved surface of the whirlpool is $y = C^3/x^2$, where y is the depth of any point of the curve below the level of the fluid taken at any part far away from the whirlpool, where there is no perceptible depression; x the distance of the point from the axis of revolution; and C a constant quantity.

Every point of the surface of the fluid moves with the velocity which a heavy body would attain in falling from the level of the surface far away from the whirlpool to the level of the point. Also every point in the interior of the revolving mass moves with the velocity of the point on the surface vertically above itself; and it follows, that the velocities of points at various distances from the centre are inversely proportional to the distances.

It follows also that the velocity of each point in the mass, is the greatest that is possible without an increase of the velocity of every other point revolving further from the centre.

He was led from these and other properties of whirling fluids, to find that the efficiency of centrifugal pumps for water, and of fans for causing blasts of air, may be greatly increased by the provision, outside of the circumference of the wheel, of a space in which the fluid may continue to revolve without any interruption after it has left the wheel. He mentioned also, that an apparatus termed a "diffuser," and involving the same principle, has recently been applied with good results, in turbines of great power constructed in America.

2. On the Vortex Water-Wheel.

[From the *Report of the British Association*, Belfast, 1852, page 317. A Communication ordered to be printed among the Reports.]

NUMBERLESS are the varieties, both of principle and of construction, in the mechanisms by which motive power may be obtained from falls of water. The chief modes of action of the water are, however, reducible to three, as follows:—First, The water may act directly by its weight on a part of the mechanism which descends while loaded with water, and ascends while free from load. The most prominent example of the application of this mode is afforded by the ordinary bucket water-wheel. Secondly, The water may act by fluid pressure, and drive before it some yielding part of a vessel by which it is confined. This is the mode in which the water acts in the water-pressure engine, analogous to the ordinary high-pressure steam-engine. Thirdly, The water, having been brought to its place of action subject to the pressure due to the height of fall, may be allowed to issue through small orifices with a high velocity, its inertia being one of the forces essentially involved in the communication of the power to the moving part of the mechanism. Throughout the general class of water-wheels called Turbines, which is of wide extent, the water acts according to some of the variations of which this third mode is susceptible. The name Turbine is derived from the Latin word *turbo*, a top, because the wheels to which it is applied almost all spin round a vertical axis, and so bear some considerable resemblance to the top. In our own country, and more especially on the Continent, turbines have attracted much attention, and many forms of them have been made known by published descriptions. The subject of the present communication is a new water-wheel, which belongs to the same general class, and which has recently been invented and brought successfully into use by the author.

In this machine the moving wheel is placed within a chamber of a nearly circular form. The water is injected into the chamber tangentially at the circumference, and thus it receives a rapid motion of rotation. Retaining this motion it passes onwards towards the centre, where alone it is free to make its exit. The wheel, which is placed within the chamber, and which almost entirely fills it, is divided by thin partitions into a great number of radiating passages. Through these passages the water must flow on its course towards the centre; and in doing so it imparts its own rotatory motion to the wheel. The whirlpool of water acting within the wheel-chamber, being one principal feature of this turbine, leads to the name *Vortex* as a suitable designation for the machine as a whole.

The vortex admits of several modes of construction, but the two principal forms are the one adapted for high falls and the one for low falls. The former may be called the High-pressure Vortex, and the latter the Low-pressure Vortex*. Examples of these two kinds, in operation at two mills near Belfast, are delineated in figs. 1 and 2, with merely a few unimportant deviations from the actual constructions.

Figs. 1 and 2 are respectively a vertical section, and a plan of a vortex of the high-pressure kind in use at the Low Lodge Mill near Belfast, for grinding Indian corn†. In these figures AA is the water-wheel. It is fixed on the upright shaft, B, which conveys away the power to the machinery to be driven. The water-wheel occupies the central part of the upper division of a strong cast-iron

---

* These terms correspond to Hochdruckturbine, and Niederdruckturbine, used in Germany to express the like distinction in turbines.

† This vortex was only in course of erection at the time of the meeting of the British Association in Belfast. The water-wheel itself, removed from its case, being light and of small dimensions, was exhibited in Section G. It is composed chiefly of thick-tinned iron plates united by soft solder.

case, *CC*; and the part occupied by the wheel is called the *wheel-chamber*. *DD* is the lower division of the case, and is called the *supply chamber*. It receives the water directly from the supply pipe, of which the lower extremity is shown at *E*, and delivers it into the outer part of the upper division, by four large openings, *F*, in the partition between the two divisions. The outer part of the upper division is called the *guide-blade chamber*, from its

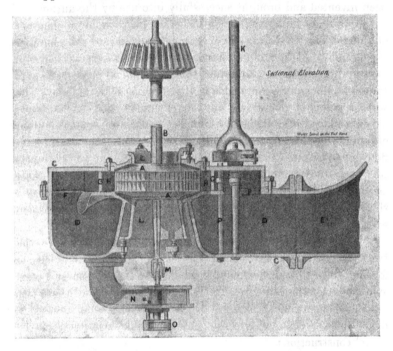

Fig. 1.

containing four guide-blades, *G*, which direct the water tangentially into the wheel-chamber. Immediately after being injected into the wheel-chamber the water is received by the curved radiating passages of the wheel, which are partly seen in fig. 2, at a place where both the cover of the wheel-chamber and the upper plate of the wheel are broken away for the purpose of exposing the interior to view. The water, on reaching the inner ends of these curved passages, having already done its work, is allowed to make its exit by two large central orifices, shown distinctly on the figures at the letters *L, L*; the one leading upwards and the other downwards. It then simply flows quietly away; for, the vortex being submerged under the surface of the water in the tail race, the water on being discharged wastes no part of the fall by a further descent. At the central orifices, close

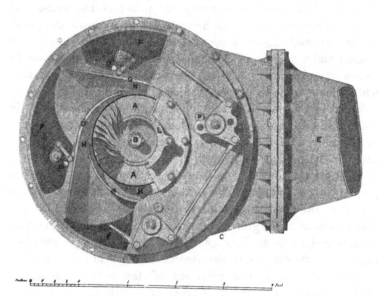

Fig. 2.

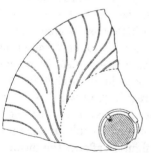

Part of the Wheel on a larger scale, to shew the form of the vanes more accurately

joints between the case and the wheel, to prevent the escape of water otherwise than through the wheel itself, are made by means of two annular pieces, *L, L,* called *joint-rings*, fitting to the central orifices of the case, and capable of being adjusted, by means of studs and nuts, so as to come close to the wheel without impeding its motion by friction. The four openings, H, H, figs. 1 and 2, through which the water flows into the wheel-chamber, each situated between the point or edge of one guide-blade and the middle of the next, determine, by their width, the quantity of water admitted, and consequently the power of the wheel. To render this power capable of being varied at pleasure, the guide-blades are made movable round gudgeons or centres near their points; and a spindle, K, is connected with the guide-blades by means of links, cranks, &c. (see the figures) in such a way that, when the spindle is moved, the four entrance orifices are all enlarged or contracted alike. This spindle, K, for working the guide-blades is itself worked by a handle in a convenient position in the mill; and the motion is communicated from the handle through the medium of a worm and sector, which not only serve to multiply the force of the man's hand, but also to prevent the guide-blades from being liable to the accident of slapping suddenly shut from the force of the water constantly pressing them inwards. The gudgeons of the guide-blades, seen in fig. 2 as small circles, are sunk in sockets in the floor and roof of the guide-blade chamber; and so they do not in any way obstruct the flow of the water.

M, in fig. 1, is the pivot-box of the upright shaft. It contains, fixed within it, an inverted brass cup, shown distinctly on the figure; and the cup revolves on an upright pin, or pivot, with a steel top. The pin is held stationary in a bridge, N, which is itself attached to the bottom of the vortex-case. For adjusting the pin as to height, a little cross bridge, O, is made to bear it up, and is capable of being raised or lowered by screws and nuts shown distinctly on the figure. Also, for preventing the pin from gradually becoming loose in its socket in the large bridge, two pinching-screws are required, of which one is to be seen in the figure. A small pipe, fixed at its lower end into the centre of the inverted brass cup, and sunk in an upright groove in the vortex-shaft (see the figures), affords the means of supplying oil to the rubbing surfaces, over which the oil is spread by a radial groove in the brass. A cavity, shown in the figures, is provided at the lower part of the cup, for the purpose of preventing the oil from being rapidly washed away by the water*.

* Great stress has been laid on the supposed necessity for oiling the pivots of turbines by continental engineers and authors. The author of the present

Four tie-bolts, marked P, bind the top and bottom of the case together, so as to prevent the pressure of the water from causing the top to spring up, and so occasioning leakage at the guide-blades or joint-rings.

The height of the fall for this vortex is about 37 feet, and the standard or medium quantity of water, for which the dimensions of the various parts of the wheel and case are calculated, is 540 cubic feet per minute. With this fall and water supply the estimated power is 28 horse-power, the efficiency being taken at 75 per cent. The proper speed of the wheel, calculated in accordance with its diameter and the velocity of the water entering its chamber, is 355 revolutions per minute. The diameter of the wheel is 22⅝ inches, and the extreme diameter of the case is 4 feet 8 inches.

A low-pressure vortex, constructed for another mill near Belfast, is represented in vertical section and plan in figs. 3 and 4. This is essentially the same in principle as the vortex already described, but it differs in the material of which the case is constructed, and in the manner in which the water is led to the guide-blade chamber. In this the case is almost entirely of wood; and, for simplicity, the drawings represent it as if made of wood alone, though in reality, to suit the other arrangements of the mill, brick-work, in certain parts, was substituted for the wood. The water flows with a free upper surface, W, W, into this wooden case, which consists chiefly of two wooden tanks, AA and BB, one within the other. The water-wheel chamber and the guide-blade chamber are situated in the open space between the bottom of the outer and that of the inner tank, and will be readily distinguished by reference to the figures. The water of the head race, having been led all round the outer tank in the space CC, flows inwards over its edge, and passes downwards by the space DD, between the

communication has thus been led to endeavour to find and adopt the best means for oiling pivots working under water. The oiling, however, is a source of much trouble; and he has found in the course of his experience, that pivots of the kind described above, made with brass working on hard steel, and with a radial groove in the brass suitable for spreading water over the rubbing surfaces, will last well without any oil being supplied. The rapid destruction, which is commonly reported as having been of frequent occurrence in turbine pivots, he believes may in many cases have arisen from the employment of an inverted cup like a diving-bell as one of the rubbing parts, without any provision for the escape of air from the cup. It is evident that a pivot of this kind, although under water, might be perfectly dry at the rubbing surfaces.

sides of the two tanks.   It then passes through the guide-blade
chamber and the water-wheel, just in the same way as was
explained in respect to the high-pressure vortex already described;
and in this one likewise it makes its exit by two central orifices,
the one discharging upwards and the other downwards.   The part
of the water which passes downwards flows away at once to the
tail race, and that which passes upwards into the space E within

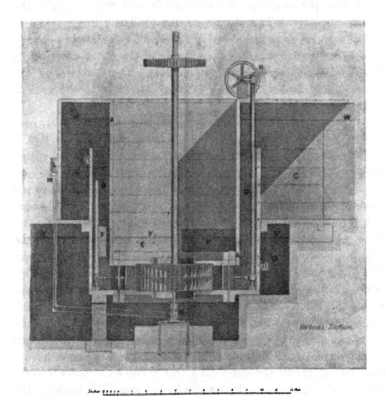

Fig. 3.

the innermost tank, finds a free escape to the tail race through
boxes and other channels, F and G, provided for that purpose.
The wheel is completely submerged under the surface of the water
in the tail race, which is represented at its ordinary level at YYY,
fig. 3, although in floods it may rise to a much greater height.
The power of the wheel is regulated in a similar way to that
already described in reference to the high-pressure vortex.   In

this case, however, as will be seen by the figures, the guide-blades
are not linked together, but each is provided with a hand-wheel,
H, by which motion is communicated to itself alone.

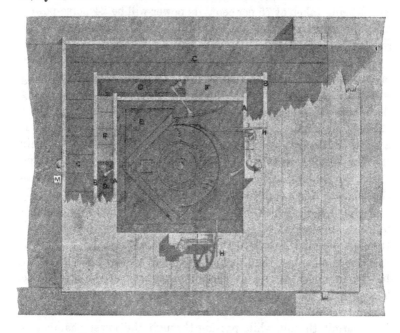

Fig. 4.

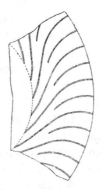

*Part of the Wheel
drawn on a larger
scale to shew the cur-
vature of the vanes
more correctly*

In this vortex, the fall being taken at 7 feet, the calculated quantity of water admitted, at the standard opening of the guide-blades, is 2460 cubic feet per minute. Then, the efficiency of the wheel being taken at 75 per cent., its power will be 24 horse-power. Also the speed at which the wheel is calculated to revolve is 48 revolutions per minute.

In connexion with the pivot of this wheel arrangements are made which provide for the perfect lubrication of the rubbing surfaces with clean oil. The lower end of the upright revolving shaft enters a stationary pivot-box, K, through an opening made oil-tight by hemp and leather packing. Within the box there is a small stationary steel plate on which the shaft revolves. Within the box, also, there are two oil-chambers, one situated above and round the rubbing surface of this plate, and the other underneath the plate. A constant circulation of the oil is maintained by centrifugal force, which causes it to pass from the lower chamber upwards through a central orifice in the steel plate, then outwards through a radial groove in the bottom of the revolving shaft to the upper chamber, then downwards back to the lower chamber, by one or more grooves at the circumference of the steel plate. The purpose intended to be served by the provision of the lower chamber combined with the passages for the circulation of the oil, is to permit the oil, while passing through the lower chamber, to deposit any grit or any worn metal which it may contain, so that it may be maintained clean and may be washed over the upper surface of the steel plate at every revolution of the radial groove in the bottom of the shaft. A pipe leading from an oil cistern, L, in an accessible situation conducts oil to the upper chamber of the pivot-box; and another pipe leaves the lower chamber, and terminates, at its upper end, in a stop-cock, M. This arrangement allows a flow of oil to be obtained at pleasure from the cistern, down by the one pipe, then through the pivot-box, and then up by the other pipe, and out by the cock. Thus, if any stoppage were to occur in the pipes, it could be at once detected; or if water or air were contained in the pivot-box after the first erection, or at any other time, the water could be removed by the pipe leading to the stop-cock, or the air would of itself escape by the pipe leading to the cistern, which, as well as the other pipe, has a continuous ascent from the pivot-box. Certainty may consequently be attained that the pivot really works in clean oil.

The author was led to adopt the pivot-box closed round the shaft with oil-tight stuffing, from having learned of that arrangement having been successfully employed by Köchlin, an engineer of Mühlhausen. As to the other parts of the arrangements just described, he believes the settling chamber with the circulation of oil to be new, and he regards this part of the arrangements as being useful also for pivots working not under water. In respect to the materials selected for the rubbing parts, however, he thinks it necessary to state that some doubts have arisen as to the suitableness of wrought iron to work on steel even when perfectly lubricated; and he would, therefore, recommend that a small piece of brass should be fixed into the bottom of the shaft, all parts being made to work in the manner already explained.

The two examples which have now been described of vortex water-wheels adapted for very distinct circumstances, will serve to indicate the principal features in the structural arrangements of these new machines in general. Respecting their principles of action some further explanations will next be given. In these machines the velocity of the circumference is made the same as the velocity of the entering water, and thus there is no impact between the water and the wheel; but, on the contrary, the water enters the radiating conduits of the wheel gently, that is to say, with scarcely any motion in relation to their mouths. In order to attain the equalization of these velocities, it is necessary that the circumference of the wheel should move with the velocity which a heavy body would attain in falling through a vertical space equal to half the vertical fall of the water, or in other words, with the velocity *due* to half the fall; and that the orifices through which the water is injected into the wheel-chamber should be conjointly of such area that when all the water required is flowing through them, it also may have a velocity due to half the fall. Thus one-half only of the fall is employed in producing velocity in the water; and, therefore, the other half still remains acting on the water within the wheel-chamber at the circumference of the wheel in the condition of fluid pressure. Now, with the velocity already assigned to the wheel, it is found that this fluid pressure is exactly that which is requisite to overcome the centrifugal force of the water in the wheel, and to bring the water to a state of rest at its exit, the mechanical work due to both halves of the fall being transferred to the wheel during the combined action of the moving water and the moving wheel. In the foregoing statements, the effects of fluid friction, and of some other modifying influences, are, for simplicity, left out of consideration; but in the practical application of the principles, the skill and judgement of the designer must be exercised in taking all such elements as far as possible into account. To aid in this, some practical rules, to which the author as yet closely adheres, were made out by him previously to the date of his patent. These are to be found in the specification of the patent, published in the *Mechanics' Magazine* for Jan. 18 and Jan. 25, 1851 (London)*.

* From the *Mechanics' Magazine*, 18 January, 1851, pp. 45, 46.

[According to the best calculations I have been able to make as to the principal dimensions of the wheel and case, and the velocity of the wheel, these should be as follows:

Let Q be the quantity of water in cubic feet per minute, and H the height of fall in feet.

Then, Total area of entrance orifice or orifices

$$= \cdot 00329 \, \frac{Q}{\sqrt{H}} \text{ square feet.}$$

Diameter of each central orifice of the movable wheel

$$= \cdot 0915 \, \sqrt{\frac{Q}{\sqrt{H}}} \text{ feet.}$$

Diameter of the wheel, if the vanes be straight $= 3\cdot5$ times the diameter of each central orifice of the wheel.

$$[= \cdot 32 \, \sqrt{\frac{Q}{\sqrt{H}}} \text{ feet.}]$$

Diameter of the wheel, if the vanes be curved $=$ twice the diameter of each central orifice of the wheel.

$$[= \cdot 183 \, \sqrt{\frac{Q}{\sqrt{H}}} \text{ feet.}]$$

Revolutions per minute if the vanes be straight

$$= 316 \, \sqrt{\frac{H\sqrt{H}}{Q}}.$$

Revolutions per minute if the vanes be curved

$$= 553 \, \sqrt{\frac{H\sqrt{H}}{Q}}.$$

Again; to determine the angle which the inner end of each vane should make with the radius passing through it:

Let $c =$ circumference of each central orifice in feet;

$d =$ distance in feet between the top and the bottom plate of the wheel, minus the thickness of the plate for attaching the vanes to the shaft;

$t =$ thickness of each vane in feet (or fractions of a foot);

$n =$ number of vanes which terminate at the circumference of the central orifice;

$v =$ velocity of the inner end of each vane in feet per minute.

In respect to the numerous modifications of construction and arrangement which are admissible in the Vortex, while the leading principles of action are retained, it may be sufficient here merely to advert—first, to the use (as explained in the specification of the patent) of straight instead of curved radiating passages in the wheel; secondly, to the employment, for simplicity, of invariable entrance orifices, or of fixed instead of movable guide-blades; and lastly, to the placing of the wheel at any height, less than about thirty feet, above the water in the tail race, combined with the employment of suction pipes descending from the central discharge

Also let C, fig. 5, be the centre of the wheel; DSG part of its outer circumference; NFB part of the circumference of one of its central orifices; and let F be the termination of one vane, and K that of the next which reaches to the circumference of the central orifices.

Make TH perpendicular to FC, and such that

$$FT : TH :: \frac{Q}{cd} : v.$$

Then FH would be the direction of the inner end of the vane passing through F, if the vanes were infinitely thin. To correct for the thickness make

$$FT : FO :: \frac{Q}{cd} : \frac{Q}{cd - \dfrac{ntd}{\cos HFT}}.$$

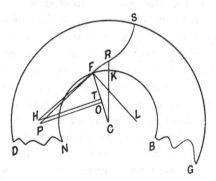

Fig. 5.

Draw OP equal and parallel to TH. FP is the true direction of the end of the vane.

Again; draw FL perpendicular to FP, and a little longer than FC. It may be about $=1·2$ FC. From L, as centre, describe an arc of a circle FR, terminating at R in the continuation of CK. FR is the inner portion of the vane, and in forming the remaining portion, all that need be attended to, is to give it a gentle curvature, and to make a short portion of it at S be in the direction of a radius passing through S.]

orifices, and terminating in the water of the tail race, so as to render available the part of the fall below the wheel.

In relation to the action of turbines in general, the chief and most commonly recognized conditions, of which the accomplishment is to be aimed at, are that the water should flow through the whole machine with the least possible resistance, and that it should enter the moving wheel without shock, and be discharged from it with only a very inconsiderable velocity. The vortex is in a remarkable degree adapted for the fulfilment of these conditions. The water moving centripetally (instead of centrifugally, which is more usual in turbines) enters at the period of its greatest velocity (that is, just after passing the injection orifices) into the most rapidly moving part of the wheel, the circumference; and, at the period when it ought to be as far as possible deprived of velocity, it passes away by the central part of the wheel, the part which has the least motion. Thus in each case, that of the entrance and that of the discharge, there is an accordance between the velocities of the moving mechanism and the proper velocities of the water.

The principle of injection from without inwards, adopted in the vortex, affords another important advantage in comparison with turbines having the contrary motion of the water; as it allows ample room, in the space outside of the wheel, for large and well-formed injection channels, in which the water can be made very gradually and regularly to converge to the most contracted parts, where it is to have its greatest velocity. It is as a concomitant also of the same principle, that the very simple and advantageous mode of regulating the power of the wheel by the movable guide-blades already described can be introduced. This mode, it is to be observed, while giving great variation to the areas of the entrance orifices, retains at all times very suitable forms for the converging water channels.

Another adaptation in the vortex is to be remarked as being highly beneficial, that namely according to which, by the balancing of the contrary fluid pressures due to half the head of water and to the centrifugal force of the water in the wheel, combined with the pressure due to the ejection of the water backwards from the inner ends of the vanes of the wheel when they are curved, only one-half of the work due to the fall is spent in communicating *vis viva* to the water, to be afterwards taken from it during its passage through the wheel; the remainder of the work being communicated through the fluid pressure to the wheel, without any intermediate generation of *vis viva*. Thus the velocity of the water, where it moves fastest in the machine, is kept comparatively low; not exceeding that due to half the height of the fall, while in other turbines the water usually requires to act at much higher velocities. In many of them it attains at two successive times the velocity due to the whole fall. The much smaller amount of action, or agitation, with which the water in the vortex performs its work, causes a material saving of power by diminishing the loss necessarily occasioned by fluid friction.

In the Vortex, further, a very favourable influence on the regularity of the motion proceeds from the centrifugal force of the water, which, on any increase of the velocity of the wheel, increases, and so checks the water supply; and on any diminution of the velocity of the wheel, diminishes, and so admits the water more freely; thus counteracting, in a great degree, the irregularities of speed arising from variations in the work to be performed. When the work is subject to great variations, as for instance in saw-mills, in bleaching works, or in forges, great inconvenience often arises with the ordinary bucket water-wheels and with turbines which discharge at the circumference, from their running too quickly when any considerable diminution occurs in the resistance to their motion.

The first vortex which was constructed on the large scale was made in Glasgow, to drive a new beetling-mill of Messrs C. Hunter and Co., of Dunadry, in County Antrim. It was the only one in action at the time of the Meeting of the British Association in Belfast; but the two which have been particularly described in the present article, and one for an unusually high fall, 100 feet, have since been completed and brought into operation. There are also several others in progress; of which it may be sufficient to particularize one of great dimensions and power, for a new flax-mill at Ballyshannon in the West of Ireland. It is calculated for working at 150 horse-power, on a fall of 14 feet, and it is to be impelled by the water of the River Erne. This great river has an ample reservoir in the Lough of the same name; so that the water of wet weather is long retained, and continues to supply the river abundantly even in the dryest weather. The lake has also the effect of causing the floods to be of long duration, and the vortex will consequently be, through a considerable part of the year, and for long periods at a time, deeply submerged under back-water. The water of the tail race will frequently be 7 feet above its ordinary summer level; but as the water of the head race will also rise to such a height as to maintain a sufficient difference of levels, the action of the wheel will not be deranged or impeded by the floods. These circumstances have had a material influence in leading to the adoption in the present case of this new wheel in preference to the old breast or undershot wheels.

3. ON A CENTRIFUGAL PUMP AND WINDMILL ERECTED FOR DRAINAGE AND IRRIGATION IN JAMAICA.

[From the *British Association Report*, Section G, Glasgow, 1855, page 210.]

IN this paper Mr Thomson gave explanations, with the aid of large drawings, of a centrifugal pump recently constructed, embodying the improvement of an exterior whirlpool which he had first made public in the Mechanical Section at the Belfast Meeting in 1852*. He also described a windmill, with its framing of very simple construction, which had been specially designed for working the pump. The apparatus was prepared for purposes of drainage and irrigation in Jamaica, the costliness of fuel and the habitual use of windmills in that island having led to the selection of the windmill in this case as the source of power. The whole apparatus was constructed in Glasgow and afterwards erected and brought into action at its destination.

4. ON A CENTRIFUGAL PUMP WITH EXTERIOR WHIRLPOOL, CONSTRUCTED FOR DRAINING LAND.

[From the *Proceedings of the Institution of Engineers in Scotland*, October 27, 1858.]

THE centrifugal pump which forms the subject of the present communication has been designed and constructed in the course of this year, for Messrs James Ewing & Co. of Glasgow, for the drainage of lands belonging to them in Demerara.

The cultivated district of Demerara in which the lands referred

* [*Supra*, p. 2.]

to are situated, forms a narrow strip or belt of low ground extending along the sea-coast. It is about four or five miles wide, as
measured from the sea back towards the interior. It lies generally
below the level of high tide; and is protected in front against the
waters of the sea, and in rear against the waters of the Savannahs,
or inland swamps, by embankments. The district was originally
embanked and reclaimed by the Dutch, and was taken from them
by the English in 1796, along with other parts of what now forms
British Guiana. The drainage is ordinarily effected by valves or
kokers, as they are designated, which discharge to the sea when
the tide is low, but hinder the entrance of the tidal water when
it rises. Of late years banks of soft and shifting mud, or slob,
have been gathering and increasing for great distances out to sea,
for miles perhaps, in front of some of the estates. These banks,
by stopping the drainage outlets to the sea, have caused several
estates to be swamped, and have been threatening like destruction
to others. Some estates, it is understood, have been saved, mainly
by the employment of centrifugal pumps driven by steam power,
which keep up the drainage when the slob rises so high as to stop
the ordinary flow to the sea by gravitation through the valves or
kokers; and which, by keeping up a run of water over the slob,
tend, as is believed, in a material degree, to keep open the outlet
channels through the slob, or to open them again when choked
up by the shifting of the mud banks. Some of the estates of
Messrs James Ewing & Co., after having been seriously threatened
for several years by the approach of mud banks to their front,
were last year partially injured by deficient drainage; and thus
the proprietors were led to determine on speedily procuring
a centrifugal pump, to be driven by a steam-engine already
existing on their land, as the driving power for one of their sugar
mills. They engaged the writer to prepare the designs according
to an improved arrangement, which had been devised by him for
increasing the work done by a given power, the main feature
of which is the employment of an exterior whirlpool round the
circumference of the pump wheel. The exterior whirlpool had
been previously introduced in a centrifugal pump designed by
him in 1853, for Messrs James Ewing & Co., for drainage of
lands on an estate of theirs in Jamaica, and constructed by
Messrs W. & A. M'Onie & Co. of Glasgow. It had also more
recently been introduced in a centrifugal pump constructed

according to his designs by Messrs Robert Napier & Sons, for
a steamer for the Great West of Scotland Fishery Company*.
The nature of this modification in centrifugal pumps, consisting
of the introduction of the exterior whirlpool, may be understood
from the following explanations:

In centrifugal pumps for water, or fans for air, when doing
actual work in lifting the fluid, or in forcing it against a pressure,
the fluid has necessarily a considerable tangential velocity on
leaving the circumference of the wheel. This velocity in wheels
in which the vanes or blades are straight and radial, is the
same as that of the circumference of the wheel; in others, in
which the vanes are curved backwards in receding from the
centre, it is somewhat less; but in all cases it is so great that the
water on leaving the wheel carries away, in its energy of motion,
a large and important part of the work applied to the wheel by
the steam-engine, or other prime mover. This energy of motion
in centrifugal pumps and fans, as ordinarily constructed, is mainly
consumed in friction and eddies in the discharge pipe, which
receives the water or air directly from the circumference of the
wheel. The object of the introduction of the exterior whirlpool is
to prevent this waste, and to apply usefully, towards increasing
the pumping power, the energy of the motion inherent in the
fluid on its leaving the circumference of the wheel. For this
purpose there is provided, round the circumference of the wheel,
an exterior chamber, in which the fluid is left free to revolve, in
virtue of the motion it had on leaving the wheel. This chamber
may be called the *exterior whirlpool chamber*, and its diameter is
ordinarily made about double that of the wheel. The fluid
revolving in it assumes the condition of a vortex or whirlpool,
which has been designated by the writer as the Vortex or
Whirlpool of free mobility. The principles of the motion of this
kind of whirlpool, and of its application in improving the action
of centrifugal pumps and blowing fans, was first brought forward
at the Belfast Meeting of the British Association in 1852, in a
paper read in the Mechanical Section, of which an abstract is to
be found in the Report of the Meeting†. The chief properties for
which this whirlpool is remarkable, are, that its particles move
with velocities inversely proportional to their distances from the

* See *Transactions of the Institution of Engineers in Scotland*, Vol. i. p. 90.
† [*Supra*, p. 1.]

centre or axis of rotation, and that each particle composing it is free to move to any position within the whirlpool, without interfering with the general motion of the other particles, as each one in moving towards or from the centre assumes of itself, subject simply to the laws of motion under a central force, the velocity due to its position in the whirlpool. It is also a property of this whirlpool, that, for any equal particles, in whatever situations in it they may momentarily be, the sum of the energies corresponding to velocity, to pressure, and to height, is constant. Thus it arises that, in the exterior whirlpool in the pump for raising water, as each particle of the water in moving from the centre, gives up its velocity according to the law of motion already stated, it either actually ascends or becomes subject to a pressure capable of causing it to ascend, through a height corresponding to the energy deducted from its motion. It follows, therefore, that through the medium of the exterior whirlpool a decided increase in the working efficiency of centrifugal pumps is attainable; the work contained in the rapid motion of the water leaving the wheel, which in centrifugal pumps as ordinarily constructed is wasted, being, according to the new arrangement, usefully employed in increasing the pumping power of the machine.

To arrive by a simpler method, than that just given, at a general idea of the mode of action of the exterior whirlpool in improving the efficiency of the centrifugal pump, it is only necessary to consider that the mass of water revolving in the whirlpool chamber, round the circumference of the wheel, must necessarily exert a centrifugal force, and that this centrifugal force may readily be supposed to add itself to the outward force generated within the wheel; or, in other words, to go to increase the pumping power of the wheel. The outward force generated within the wheel is to be understood as being produced entirely by the medium of centrifugal force, if the vanes of the wheel be straight and radial; but if they be curved, as is more commonly the case, the outward force is partly produced through the medium of centrifugal force, and partly applied by the vanes to the water, as a radial component of the oblique pressure, which, in consequence of their obliquity to the radius, they apply to the water as it moves outwards along them. On this subject it is well to observe, that while the quantity of water made to pass through a given pump with curved vanes is perfectly variable

at pleasure, the smaller the quantity becomes the more nearly
will the force generated within the wheel for impelling the water
outwards, become purely centrifugal force, and the more nearly
will the pump become what the name ordinarily given to it would
seem to indicate—a purely centrifugal pump.   When, however, a
centrifugal pump with vanes curved backwards in such forms as
are ordinarily used in well-constructed examples of the machine,
is driven at a speed considerably above that requisite merely to
overcome the pressure of the water, and cause lifting or propulsion
to commence, the radial component of the force applied to the
water by the vanes will become considerable, and the water
leaving the circumference of the wheel will have a velocity less
than that of the circumference of the wheel, in a degree having
some real importance in practice.   It has, indeed, been promulgated in respect to some centrifugal pumps with curved vanes
brought into use in the last six or eight years, that they are
capable of propelling the water from the centre to the circumference, along the line of a stationary radius, and of
discharging the water at the circumference without rotatory
velocity.   This supposition needs only to be mentioned as being
fallacious.   A motion approximately along a stationary radius
may no doubt have been attained when the pump was made to
impel a fluid against no resistance; but when a pump is actually
doing work in raising water, the rotatory motion of the water on
leaving the circumference of the wheel is unavoidable.   In practice
it involves, in ordinary pumps, an important loss, the loss being
commonly about as much as the entire power usefully applied in
overcoming the gravitation of the water; and the efficiency of the
pump comes out, therefore, when friction and other further losses
are taken into account, in ordinary circumstances, considerably
less than fifty per cent. of the applied power.

From calculations relative to the various sources of loss
occurring in centrifugal pumps of the improved kind with the
exterior whirlpool, according to the design carried out for
Messrs James Ewing & Co.—the writer thinks the efficiency of
such pumps may be fairly estimated at about seventy per cent.
Now the steam-engine, already existing on the estate, for which
the pump forming the special subject of the present paper was
required, is estimated as being capable of applying 25 horse-
power (of 33,000 foot-pounds per minute) to the driving of the

pump. The lift of the water is intended to be variable, according
to the state of the weather, and is to range ordinarily from
2½ feet to 5 feet. Then for a 2½ feet lift, and an assumed
efficiency of the pump of seventy per cent., the quantity of water
lifted comes to be 3700 cubic feet per minute. And for a lift
of 5 feet, the quantity would be about half as much, or, say 1850
cubic feet per minute.

The details of the constructive arrangements of the centrifugal
pump are shown in figs. 1 and 2, prepared from drawings supplied
to the Institution by Messrs W. & A. M'Onie & Co., the engineers
by whom the execution of the work has been carried out. In
reference to them, the writer here wishes to recognize the very
efficient aid they afforded him during the progress of the work in
arranging many of the practical details of the construction. The
work was required to be designed and executed in great haste,
and their ready co-operation tended very materially to facilitate
the business of the undertaking. Fig. 1 is an elevation, and
fig. 2 is a plan of the pump—one-half of each figure being in
section—fig. 1 as through the line a b in fig. 2, and fig. 2 as
through the line c d in fig. 1. In addition to the description
already given of the principles and mode of action of pumps of
this kind, attention may be directed to the helical outlet chamber,
B C, which is arranged for receiving the water from all parts of the
circumference of the whirlpool chamber, except the part from
which an immediate outlet into the discharge pipe is available.
This helical chamber has a gradually increasing area to meet the
regular accessions of water which it receives from its commencement at its smaller extremity, B, forward to its termination at C
in the outlet pipe. This helical outlet chamber is arranged with
a large part of its capacity lying nearer to the axis of rotation
than the edge of the whirlpool chamber cover, D. This arrangement is made with the special object in view of preventing
eddying water that will exist in the outlet pipe at times when
the pump is raising a small quantity of water through a high lift,
from being able to return to the circumference of the whirlpool,
with the risk of breaking through the water of the whirlpool, and,
as it were, falling in towards the centre in consequence of having
less rotatory motion, and, therefore, less centrifugal force than the
water of the whirlpool. If any such return water were to
penetrate into the whirlpool, it would very materially damage

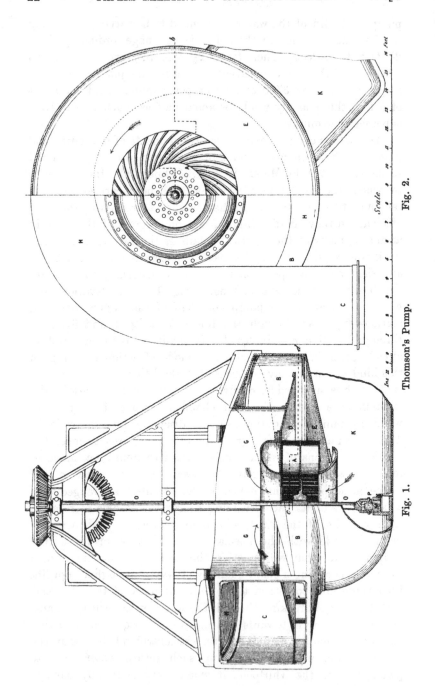

Fig. 2.

Scale

Thomson's Pump.

Fig. 1.

the action of the whirlpool, by disturbing its motion, and introducing impacts of quickly-moving against slowly-moving water.
The helical chamber, as may readily be seen by consideration of
the figures, is adapted for receiving any such eddying water, and
hindering it from getting access to the whirlpool; as the most
rapidly moving water coming direct from the whirlpool will tend
always to fly outwards, and will occupy the outer parts of the
helical chamber, leaving the inner parts of the same chamber for
the dead or eddying water, when from any of the varying circumstances of the height of lift and power applied, such eddying or
dead water exists.

The writer would further direct attention to the peculiar
adaptation of the forms selected for the various parts of the case
for the attainment of a high degree of stiffness for preventing the
top and bottom of the case from changing their distance apart
under varying circumstances of the internal water pressure. It is
to be observed, on this point, that no tie-bolts through the case
for holding the top and bottom together are admissible, as any
such tie-bolts would interrupt the revolving motion of the water.
Yet, at the same time, there ought to be as close a fit as possible,
without rubbing, between the joint-rings of the wheel and case at
the central orifices, in order to prevent leakage and waste of the
water pumped. The conical forms of the bottom, E, and top, D, of
the whirlpool chamber, and the cylindrical vertical walls, G, with
the somewhat conical form of the roof, H, of the helical chamber,
all taken together, and, in conjunction with riveted boiler-plate
as the material, form a shell of peculiar stiffness. The chief
difficulty connected with this undertaking has had reference to
the transmission by sea and land of so bulky a thing as the
boiler-plate casing, which is 16 feet in diameter, and which could
not well be sent in parts, to be put together at the site where the
pump is to work. The casing was, of course, sent separate from
the *boat*, K, or large conduit for leading the water to the lower
central orifice, L, and the casing was shipped on deck, being too
large to enter the hold of the vessel by which it was transmitted.
Accounts have been received of the safe landing in the colony of
these bulky parts; and news is soon expected of further progress
having been made towards their erection at the site where they
are intended to operate.

It may be mentioned that the pivot, M, on which the wheel, A, turns, is made of lignum vitae. It is fixed standing upright in the bottom of the boat, K; and in the bottom of the shaft, O, there is fixed a brass socket, P, which turns on the pivot. Arrangements are made for spreading an ample supply of water over the rubbing surfaces of the wood and brass, for lubrication. Bearings of lignum vitae have of late years been found to be much superior in durability, when working under water, to any kinds of metallic pivots. Lignum vitae, from this reason, has recently come much into use for the bearings of propeller shafts, where the shafts pass through the sterns of steamers. Lignum vitae pivots have likewise come much into use for turbines in America, and the writer is now in the habit of introducing them in his vortex turbines, in preference to all other kinds of pivots.

In conclusion, the writer considers centrifugal pumps, constructed on the principles which have been described, to be capable of affording the best available means of raising large quantities of water through low lifts, or through lifts of such moderate height as may not require the wheel to revolve at an inconveniently high speed. For drainage of fens, and for raising the sewerage of towns through moderate lifts, they are peculiarly suitable. They are much less liable than common pumps to be choked or injured by the entrance of solid materials; and, having no working valves or pistons, they are remarkably free from injury by wearing; whilst, as regards economy of power in the applications referred to, they will, the writer believes, be found decidedly superior to pumps of the ordinary kinds, constructed with valves and pistons or plungers.

5. ON THE FRICTION BREAK DYNAMOMETER.

[From the *Report of the British Association*, Section G, Glasgow, 1855, p. 209.]

IN this paper Mr Thomson explained the nature and principles of the Friction Break Dynamometer, characterizing it as a highly valuable apparatus for the measurement of mechanical power. He turned special attention to matters having important bearings on its successful employment in practice, and to the consideration of which he had been led by experience in its use on the large scale.

The chief difficulties to be contended with, he stated as follows :

1. The heat generated in the consumption of the mechanical power.

2. Vibration or even entire instability of the arm of the break due to ovalness or other imperfections of the friction drum.

3. Tremor of the driving shaft occasioned by alternate sticking and slipping of the drum in its friction blocks, instead of steady slipping.

In regard to these matters he made statements and explanations to the following effect :—Unless the drum be very large with reference to the power to be consumed by it, the heat generated by the friction usually requires to be carried off by streams of water carefully distributed over the drum. On the proper distribution of the water, and its regularity of supply, much of the success of the apparatus depends, since great irregularities in the friction are liable to result from imperfect arrangements for the water supply.

In the practical employment of the friction break dynamometer it is often necessary to form the drum in two halves, in order that it may be got into its place on the driving shaft. This arrangement, however, is often a source of great detriment to the working of the apparatus, on account of the difficulty or impossibility of bringing the two halves of the divided drum so correctly together as to form a sufficiently perfect cylindrical surface for producing a uniform friction. In cases therefore in which the dividing of the drum cannot be avoided, the greatest possible care ought to be taken in effecting a correct and secure union of the two parts. He had on some occasions diminished very materially the inconvenience arising from the vibrations of the arm, by connecting a spring balance with the scale for bearing the weights, in such a way as to make the arm tend to stable equilibrium in the position intended for it when working.

It very frequently happens that, from no clearly apparent or controllable cause, a violent torsional tremor occurs in the friction drum and driving shaft; while through some very slight change of circumstances, such as a change in the heat of the drum, or in the mode of application of the water, or in the speed of revolution of the shaft, steadiness of motion may be instantly restored, and perhaps soon again destroyed. The origin of the tremor he attributes to one of the known laws of the friction and cohesion of bodies; namely, that the force necessary to overcome the cohesion before sliding has commenced, is usually more than the force necessary to overcome the friction of the sliding motion. The evil liable to arise from the tremor he had found to be very great, the danger to life of the by-standers in such experiments being sometimes considerable. He had himself witnessed a case in which a violent tremor occurred in the testing of a powerful water-wheel; and, on the conclusion of the experiments, the working shaft of the wheel was found to be split and twisted.

Notwithstanding the difficulties occasionally arising in the use of the friction dynamometer, however, its remarkable efficiency, when not marred by such occurrences, and the certainty of its indications when working properly, render it a most valuable apparatus for practical use in many important and delicate cases often arising for decision. It is therefore a mechanism in which improvements are much to be desired; and also in which, he is of opinion, they are likely to be found.

6. On a Jet Pump, or Apparatus for drawing up Water by the Power of a Jet.

[From the *British Association Report*, Section G, Belfast, 1852, page 130.]

THE purpose for which the author has designed this new pump, is to clear the water out of the pits of submerged water-wheels, when access to them is required for inspection or repairs. This pump may also be used for raising water in other cases where an abundant fall of water is available; as, for instance, for draining a marsh in the neighbourhood of a waterfall. Its action depends on two principles. One of these is the same as that of the steam blast used in locomotive engines, and in the ventilation of mines. The other is that of the increased flow of water from a pipe, produced by giving a gradually widening form to its discharging extremity.

A sketch of the apparatus is given in the accompanying figure, where A is a pipe which supplies the water to the nozzle B for the jet, and C is a pipe which receives, at its narrow end, the jet from the nozzle, and on account of its gradually widening form, causes a suction capable of raising water by the pipe *D*.

The various principles brought into action in this apparatus have, as was stated by the author, been long known in hydro-dynamics; but their combination in this form for use he believed to be new. A rush of water had been used previously in a

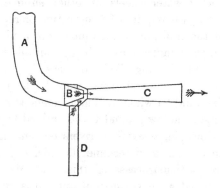

somewhat similar way in Italy to draw up and carry off the water of a marsh. In respect to the method there employed he had not been able to obtain full information; but the description of it he had received led him to suppose that it was not so efficacious as the method which formed the subject of his communication to the Meeting.

7. On an Experimental Apparatus constructed to deter-mine the Efficiency of the Jet Pump; and a Series of Results obtained.

[From the *Report of the British Association*, Section G, Hull, 1853, page 130.]

Mr Thomson had last year given, at the Mechanical Section, an account of a very simple machine which he had contrived for the purpose of raising water from beneath the lowest available level of discharge, by means of a supply of other water coming from a higher level. This machine he designated a Jet Pump, because it raised water by the action of a jet. A drawing and an explanation of it, in its original form, are to be found in the Report of the Transactions of the Mechanical Section for last year. The machine is remarkably simple and free from liability to derangement, having no valves, pistons, or other moving mechanisms. It consists indeed only of pipes with an internal jet, and is capable of working properly when left entirely to itself without the care of an attendant. It had at first been intended chiefly for one especial purpose, namely, to empty the pits of his own patent vortex water-wheels, or other submerged turbines, when access to them is required for inspection or repairs. During the progress of the trials, however, which were made of it for this purpose, it soon gave indications of being suitable for much more extensive uses, and of being likely to prove, in certain cases, an advantageous machine for draining swampy lands or shallow lakes. The cases of this kind for which its employment was contemplated are those in which the low ground to be drained happens to have, adjacent to its margin, streams or rivers descending from higher ground. With a view to determine its efficiency and its applicability in any particular cases of this kind, Mr Thomson had recently constructed an experimental apparatus in which a jet pump could be made to act subject to great variations in the ratio of the height of lift to the height of fall; and which was suited for indicating accurately the quantity of water lifted, and the height of the lift, corresponding to each quantity of water allowed to fall through any given distance within the working range of the apparatus. The results obtained give higher efficiencies than had been anticipated previously to the experiments, and remove all doubt as to the quantity of water which can be raised in any ordinary cases of its employment for the drainage of swampy land. They give, in fact, when taken in conjunction with known laws of the flow of fluids through orifices, the means of calculating, with full confidence, the requisite dimensions and proportions of a machine for the performance of a stated amount of work in the raising of water.

With respect to the nature of the experiments, a statement of a few principal points will here suffice, as the form and construction of the apparatus cannot be minutely explained in the absence of the drawings, which were exhibited in the Section by the author.

One of the chief difficulties anticipated by Mr Thomson, in his attempts to find good modes of experimenting, had reference to the determination of the quantity of water lifted by the pump from the low level to the discharge level, and of that let down from the high level to the discharge level. A very simple and effective mode of obviating the difficulty was devised, as follows:—An apparatus was arranged, not for measuring the absolute quantities of water lifted and let down in various experiments, but only the ratios of these quantities to one another; and it is to be observed, that, for the purpose in view, precision in the ratios of the quantities, rather than in their absolute amounts, was to be desired.

A vessel or cistern was provided and set up at a level above the jet pump; and all the water to supply the pump, as well as all the water to be lifted by it, was made to pass through the cistern, and to issue by a slit in its bottom, about one foot long, and of a width which could be varied at pleasure, but was usually about one-quarter of an inch. The water thus issued in the form of a thin sheet one foot wide and about one-quarter of an inch thick descending vertically. Out of this sheet of water any portion desired could be taken and conveyed away by means of a small movable wedge-shaped vessel made to slide in below the slit of the cistern. The water thus abstracted was conveyed down to the low level to be lifted by the jet pump, and the remainder was used for supplying the power in the jet pump. By observation of the width of the portion of water abstracted from the sheet, and comparison of that with the width of the whole sheet, the ratio of the two quantities was determined.

The absolute quantities of water could be varied at pleasure, without any alteration of their ratio, by increasing or diminishing the depth of water in the supply cistern from which the sheet of water issued. Thus the absolute quantity was adjusted so as to make either the high water supplying the power, or the low water lifted by the pump, stand at any desired level while the pump was in continuous action. The one of those two levels being thus fixed, the other, after some fluctuations, soon adjusted itself to its permanent height, and the two permanent levels were then observed, and so one experiment was completed. The series of experiments was made by successively cutting out various portions of the sheet of water to be conveyed to the low level; and then observing the height of lift, and the height of fall, which corresponded to each ratio of the quantity sent to be lifted, to the quantity sent to supply the power.

The following table gives a summary of the chief practical

results obtained. It is derived from two sets of experiments made on a jet pump with a jet of seven-eighths of an inch diameter. The height of fall varied from twenty-one inches to twenty-eight and three-fourths inches, and the height of lift from six inches to thirty-six and a half inches.

Ratio of lift to the fall	Quantity of water lifted if the water supplying the power is 100	Mechanical work performed in the raising of water if the work due to the fall is 100	Ratio of lift to the fall	Quantity of water lifted if the water supplying the power is 100	Mechanical work performed in the raising of water if the work due to the fall is 100
0·2	51	10·3	1·0	18	18·0
0·3	44	13·2	1·1	16	17·9
0·4	37	14·8	1·2	15	17·6
0·5	33	16·3	1·3	13	17·3
0·6	29	17·3	1·4	11·5	16·3
0·7	26	18·1	1·5	10·2	15·3
0·8	23	18·2	1·6	9·0	14·2
0·9	20	18·1	1·7	7·7	13·2

8. ON THE JET PUMP AND ITS APPLICATION TO THE DRAINAGE OF SWAMPY LANDS AND SHALLOW LAKES.

[From MS. notes of a paper read at the Royal Dublin Society, 1st June, 1855.]

IN this paper I propose to explain a novel application of a set of Jet Pumps to the drainage of swampy lands or shallow lakes.

The Jet Pump is a new machine which I brought under the notice of the Mechanical Section of the British Association at Belfast in 1852*. At that time I intended it chiefly for use in draining the water from the pits of turbine water-wheels when access to them is required for inspection or repairs, and about the same time I got one introduced at Strabane with very advantageous results for clearing an excessive amount of soakage or spring water from a pit in which the building of foundations for one of my improved turbines was in progress. At that time I made the jet act horizontally.

* [See *supra*, p. 26.]

I subsequently observed that the fall from discharge end
might be saved by addition of a descending pipe dipping below
lowest water level.

This constituted a material improvement and led me at once
to the form in which I now intend that the machine should be
generally used, with the jet acting vertically and the discharge
end of the pipe dipping under the water level in the course
prepared for leading the whole water away from the machine.

A preliminary idea of the principle of action in the jet pump
may probably be best communicated by reference to the jet blast
of locomotive steam engines, which to many must be already
familiar. The waste steam after having left the working cylinders
in the locomotive engine is directed vertically upwards at the
bottom of the chimney in its centre. It mixes with the smoke
and gaseous matters contained in the chimney and dashes them
upwards and thereby produces the draught required for brisk
combustion in the furnace.

Now if we imagine a jet of water to be substituted instead of
the steam as the impelling power, and water also to be substituted
for the products of combustion as the substance to be acted on,
we derive a primary conception of the mode of action of the jet
pump.

For the attainment of a great increase of power, however, and
of a high degree of perfection, an additional principle is introduced.
It consists in causing the water to flow, from its point of greatest
rapidity, forwards, through a gradually enlarging or trumpet-
shaped pipe :—a principle which has the effect of applying usefully
the mechanical work or energy of motion retained in the momentum
of each portion of the united stream proceeding from the jet and
from the low source, in the propulsion of those portions of the
streams which are following.

The principle according to which, when water enters rapidly
at the small end of a gradually widening pipe and flows forwards
to the wide end, each portion of the water produces a suction
impelling forwards all the portions following it, is not new. It
was known to the ancient Romans and was even used in some
cases among them as a fraud, in fact as a means of cheating the
Water Commissioners of those days ; the water which could flow
through a pipe of a certain size being first agreed to be paid for and a
trumpet pipe being afterwards added to produce an increased flow.

The effect of the trumpet pipe will be readily understood by considering that, when the pipe is flowing full of water or free from air, the velocity of the water must diminish just as the sectional area of the pipe increases. But the water, on account of its inertia, tends to retain the velocity it has, and, as no air can enter the pipe, the water tends to dash forwards leaving a vacuum behind it; but, as the pressure of the external air prevents this, instead of the vacuum being formed the suction is produced tending to draw forwards the water which is behind. The very powerful suction which the widening pipe is capable of producing, is strikingly exemplified in some of the experiments which I have made with the jet pump, and in which a force of suction has been produced to the extent of $1\frac{1}{2}$ times the height of the fall, thus causing the water to flow through the jet with the velocity due to $2\frac{1}{2}$ times its real fall.

Having now explained the construction of the Simple Jet Pump and the leading principles on which its action depends, I proceed to show how a number of jet pumps can be so combined as to form an advantageous machine adapted for the drainage of swampy lands.

*   *   *   *   *   *   *

For the purpose of determining the efficiency of the jet pump under various circumstances of height of fall and of lift, with especial reference to obtaining data for ascertaining what amount of work can be performed in the raising of water in any of the particular cases in which the employment of the machine may be contemplated, I, some time since, constructed and tried the experimental apparatus represented in this drawing [shown to the Meeting].

These results which I obtained with this apparatus I brought before the Mechanical Section of the British Association in a second communication made to it, at the Meeting at Hull in 1853*: but I did not then divulge the principle of the combination of jet pumps which now forms the principal object of my communication to you.

I have now said enough to explain the applicability of the jet pump to drainage and also to indicate the natural circumstances requisite for its employment. It remains to consider the comparative merits of this and of other systems for effecting the same

* [See *supra*, p. 27.]

objects. It may naturally be asked, if a fall of water be available, why not apply it to drive a common water-wheel or a turbine, either of which might be made to raise the water by common pumps, by centrifugal pumps, or by the ordinary scoop wheels at present much used in the drainage of lands in Holland?

There is no doubt that the efficiency, in respect to economy of the water power, of the machines I have just named could readily be made to surpass in many cases that attainable by the jet pump. There are however numerous practical advantages connected with the jet pump which will commonly be of greater consequence than any slight saving of the water; and this more particularly in cases in which the stream supplies more power than is required for the drainage of the swamp.

In the first place the first cost of the system of jet pumps is greatly less than that of any other kind of machinery capable of applying the power of the fall for raising the drainage water. This is a very important advantage, because the cost of any kind of water-wheels with pumping machinery is such as to prove a serious obstacle to their employment. Next the cost due to wear and tear in the jet pump is practically nothing: as there are no moving parts in the machine, the motion being confined entirely to the water itself; and the most important parts of the machine may be made of copper so that they may be free from liability to corrosion. In the jet pump in particular there are no valves liable to stick or leak: no pistons requiring to be packed or to have leathering renewed. In fact the jet pump is capable of working on from week to week or even from year to year without requiring to be touched for any kind of adjustment or repair. The entire skilled labour and superintendence required for any system of water-wheels and lifting machinery is saved by the jet pump, and no additional labour is required in other respects for the working of the jet pump beyond what is equally necessary for the other kinds of machinery. The labour actually required is chiefly that of keeping the water courses and banks in order, and of cleansing the gratings or strainers which must be used for the removal of sticks and other floating bodies from the water. This work is of a kind that can be performed by the ordinary agricultural labourers required for the cultivation of the swamp when drained.

Lastly it is to be particularly observed that while the power of water is peculiarly well suited for raising the water from low lands because the same wet weather which causes an increase of the quantity of water to be raised supplies also an increased power for the performance of the work: the combination of jet pumps in a series affords a remarkably simple self acting arrangement for increasing the power of the working apparatus just in proportion as the available power and the work to be performed are increased.

9. On a New Application of the Jet Pump by Steam as Auxiliary to a Large Drainage Centrifugal Pump.

[Read before the Belfast Natural History and Philosophical Society, December 21, 1864. No report is preserved except the manuscript of this introduction to the paper.]

In the early days of attempts at railway locomotion the puffing of the steam off directly from a waste steam pipe was found to be noisy and annoying and to cause danger by frightening horses. In order to mitigate these evils, it was thought advisable to throw the waste steam into the chimney from whence it might pass off into the air with less violence in its puffs. Soon however an astonishing result followed on the introduction of this new arrangement. An engine was found to be greatly improved in its power and speed, the fire being at the same time observed to burn much more briskly and the steam being generated more quickly. This simple chance proved to be, more perhaps than any other single event, or than any of the contrivances designed with high efforts of ingenuity and skill, the turning point to success in the railway system. The improvement in the engine referred to was found to be due to the end of the pipe throwing the steam into the chimney having happened to be directed upwards. The steam blown upwards with force, and entangling itself with the smoke, dashed the smoke forwards along with itself, and so caused a great rush of air through the fire to the lower part of the chimney where the partial void was thus produced. The jet blast in the chimney has ever since been universally adopted in railway engines as an essential stimulant to their energetic action.

Many years ago, on my having occasion to wish for a simple means for raising water from pits beside which other water happened to be in store at higher levels, it occurred to me that the principle used in the locomotive steam engine chimney might be applied to effect the desired object. Following up this idea and seeking to apply it to the best advantage under the new proposed circumstances, I was led to combine with it some other known hydraulic principles and so arrived at the simple and effective apparatus which I call the Jet Pump.

10. ON PRACTICAL DETAILS OF THE MEASUREMENT OF RUNNING WATER BY WEIR-BOARDS.

[From a manuscript abstract prepared by the Author, having been read at the British Association, Glasgow, 1855.]

MR THOMSON did not profess in this paper to enter into the reasoning and experiments by which formulae and tables have been made out for ascertaining the quantity of water which flows over a weir-board or through a notch in a vertical plate when the width is known and the depth from the water level to the horizontal edge of the board is observed. These reasonings and experiments had been already discussed more or less fully in many treatises, and very elaborately in some. Among the most accurate and refined experiments hitherto made for providing data for use in special cases he instanced those of MM. Poncelet and Lesbros, carried out at Metz, in the years 1827 and 1828, and detailed in a *Mémoire* read at the Academy of Sciences in November 1829. His object in the present paper for the British Association was to give explanations of the practical arrangements which in the course of his experience in the measurement of water by this method he had found most convenient, of niceties to be attended to for the attainment of accurate observations, of difficulties to be overcome, and of errors to be avoided which he knew were very frequently fallen into and often led to serious results.

He exhibited detailed drawings of gauging weirs which he had used. He stated as some of the chief matters to be attended to:

(1) That the weir-board, which dams back the water and in which the rectangular horizontal notch is cut, should present a flat vertical surface for contact with the water, and the upper surface

of the bottom of the notch should be bevelled off in such a way that the water may part at once from the edge of the flat vertical face, as if it were flowing from a perfectly thin plate instead of a board which for strength requires thickness.

(2) The pool or pond of water dammed back should be so deep that the water may lie nearly at rest in it with a level surface.

(3) The water should issue free from the edge of the board with air below the falling stream.

11. ON EXPERIMENTS ON THE MEASUREMENT OF WATER BY TRIANGULAR NOTCHES IN WEIR-BOARDS.

[From the *Report of the British Association*, Leeds, 1858 ; pp. 181—185.]

THE experiments proposed to be comprehended in the investigations to which the present interim Report relates, have for their object to determine the suitableness of triangular (or V-shaped) notches in vertical plates for the gauging of running water, instead of the rectangular notches in ordinary use. The ordinary rectangular notches, accurately experimented on as they have been, at great cost and with high scientific skill, in various countries, with the view of determining the necessary formulas and coefficients, for their application in practice, are for many purposes suitable and convenient. They are, however, but ill adapted for the measurement of very variable quantities of water, such as commonly occur to the engineer to be gauged in rivers and streams. If the rectangular notch is to be made wide enough to allow the water to pass in flood times, it must be so wide, that for long periods in moderately dry weather, the water flows so shallow over its crest that its indications cannot be relied on. To remove, in some degree, this objection, gauges for rivers or streams are sometimes formed in the best engineering practice, with a small rectangular notch cut down [at one end of the crest] below the general level of the crest of a large rectangular notch. If, now, instead of one depression being made for dry-weather use, in a crest wide enough for use in floods, we conceive of a large number of depressions extending so as to give to the crest the

appearance of a set of steps of stairs, and if we conceive the
number of such steps to become infinitely great, we are led at
once to the conception of the triangular, instead of the rectangular
notch. The principle of the triangular notch being thus arrived
at, it becomes evident that there is no necessity for having one
side of the notch vertical and the other slanting; but that, as
may in many cases prove more convenient, both sides may be
made slanting, and their slopes may be alike. It is then to be
observed, that by the use of the triangular notch with proper
formulas and coefficients derivable by due union of theory and
experiments, quantities of running water, from the smallest to the
greatest, may be accurately gauged by their flow through the same
notch. The reason of this is obvious, from considering that in the
triangular notch, when the quantity flowing is very small, the flow
is confined to a small space admitting of accurate measurement;
and that the space for the flow of water increases as the quantity
to be measured increases, but still continues such as to admit of
accurate measurement.

Further, the ordinary rectangular notch, when applied for the
gauging of rivers, is subject to a serious objection from the difficulty,
or impossibility, of properly taking into account the influence of the
bottom of the river on the flow of the water to the notch. If it were
practicable to dam up the river so deep that the water would flow
through the notch as if coming from a reservoir of still water, the
difficulty would not arise. This, however, can seldom be done in
practice; and although the bottom of the river may be so far below
the crest as to produce but little effect on the flow of the water
when the quantity flowing is small, yet when the quantity becomes
great, the " Velocity of Approach " comes to have a very material
influence on the flow of the water, but an influence which it
is usually difficult, if not impracticable, to ascertain with satisfactory accuracy. In the notches now proposed, of triangular
form, the influence of the bottom may be rendered definite, and
such as to affect alike (or at least by some law that may be readily
determined by experiments) the flow of the water when very
small, or when very great, in the same notch. The method by
which I propose that this may be effected, consists in carrying out
a floor starting exactly from the vertex of the notch, and extending
both up stream and laterally so as to form a bottom to the channel
of approach, which will both be smooth, and will serve as the

lower bounding surface of a passage of approach unchanging in form while increasing in magnitude at the places at least which are adjacent to the vertex of the notch. The floor may either be perfectly level, or may consist of two planes whose intersection would start from the vertex of the notch, and, as seen in plan, would pass up stream perpendicularly to the direction of the weir board, the two planes slanting upwards from their intersection more gently than the sides of the notch. The level floor, although theoretically not quite so perfect as the floor of two planes, would probably, for most practical purposes, prove the more convenient arrangement. With reference to the use of the floor, it may be said, in short, that by a due arrangement of the notch and the floor, a discharge orifice and channel of approach may be produced, of which (the upper surface of the water being considered as the top of the channel and orifice) the form will be unchanged or but little changed, with variations of the quantity flowing;—very much less certainly than is the case with rectangular notches.

The laws regulating the quantities of water flowing in such orifices as have now been described, come naturally next to be considered. Without, however, in the present interim Report, attempting to enter on a detailed discussion of theoretical considerations on this subject, I shall here merely advert briefly to the principal results and methods of reasoning.

By theory I have been led to anticipate that the quantity flowing in a given notch should be proportional, or very nearly so, to the $\frac{5}{2}$ power of the lineal dimensions of the cross section of the issuing jet, or to the $\frac{5}{2}$ power of the head of water over the vertex of the notch. This head is to be understood, in the case of water flowing from a still reservoir, as being measured vertically from the level of the water surface down to the vertex of the notch; or, in the case of water flowing to the notch, with a considerable velocity of approach over a floor arranged as above prescribed, the head is to be considered as being measured vertically from the water surface where the motion is nearly stopped by the weir board, at a place near the board, but as far as may be found practicable from the centre of the notch. The law here enunciated, to the effect that the quantity flowing should be proportional to the $\frac{5}{2}$ power of the head, I consider should hold good rigidly in reference to water flowing by a triangular notch in a thin vertical plate from a large and deep reservoir of still water, if the water were a perfect fluid, free from viscidity and friction, and free from capillary attraction at its surface, and from any other slight disturbing causes that may have minute influences on the flow, the flow being supposed to be that due simply to gravitation resisted by the inertia of the fluid. The like may be said of water flowing from triangular notches with shallow channels of approach, having floors as described above, when due attention is given to make the passages of approach so as really to remain unchanged in form for a sufficient distance from the notch, while increasing in magnitude as the flow increases (such being supposed, according to my theory, to be possible); and if due attention be paid to measuring the heads in all cases in positions similarly situated with reference to the varying dimensions of the issuing streams.

In illustration of these statements, or suppositions, I would merely say that if two triangular notches, similar in form, have water flowing in them at different depths but with similar passages of approach, the cross sections of the two jets at the notches may be similarly divided into the same number of elements of area; and that the areas of the corresponding elements will be proportional to the squares of the lineal dimensions of the cross sections, or, as from various considerations may readily be assumed, proportional to the squares of the heads; also the velocities of the water in the corresponding elements may be taken as proportional to the square roots of the lineal dimensions, or to the square roots of the heads. From these considerations, supported by numerous others, it appears that the quantities flowing should be proportional to the products of the squares of the heads into their square roots, or to the $\frac{5}{2}$ powers, as already stated.

The friction of the fluid on the solid bounding surfaces of the passages of approach where the water moves rapidly adjacent to the notch, may readily be assumed, from all previous experience in similar subjects, not to have a very important influence even on the absolute amount of the flow of the water; and if we assume (as is known to be nearly the case for high velocities, such as occur in notches used for practical purposes, unless unusually small) that the tangential force of friction of the fluid, per unit of area of surface flowed along, is proportional to the square of the velocity of flow, it follows by theory that the friction, although slightly influencing the absolute amount of the flow, will not, according to that assumption, at all interfere with its proportionality to the $\frac{5}{2}$ power of the head, and this condition will very nearly hold good if the assumption is very nearly correct.

How closely the theory thus briefly sketched may be found to agree with the actual flow of water will be a subject for experimental investigation; and whatever may be the result in this respect, the main object must be to obtain, for a moderate number of triangular notches of different forms, and both with and without floors at the passage of approach, the necessary coefficients for the various forms of notches and approaches selected, and for various depths in any one of them, so as to allow of water being gauged for practical purposes, when in future convenient, by means of similarly-formed notches and approaches. The utility of the proposed system of gauging, it is to be particularly observed, will not depend on a perfectly close agreement of the theory described with the experiments, because in respect to any given form of notch and approach, a table of experimental coefficients for various depths, or an empirical formula slightly modified from the theoretical one, will serve all purposes. To one evident simplification in the proposed system of gauging, as compared with that by rectangular notches, I would here advert, namely, that in the proposed system, when once the form of the notch and channel of approach is fixed for gauging any set of streams, the quantity flowing comes to be treated as a function of only one variable, namely, the measured head of water; while in the rectangular notches it is practically treated as a function of at least two variables, namely, the head of water and the horizontal width of the notch; because in practice it would be inconvenient, if not impossible, to select any single width of notch, or any moderate number of widths of notches, for general use, for very varied quantities of water. It is commonly also a function of a third variable, very difficult to be taken into account, namely, the depth from the crest of the notch down to the bottom of the channel of approach; which depth must vary in its influence with all the varying ratios between it and the other two quantities of which the flow is a function.

The proposed system of gauging also gives facilities for taking another element into account which often arises in practice; namely, the influence of back water on the flow of the water in the gauge, when, as frequently occurs in rivers, it is found

impracticable to dam the river up sufficiently to give it a clear
overfall, free from the back or tail water. For any given ratio of
the height of the tail water above the vertex of the notch, to the
height of the head water above the vertex of the notch, I would
anticipate that the quantities flowing would still be, approximately
at least, proportional to the $\frac{5}{2}$ power of the head as before; and
a set of coefficients would have to be determined experimentally
for different ratios of the height of the head water to the height
of the tail water above the vertex of the notch.

With the aid of the grant placed at my disposal by the
Association at last year's meeting, for the purposes of these
researches, I have got an experimental apparatus constructed and
fitted up at a place a few miles distant from Belfast, in Carr's
Glen, on the grounds of Mr Neeson, who has kindly afforded me
all the necessary facilities regarding the water supply and the site
for the experiments; and I have got some preliminary experiments
made on a right-angled notch in a vertical plane surface, the sides
of the notch making angles of 45° with the horizon, and the flow
being from a deep and wide pool of quiet water, and the water
thus approaching the notch uninfluenced by any floor or bottom.
The principal set of experiments as yet made were on quantities
of water varying from about 2 to 10 cubic feet per minute; and
the depths or heads of water varied from 2 to 4 inches in the
right-angled notch.   From these experiments I derive the formula
$Q = \cdot 317 H^{\frac{5}{2}}$, where Q is the quantity of water in cubic feet per
minute, and H the head as measured, vertically, in inches, from
the still water level of the pool, down to the vertex of the notch.
This formula is submitted, at present temporarily, as being
accurate enough for use for ordinary practical purposes for the
measurement of water by notches similar to the one experimented
on, and for quantities of water limited to nearly the same range
as those in the experiments; but as being, of course, subject
to amendment by more perfect experiments extending through
a wider range of quantities of water.

Out of the grant of £10 from the Association for these
experiments, the amount for which I have hitherto had to apply
to the Treasurer as having been expended in them is £8. 0s. 4d.;
which leaves a balance remaining of £1. 19s. 8d.

It will be readily observed, that the experimental investigations
indicated in the foregoing report as desirable, are such as would

require for their completion, and extension to large flows of water, a great expenditure both of time and money, like as has already been the case with researches on the flow of water in rectangular notches. All that I can myself for the present propose to attempt, is to open up the subject with experiments on moderately small flows of water; and with this view, I would be glad to be aided, by a further grant from the Association, in continuing experiments of the kinds already undertaken.

12. ON EXPERIMENTS ON THE GAUGING OF WATER BY TRIANGULAR NOTCHES.

[From the *Report of the British Association*, Manchester, 1861 ; pp. 151—158.]

IN 1858 I presented to the Association an interim Report on the new method which I had proposed for the gauging of flowing water by triangular (or V-shaped) notches, in vertical plates, instead of the rectangular notches, with level bottom and upright sides, in ordinary use. I there pointed out that the ordinary rectangular notches, although for many purposes suitable and convenient, are but ill adapted for the measurement of very variable quantities of water, such as commonly occur to the engineer to be gauged in rivers and streams; because, if the rectangular notch be made wide enough to allow the water to pass through it in flood times, it must be so wide that for long periods, in moderately dry weather, the water flows so shallow over its crest, that its indications cannot be relied on. I showed that this objection would be removed by the employment of triangular notches, because, in them, when the quantity flowing is small, the flow is confined to a narrow and shallow space, admitting of accurate measurement; and as the quantity flowing increases, the width and depth of the space occupied in the notch increase both in the same ratio, and the space remains of the same form as before, though increased in magnitude. I proposed that in cases in which it might not be convenient to form a deep pool of quiet water at the up-stream side of the weir-board, the bottom of the channel of approach, when the triangular notch is used, may be formed as a level floor, starting exactly from the vertex of the notch, and extending both up stream and laterally so far as that the water entering on it at its margin may be practically considered as still water, of which the height of the surface above the vertex of the notch may be measured in order to determine the quantity flowing. I indicated theoretic considerations which led to the anticipation that in the triangular notch, both without and with the floor, the quantity flowing would be proportional, or very nearly so, to the $\frac{5}{2}$ power of the height of the still-water surface above the vertex of the notch. As the result of moderately accurate experiments which I had at that time been able to make on the flow in a right-angled notch, without floor, I gave the formula $Q = 0.317 H^{\frac{5}{2}}$, where Q is the quantity of water in cubic feet per minute, and H the head of water, as measured vertically, in inches, from the still-water level of the pool down to the vertex of the notch. This formula I submitted at that time temporarily, as being accurate enough for use for many ordinary practical purposes for the measurement of water by notches similar to the one experimented on, and for quantities of water limited to nearly the same range as those in the experiments (from about two to ten cubic feet per minute), but as being subject to amendment by future experiments which might be of greater accuracy, and might extend over a wider range of quantities of water.

Having been requested by the General Committee of the Association to continue my experiments on this subject, with a grant placed at my disposal for the purpose, I have, in the course of last summer and of the present summer, devoted much time to the carrying out of more extended and more accurate experiments. The results which I have now obtained are highly satisfactory. I am confident of their being very accurate. I find them to be in close accordance with the law which had been indicated by theoretical considerations; and I am satisfied that the new system of gauging, now by these experiments made completely ready for general application, will prove to be of great practical utility, and will afford, for a large class of cases, important advantages over the ordinary method—for such cases, especially, as the very varying flows of rivers and streams.

The experiments were made in the open air, in a field adjacent to a corn-mill belonging to Mr Henry Neeson, in Carr's Glen, near Belfast. The water-supply was obtained from the course leading

to the water-wheel of the mill, and means were arranged to allow
of a regulated supply, variable at pleasure, being drawn from that
course to flow into a pond, in one side of which the weir-board
with the experimental notch was inserted. The inflowing stream
was so screened from the part of the pond next the gauge-notch,
as to prevent any sensible agitation being propagated from it to
the notch, or to the place where the water level
was measured. For measuring the water level,
a vertical slide-wand of wood was used, with
the bottom end cut to the form of a hook (as
shown in the marginal figure), the point of
which was a small level surface of about one-
eighth of an inch square. This point of the
hook, by being brought up to the surface of the
water from below, gave a very accurate means
for determining the water level, or its rise or
fall, which could be read off by an index mark
near the top of the wand, sliding in contact
with the edge of a scale of inches on a fixed
framing which carried the wand.

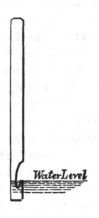

By other experimenters a sharp-pointed hook, like a fishing-
hook, has sometimes, especially of late, been used for the same
purpose, and such a hook affords very accurate indications. The
result of my experience, however, leads me to incline to prefer
something larger than the sharp-pointed hook, and capable of
producing an effect on the water surface more easily seen than
that of a sharp-pointed hook ; and on the whole I would recommend
a level line like a knife-edge, which might be from one-eighth to
half an inch long, in preference either to a blunt point with level
top or a sharp point. The blunt point which I used was so small,
however, as to suit very perfectly. If the point be too large,
it holds the water up too much on its top as the water in the pond
descends, and makes too deep a pit in the surface as the water
ascends and begins to flow over it. The knife-edge would be free
from this kind of action, and would, I conceive, serve every
purpose, perfectly, except when the water has a sensible velocity
of flow past the hook, and in that case, perhaps, the sharp point,
like that of a fishing-hook, might be best.

To afford the means for keeping the water surface during
an ex eriment exactly at a constant level, as indicated by the

point of the wooden hook, a small outlet waste-sluice was fitted
in the weir-board. The quantity of water admitted to the pond
was always adjusted so as to be slightly in excess of that required
to maintain the water level in the pond at the height at which the
hook was fixed for that experiment. Then a person lying down,
so as to get a close view of the contact of the water surface with
the point of the hook, worked this little waste or regulating
sluice, so as to maintain the water level constantly coincident with
the point of the hook.

The water issuing from the experimental notch was caught in
a long trough, which conveyed it forward with slight declivity, so
as to be about seven or eight feet above the ground further down
the hill-side, where two large measuring-barrels were placed side
by side at about six feet distance apart from centre to centre.
Across and underneath the end of the long trough just mentioned,
a tilting-trough six feet long was placed, and it was connected at
its middle with the end of the long trough by a leather flexible
joint, in such a way that it would receive the whole of the water
without loss, and convey it at pleasure to either of the barrels,
according as it was tilted to one side or the other.

Each barrel had a valve in the bottom, covering an aperture
six inches square, and the valve could be opened at pleasure, and
was capable of emptying the barrel very speedily. The capacity
of the two barrels jointly was about 230 gallons, and their content
up to marks fixed near the top for the purpose of the experiments
was accurately ascertained by gaugings repeated several times
with two- or four-gallon measures with narrow necks.

By tilting the small trough so as to deliver the water
alternately into the one barrel and the other, and emptying each
barrel by its valve while the other was filling, the process of
measuring the flowing water could be accurately carried on for as
long time as might be desired. With this apparatus, quantities
of water up to about 38 cubic feet per minute could be measured
with very satisfactory accuracy.

The experiments of which I have now to report the results
were made on two widths of notches in vertical plane surfaces.
The notches were accurately formed in thin sheet iron, and were
fixed so as to present next the water in the pond a plane surface,
continuous with that of the weir-board.

The one notch was right-angled, with its sides sloping at

45° with the horizon, so that its horizontal width was twice its depth. The other notch had its sides each sloping two horizontal to one vertical, so that its horizontal width was four times its depth.

In each case experiments were made both on the simple notch without a floor, and on the same notch with a level floor starting from its vertex, and extending for a considerable distance both up stream and laterally. The floor extended about two feet on each side of the centre of the notch, and about two and a half feet in the direction up stream, and this size was sufficient to allow the water to enter on it with only a very slow motion—so slow as to be quite unimportant. The height of the water surface above the vertex of the notch was measured by the sliding hook at a place outside the floor, where the water of the pond was deep and still.

The principal results of the experiments on the flow of the water in the right-angled notch without floor are briefly given in the annexed table:

H [Ins.]	Q [Cubic ft. per min.]	c
7	39·69	·3061
6	26·87	·3048
5	17·07	·3053
4	9·819	·3068
3	4·780	·3067
2	1·748	·3088

the quantity of water given in column 2 for each height of 2, 3, 4, 5, 6, and 7 inches being the average obtained from numerous experiments comprised in two series, one made in 1860, and the other made in 1861, as a check on the former set, and with a view to the attainment of greater certainty on one or two points of slight doubt. The second set was quite independent of the first, the various adjustments and gaugings being made entirely anew. The two sets agreed very closely, and I present an average of the two sets in the table as being probably a little more nearly true than either of them separately. The third column contains the values of the coefficient c, calculated for the formula $Q = cH^{\frac{5}{2}}$, from the several heights and corresponding quantities of water given in the first and second columns, H being the height, as measured vertically in inches from the vertex of the notch up to the still-water surface of the pond, and Q being the corresponding quantity of water in cubic feet per minute, as ascertained by the experiments. It will be observed from this table that, while the quantity of water varies so greatly as from $1\frac{3}{4}$ cubic feet per minute to 39, the coefficient c remains almost absolutely constant; and thus the theoretic anticipation that the quantity should be proportional, or very nearly so, to the $\frac{5}{2}$ power of the depth is fully confirmed by experiment. The mean of these six values of c is ·3064; but, being inclined to give rather more weight, in the determination of the coefficient as to its amount, to some of the experiments made this year than to those of last year, I adopt ·305 as the coefficient, so that the formula for the right-angled notch without floor will be

$$Q = ·305 H^{\frac{5}{2}}.$$

My experiments on the right-angled notch with the level floor, fitted as already described, comprised the flow of water for depths of 2, 3, 4, 5, and 6 inches. They indicate no variation in the value of c for different depths of the water, but what may be attributed to the slight errors of observation. The mean value which they show for c is ·308; and as this differs so little from that in the formula for the same notch without the floor, and as the difference is within the limits of the errors of observation, and because some consecutive experiments, made without and with the floor, indicated no change of the coefficient on the insertion of the floor, I would say that the experiments prove that, with the right-angled notch, the introduction of the floor produces scarcely any increase or diminution on the quantity flowing for any given depth, but do not show what the amount of any such small increase or diminution may be, and I would give the formula

$$Q = ·305 H^{\frac{5}{2}},$$

as sufficiently accurate for use in both cases. The experiments in both cases were made with care, and are without doubt of very satisfactory accuracy; but those for the notch without the floor are, I consider, slightly the more accurate of the two sets.

The experiments with the notch with edges sloping two horizontal to one vertical showed an altered feature in the flow of the issuing vein as compared with the flow of the vein issuing from the right-angled notch. The edges of the vein, on issuing from the notch with slopes two to one, had a great tendency to cling to the outside of the iron notch and weir-board, while the portions of the vein issuing at the deeper parts of the notch would shoot out and fall clear of the weir-board. Thus, the vein of water assumed the appearance of a transparent bell, as of glass, or rather of the half of a bell closed in on one side by the weir-board and enclosing air. Some of this air was usually carried away in bubbles by the stream at bottom, and the remainder continued shut up by the bell of water, and existing under slightly less than atmospheric pressure. The diminution of pressure of the enclosed air was manifested by the sides of the bell being drawn in towards one another, and sometimes even drawn together, so as to collapse with one another at their edges which clung to the outside of the weir-board. On the full atmospheric pressure being admitted, by the insertion of a knife into the bell of falling water, the collapsed sides would instantly spring out again. The vein of water did not always form itself into the bell; and when the bell was formed, the tendency to the withdrawal of air in bubbles was not constant, but was subject to various casual influences. Now it evidently could not be supposed that the formation of the bell and the diminution of the pressure of the confined air could occur as described without producing some irregular influences on the quantity flowing through the notch for any particular depth of flow, and this circumstance must detract more or less from the value of the wider notches as means for gauging water in comparison with the right-angled notch with edges inclined at 45° with the horizon. I therefore made numerous experiments to determine what might be the amount of the ordinary or of the greatest effect due to the diminution of pressure of the air within the bell. I usually failed to meet with any perceptible alteration in the quantity flowing due to this cause, but sometimes the quantity seemed to be increased, by some small fraction, such as one, or perhaps two, per cent. On the whole, then, I do not think that this circumstance need prevent the use, for many practical purposes, of notches of any desired width for a given depth.

My experiments give as the formula for the notch, with slopes of two horizontal to one vertical and without the floor,

$$Q = 0{\cdot}636\,H^{\frac{5}{2}},$$

and for the same notch, with the horizontal floor at the level of its vertex,

$$Q = 0\text{·}628 H^{\frac{5}{2}}.$$

In all the experiments from which these formulas are derived, the bell of falling water was kept open by the insertion of a knife or strip of iron, so as to admit the atmospheric pressure to the interior. The quantity flowing at various depths was not far from being proportional to the $\frac{5}{2}$ power of the depth, but it appeared that the coefficient in the formula increased slightly for very small depths, such as 1 or 2 inches. For instance, in the notch with slopes 2 to 1 without the floor, the coefficient for the depth of 2 inches came out experimentally 0·649, instead of 0·636, which appeared to be very correctly its amount for 4 inches depth. It is possible that the deviation from proportionality to the $\frac{5}{2}$ power of the depth, which in this notch has appeared to be greater than in the right-angled notch, may be due partly to small errors in the experiments on this notch, and partly to the clinging of the falling vein of water to the outside of the notch, which would evidently produce a much greater proportionate effect on the very small flows than on great flows. The special purpose for which the wide notches have been proposed is to serve for the measurement of wide rivers or streams in cases in which it would be inconvenient or impracticable to dam them up deep enough to effect their flow through a right-angled notch. In such cases I would now further propose that, instead of a single wide notch, two, three, or more right-angled notches might be formed side by side in the same weir-board, with their vertices at the same level, as shown in the annexed figure. In cases in which this method may be selected,

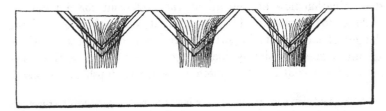

the persons using it, or making comparisons of gaugings obtained by it, will have the satisfaction of being concerned with only a single standard form of gauge-notch throughout the investigation in which they may be engaged.

By comparison of the formulas given above for the flows through the two notches experimented on, of which one is twice as wide for a given depth as the other, it will be seen that in the formula for the wider notch the coefficient ·636 is rather more than double the coefficient ·305 in the other. This indicates that as the width of a notch, considered as variable, increases from that of a right-angled notch upwards, the quantity of water flowing increases somewhat more rapidly than the width of the notch for a given depth. Now, it is to be observed that the contraction of the stream issuing from an orifice open above in a vertical plate is of two distinct kinds at different parts round the surface of the vein. One of these kinds is the contraction at the places where the water shoots off from the edges of the plate. The curved surface of the fluid leaving the plate is necessarily tangential with the surface of the plate along which the water has been flowing, as an infinite force would be required to divert any moving particle suddenly out of its previous course*. The other kind of contraction in orifices open above consists in the sinking of the upper surface, which begins gradually within the pond or reservoir, and continues after the water has passed the orifice. These two contractions come into play in very different degrees, according as the notch (whether triangular, rectangular, or with curved edges) is made deep and narrow, or wide and shallow. From considerations of the kind here briefly touched upon, I would not be disposed to expect theoretically that the coefficient c for the formula for V-shaped notches should be at all truly proportional to the horizontal width of the orifice for a given depth; and the experimental results last referred to are in accordance with this supposition. I would, however, think that, from the experimental determination now arrived at, of the coefficient for a notch so wide as four times its depth, we might very safely, or without danger of falling into important error, pass on to notches wider in any degree, by simply increasing the coefficient in the same ratio as the width of the notch for a given depth is increased.

---

* This condition appears not to have been generally noticed by experimenters and writers on hydrodynamics. Even MM. Poncelet and Lesbros, in their de-lineations of the forms of veins of water issuing from orifices in thin plates, after elaborate measurements of those forms, represent the surface of the fluid as making a sharp angle with the plate in leaving its edge.

APPENDIX—April, 1862.

With reference to the comparison made, in the concluding sentences of the foregoing Report, between the quantities of water which, for any given depth of flow, are discharged by notches of different widths, and to the opinion there expressed, that we might, without danger of falling into important error, pass from the experimental determination of the coefficient for a notch so wide as four times its depth, to the employment of notches wider in any degree, by simply increasing the coefficient in the same ratio as the width of the notch for a given depth is increased, I now wish to add an investigation since made, which confirms that opinion, and extends the determination of the discharge beyond the notches experimented on, to notches of any widths great in proportion to their depths. This investigation is founded on the formula for the flow of water in rectangular notches obtained from elaborate and careful experiments made on a very large scale by Mr James B. Francis, in his capacity as engineer to the Water-power Corporation at Lowell, Massachusetts, and described in a work by him, entitled *Lowell Hydraulic Experiments*, Boston, 1855[*]. That formula, for either the case in which there are no end-contractions of the vein, or for that in which the length of the weir is great in proportion to the depth of the water over its crest, and the flow over a portion of its length not extending to either end is alone considered, is

$$Q_1 = 3 \cdot 33 \, L_1 H_1^{\frac{3}{2}} \quad \dots\dots\dots\dots\dots\dots\dots(1),$$

where L_1 = length of the weir over which the water flows, without end-contractions; or length of any part of the weir not extending to the ends, in feet: H_1 = height of the surface-level of the impounded water, measured vertically from the crest of the weir, in feet: and Q_1 = discharge in cubic feet per second over the length L_1 of the weir.

It is to be understood that, in cases to which this formula is applicable, the weir has a vertical face on the upstream side, terminating at top in a level crest; and the water, on leaving the crest, is discharged through the air, as if the weir were a vertical thin plate.

To apply this to the case of a very wide triangular notch :—

---

[*] The formula is to be found at page 133 of that work.

Let ABC be the crest of the notch, and AC the water level in the impounded pool. Let the slopes of the crest be each m horizontal to 1 vertical; or, what is the same, let the cotangent of the

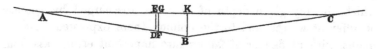

inclination of each side of the crest to the horizon be $= m$. Let AE, a variable length, $= x$. Then $ED = \dfrac{x}{m}$. Let EG be an infinitely small element of the horizontal length or width from A to C. Then EG may be denoted by dx. Let $q =$ quantity in cubic feet per second flowing under the length x, that is, under AE in the figure. Then dq will be the quantity discharged per second between ED and GF.

Then, by the Lowell formula just cited, we have

$$dq = 3 \cdot 33\, dx \left(\frac{x}{m}\right)^{\frac{3}{2}};$$

whence, by integrating, we get

$$q = 3 \cdot 33 \left(\frac{1}{m}\right)^{\frac{3}{2}} \cdot \frac{2}{5} x^{\frac{5}{2}} + C,$$

in which the constant quantity is to be put $= 0$, because when $x = 0$, q also $= 0$. Hence we have

$$q = \frac{2}{5} \times 3 \cdot 33 \left(\frac{1}{m}\right)^{\frac{3}{2}} x^{\frac{5}{2}} \ \dots\dots\dots\dots\dots(2).$$

Let now $H_2 =$ height in feet from the vertex of the notch up to the level surface of the impounded water $= BK$ in the figure. Then $AK = mH_2$. Let also $Q_2 =$ the discharge per second in the whole triangular notch $=$ twice the quantity discharged under AK. Then by formula (2) we get

$$Q_2 = \frac{4}{5} \times 3 \cdot 33 \times \left(\frac{1}{m}\right)^{\frac{3}{2}} (mH_2)^{\frac{5}{2}},$$

or

$$Q_2 = 2 \cdot 664\, m H_2^{\frac{5}{2}} \ \dots\dots\dots\dots\dots(3).$$

To bring the notation to correspond with that used in the fore-going Report, let $Q =$ the quantity of water in cubic feet per minute, and $H =$ the height of the water level above the vertex in inches.

Then $Q_2 = \dfrac{Q}{60}$ and $H_2 = \dfrac{H}{12}$ and by substitution in (3), we get

$$Q = \cdot 320\, m H^{\frac{5}{2}} \quad \dots\dots\dots\dots\dots\dots(4).$$

This formula then gives, deduced from the Lowell formula, the flow in cubic feet per minute through a very wide notch in a vertical thin plate, when H is the height from the vertex of the notch up to the water level, in inches, and when the slopes of the notch are each m horizontal to 1 vertical.

As to the confidence which may be placed in this formula, I think it clear that, for the case in which the notch is so wide, or, what is the same, the slopes of its edges are so slight, that the water may flow over each infinitely small element of the length of its crest without being sensibly influenced in quantity by lateral contraction arising from the inclination of the edges, the formula may be relied on as having all the accuracy of the Lowell formula from which it has been derived; and I would suppose that when the notch is of such width as to have slopes of about four or five to one, or when it is of any greater width whatever, the deviation from accuracy in consequence of lateral contraction might safely be neglected as being practically unimportant or inappreciable.

This formula for wide notches bears very satisfactorily a comparison with the formulas obtained experimentally for narrower notches, as described in the foregoing Report. For slopes of 1 to 1 the formula was $Q = \cdot 305\, H^{\frac{5}{2}}$, and for slopes of 2 to 1 the formula was $Q = \cdot 636\, H^{\frac{5}{2}}$. To compare these with the one now deduced for any very slight slopes, we may express them thus:

For slopes of 1 to 1

$$Q = \cdot 305\, m H^{\frac{5}{2}},$$

and for slopes of 2 to 1

$$Q = \cdot 318\, m H^{\frac{5}{2}};$$

while for any very slight slopes, or for any very wide notches, the formula now deduced from the Lowell one is

$$Q = \cdot 320\, m H^{\frac{5}{2}}.$$

The very slight increase from ·318 to ·320 here shown in passing from the experimental formula for notches with slopes of 2 to 1, to notches wider in any degree—that slight change, too, being in the right direction, as is indicated by the increase from ·305 to ·318 in passing from slopes of 1 to 1, to slopes of 2 to 1—gives a verification of the concluding remarks in the foregoing Report; and this may serve to induce confidence in the application in practice of the formula now offered for wide notches.

[If two cases of liquid flow through orifices or channels are compared, in the second of which all distances parallel to one direction (say the axis of h) are contracted in the same ratio as compared with the first, it may be readily seen from the dynamical equations that, provided the motion is steady, the paths of all the elements of liquid, and also the pressures, will correspond in the two cases, when the effects of viscosity and capillary attraction are neglected. In these latter circumstances the law that the delivery of a notch is proportional to its width will therefore be exact. See also *infra*, pp. 67 *seq.* Dec. 1910.]

13. Table of Flow of Water in the Right-Angled V-Notch.

[The following MS. table has been found among Prof. Thomson's papers.]

According to the formula, $Q = ·305 h^{\frac{5}{2}}$, where Q denotes the quantity of water in cubic feet per minute, and h denotes the height in inches from the vertex of the notch up to the still water surface level.

The notch may be either *without floor* or *with floor*, as the quantities flowing in these two cases for same heights are found experimentally to be very nearly the same.

The Formula, however, and with it the Table, is reliable for more minute exactitude in the case of the *notch without floor*.

Fuller information is to be found in an Interim and Final Report to the British Association by Prof. James Thomson published in the Volumes of the Association for 1858 and 1861;

and in a paper by him on "The Flow of Water through Orifices" in the Volume of the Association for 1876.

TABLE.

h ins.	Q cub. ft. per min.	h ins.	Q cub. ft. per min.	h ins.	Q cub. ft. per min.	h ins.	Q cub. ft. per min.
½	·054	5¼	19·26	9⅞	93·5	19	480
⅝	·094	5⅜	20·43	10	96·4	19¼	496
¾	·148	5½	21·64	10¼	102·6	19½	512
⅞	·218	5⅝	22·89	10½	109·0	19¾	529
1	·305	5¾	24·18	10¾	115·6	20	546
1⅛	·409	5⅞	25·52	11	122·4	20½	580
1¼	·533	6	26·89	11¼	129·5	21	616
1⅜	·676	6⅛	28·33	11½	136·8	21½	654
1½	·840	6¼	29·79	11¾	144·4	22	692
1⅝	1·027	6⅜	31·29	12	152·1	22½	732
1¾	1·23	6½	32·85	12¼	160·2	23	774
1⅞	1·47	6⅝	34·46	12½	168·5	23½	817
2	1·73	6¾	36·10	12¾	177·0	24	861
2⅛	2·01	6⅞	37·80	13	185·8	25	953
2¼	2·32	7	39·54	13¼	194·9	26	1051
2⅜	2·66	7⅛	41·33	13½	204·2	27	1155
2½	3·01	7¼	43·17	13¾	213·8	28	1265
2⅝	3·41	7⅜	45·05	14	223·7	29	1381
2¾	3·82	7½	46·98	14¼	233·8	30	1504
2⅞	4·27	7⅝	48·97	14½	244·2	31	1632
3	4·75	7¾	51·0	14¾	254·8	32	1767
3⅛	5·27	7⅞	53·1	15	265·8	33	1908
3¼	5·81	8	55·2	15¼	277·0	34	2056
3⅜	6·38	8⅛	57·4	15½	288·5	35	2210
3½	6·99	8¼	59·6	15¾	300·3	36	2372
3⅝	7·63	8⅜	61·9	16	312·3	37	2540
3¾	8·30	8½	64·2	16¼	324·7	38	2715
3⅞	9·01	8⅝	66·6	16½	337·3	39	2897
4	9·76	8¾	69·1	16¾	350·2	40	3086
4⅛	10·54	8⅞	71·6	17	363·4	41	3283
4¼	11·35	9	74·1	17¼	376·9	42	3487
4⅜	12·21	9⅛	76·7	17½	390·8	43	3698
4½	13·10	9¼	79·4	17¾	404·9	44	3917
4⅝	14·03	9⅜	82·1	18	419	45	4143
4¾	14·99	9½	84·8	18¼	434	46	4377
4⅞	16·00	9⅝	87·7	18½	449	47	4619
5	17·05	9¾	90·5	18¾	464	48	4869
5⅛	18·13						

14. IMPROVED INVESTIGATIONS ON THE FLOW OF WATER THROUGH ORIFICES WITH OBJECTIONS TO THE MODES OF TREATMENT COMMONLY ADOPTED.

[From the *Report of the British Association*, Glasgow, 1876, pp. 243—266.]

THE methods usually put forward for treating of the flow of water out of vessels by orifices in thin plates, slightly varied though they may be in different cases, are ordinarily founded on assumptions largely alike in these different cases, and largely erroneous. The theoretical views so arrived at, and very generally promulgated, are in reality only utterly false theories based on suppositions of the flow of the water taking place in ways which are kinematically and dynamically impossible, and are at variance with observed facts of the flow, and even at variance with the facts as put forward by the advancers themselves of those theories. The admittedly erroneous results brought out through those fallacious "theories," and commonly miscalled *"theoretical results,"* are afterwards considerably amended by the introduction into the formulas so obtained of constant or variable coefficients, or otherwise, so as to be brought into some tolerable agreement with experimental results. These means of practical amendment, however, being themselves not established on any scientific principles, can at best only conduce to the attainment of useful empirical formulas, but cannot, by their application to the originally false theoretical views, come to develop any true scientific theory. A theory may, no doubt, be regarded as a good scientific theory, and as being good for practical purposes, which leaves out of account some minor features or conditions of the actual facts. In so far as it leaves any influential elements out of account, it is imperfect; but if the conditions which, for simplicity, or from want of complete knowledge of the subject, or for any other reason, are left out, be of very slight influence on the practical results in question, the theory may be regarded as a very good one, though not quite perfect. In the case, however, of the hydraulic theories now referred to, the false principles involved in the reasonings relate to the main and important conditions of the flow, and not to any mere minor considerations, the imperfections or errors of which might be of but slight importance in the development of the main principles involved, and but little influential on the results sought to be attained.

I will now proceed to give some examples or sketches of the usual methods of treating the subject.

I will first take the case of water flowing from a state of rest through an orifice in a vertical plane face. This case is ordinarily treated by supposing the orifice to be divided into an infinite number of infinitely narrow horizontal bands of area, and supposing the velocity of the water in each band to be that due, through the action of gravity, to a fall from the still-water surface-level down to that band; then multiplying that velocity by the area of the band, and treating the product as being the volume flowing per unit of time across that horizontal band or element of the area; and integrating to find the sum of all these volumes of water for all the bands, and treating this sum as being the "theoretical" volume per unit of time flowing across the whole area of the orifice. This result is commonly called the "*theoretical discharge*" per unit of time; but, as it is known not to be the actual discharge, it is then multiplied by a numerical coefficient called by some "*the coefficient of contraction*," and by others the "*coefficient of discharge*," in order to find the actual discharge per unit of time.

Thus for the case of a rectangular orifice in a vertical plane face, as in fig. 1—where WL is the level of the still-water surface,

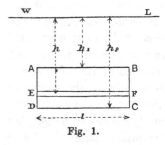

Fig. 1.

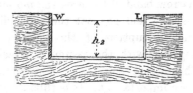

Fig. 2.

and $ABCD$ is the orifice, with two edges AB and CD level, and EF is an infinitely narrow horizontal band extending across the orifice at a depth h below the still-water surface-level, and having dh as its breadth vertically measured, while it has l, the horizontal length of the orifice, as its length, and where, as shown in the figure, the depths of the top and bottom of the orifice below WL are denoted by h_1 and h_2 respectively—if q is put to denote the so-called "theoretical" volume per unit of time, and Q the actual volume per unit of time, it is commonly stated that

$$dq = \sqrt{2gh} \, . \, ldh,$$

whence
$$q = \int_{h_1}^{h_2} \sqrt{2gh} \, . \, ldh = l\sqrt{2g} \int_{h_1}^{h_2} h^{\frac{1}{2}} dh,$$

or
$$q = \tfrac{2}{3} l \sqrt{2g} \, (h_2^{\frac{3}{2}} - h_1^{\frac{3}{2}});$$

and then when c is put to denote the so-called "*coefficient of contraction*," it is stated that the actual quantity flowing per unit of time is

$$Q = \tfrac{2}{3} cl \sqrt{2g} \, (h_2^{\frac{3}{2}} - h_1^{\frac{3}{2}}). \dots\dots\dots\dots\dots(1).$$

It is then customary to deduce from this a formula for the case of water flowing in a rectangular notch open above, as in fig. 2, by taking $h_1 = 0$, and so deriving, for the open notch, the formula

$$Q_{for\ notch} = \tfrac{2}{3} cl \sqrt{2g} \, . \, h_2^{\frac{3}{2}} \dots\dots\dots\dots(2).$$

These examples may suffice for indicating the nature of the method commonly advanced; and it may be understood that the same method with the necessary adaptations is usually given for finding the flow through circular orifices, triangular orifices, or orifices of any varied forms whatever.

Now this method is pervaded by false conceptions, and is thoroughly unscientific.

First. Throughout the horizontal extent of each infinitely narrow band of the area the motion of the water has not the same velocity, and has not the same direction at different parts; and the assumption of the velocity being the same throughout, together with the assumption tacitly implied of the direction of the motion being the same throughout, vitiates the reasoning very importantly. It is thus to be noticed at the outset that the division of the orifice into bands, infinitely narrow in height, but extending horizontally across the entire orifice, cannot lead to a satisfactory process of reasoning, and that the elements of the area to be separately considered ought to be infinitely small both in length and in breadth.

Secondly. For any element of the area of the orifice infinitely small in length and breadth it is not the velocity of the water at

it that ought to be multiplied by the area of the element to find
the volume flowing per unit of time across that element, but it is
only that the velocity's component which is normal to the plane
of the element that ought to be so multiplied.

Thirdly. Whether, for any element of the area of the orifice,
we wish to treat of the absolute velocity of the water there, or to
treat of the component of that velocity normal to the plane of
the orifice, it is a great mistake to suppose that the velocity at
the element is that due by gravity to a fall from the still-water
surface-level of the pent-up statical water down to the element.
The water throughout the area of any closed orifice in a plane
surface, with the exception of that flowing in the elements situated
immediately along the boundary of the orifice, has more than
atmospheric pressure; and hence it can be proved* that it must
have less velocity than that due to the fall from the still-water
surface-level down to the element.

The foregoing may be illustrated by consideration of the very
simple case of water flowing from a vessel through a rectangular
orifice in a vertical plane face, two sides of the rectangle being
level, and the other two vertical, and end contractions being
prevented by the insertion of two parallel guide walls or plane
faces, one at each end of the orifice, and both extending some
distance into the vessel perpendicularly to the plane of the orifice,
so that the jet of issuing water may be regarded as if it were a
portion of the flow through an orifice infinitely long in its horizontal dimensions.

Thus if the jet shown in section in fig. 3 *a* be of the kind here
referred to, while *WL* is the still-water surface-level, the so-called
"theoretical velocities" at the various depths in the orifice, which
are dealt with as if they were in directions normal to the plane
of the orifice, can be, and very commonly are, represented by the
ordinates of a parabola as is shown in fig. 3 *b*, where *BD* represents
in magnitude and direction the "theoretical velocity" at the top
of the orifice, *CE* the "theoretical velocity" at the bottom of the
orifice, and *FG* that at the level of any point *F* in the orifice—
these ordinates being each made $= \sqrt{2gh}$, where *h* is the depth
from the still-water surface down to the level of the point in the
orifice to which the ordinate belongs. Then, under the same
mode of thought, or same set of assumptions, the area of that

* Theorem I., further on [p. 64], will afford proof of this.

parabola between the upper and lower ordinates (*BD* and *CE*) will represent what is commonly taken as the "*theoretical discharge*" per unit of time through a unit of horizontal length of the orifice. But this gives an excessively untrue representation of the actual conditions of the flow. Instead of the parabola, some other curve, very different, such as the inner curve sketched in the same diagram, fig. 3 *b*, but whose exact form is unknown, would, by its ordinates, represent the velocity-components normal to the plane

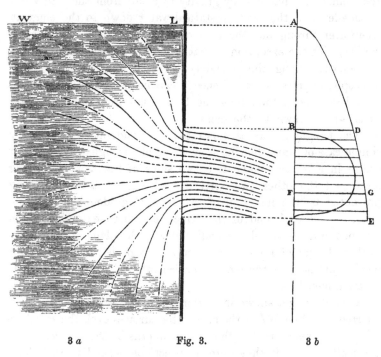

3 *a*   Fig. 3.   3 *b*

of the orifice for the various levels in the orifice, and its area would represent the real discharge in units of volume per unit of time through a unit of horizontal length of the orifice. Although the exact form of this true curve is unknown, yet we may observe that it must have its ordinates each less than the ordinate for the same level in the parabola.

The truth of this may be perceived through considerations such as the following. First, it is to be noticed that for the very top and the very bottom of the orifice, instead of the ordinates *BD* and *CE* of the parabola, the ordinates of the true curve must be each zero; because, at each of these two places, the direction of the motion is necessarily tangential to the plane of the orifice*, and so the velocity-component normal to the plane of the orifice must be zero; and that component, not the velocity itself, is what the ordinate of the true curve must represent. On the hypothesis of perfect fluidity in the water (which, throughout the present discussions and investigations, is assumed as being a close enough representation of the truth to form a basis for very good theoretical views), the velocities at top and bottom of the orifice will be those due by gravity to falls from the still-water surface-level down to the top and bottom of the orifice respectively, because at these places the water issues really into contact with the atmosphere, and consequently attains atmospheric pressure. At all intervening

---

* The assertion here made, to the effect that the directions of the stream-lines which form the external surface of the jet on its leaving the edge of the orifice must, at the edge, be tangential to the plane of the orifice when the orifice is in a plane face, or must in general be tangential to the marginal narrow band or terminal lip of the internal or water-confining face of the plate or nozzle in which the orifice is formed, can be clearly and easily proved, although, strangely, the fact has been and is still very commonly overlooked. Even MM. Poncelet and Lesbros, in their delineations of the forms of veins of water issuing from orifices in thin plates, after elaborate observations and measurements of those forms, represent the surface of the issuing fluid as making a sharp angle with the plane wetted face in leaving the edge ("Expériences Hydrauliques sur les Lois de l'Écoulement de l'Eau," a Memoir read at the Academy of Sciences in November 1829, and published in the *Mémoires, Sciences Mathématiques et Physiques*, tome III.). Other writers on Hydraulics put forward very commonly representations likewise erroneous. Weisbach, for instance, in his valuable works (*Ingenieur und Maschinen-Mechanik*, Vol. I. § 313, fig. 427, date 1846; and *Lehrbuch der theoretischen Mechanik*, 5th ed., date 1875, edited by Hermann, § 433, fig. 772), has assumed (not casually, but with deliberate care, and after experimental measurements made by himself), as the best representation which, with available knowledge of the laws of contraction of jets of water, can be given for the form of the contracting vein of water issuing from a circular orifice in a thin plate, a solid of revolution specified clearly in such a way that the water surface in leaving the plane of the plate makes an angle of about 67° with that plane, and states to the effect that that water surface is just a continuation of the paths of the stream-lines within the vessel which he represents at the margin of the orifice as crossing the plane of the orifice with converging paths making the angle already mentioned of about 67° with that plane. They ought in reality to leave the lip tangentially to the plane, and then to make a very rapid turn in a short space (or to have a very small radius of curvature) on just leaving the lip of the orifice. The prevalence of erroneous representations and notions on this subject was adverted to, and an amendment was adduced, by myself in a Report to the British Association in 1861 on the Gauging of Water by V-Notches (*Brit. Assoc. Rep. Manchester Meeting*, 1861, Part 1, p. 156). [See *supra*, p. 42.]

points in the plane of the orifice it may readily be seen, or may with great confidence be admitted, that the pressure will be in excess of the atmospheric pressure; because, neglecting for simplicity the slight and, for the present purpose, unimportant modification of the courses of the stream-lines caused by the force of gravity acting directly on the particles composing the stream-lines, as compared with the courses which the stream-lines would take if the action of gravity were removed, and the water were pressed through the orifice merely by pressure applied, as by a piston or otherwise, to the fluid in the vessel, we may say, truly enough for the present purpose, that an excess of pressure at the convex side of any stream-line is required in order that the water in the stream-line can be made to take its curved path. The mode of reasoning on this point suggested here may be obvious enough, although, for the sake of brevity, it is here not completely expressed. It follows that at all these intervening points in the plane of the orifice the absolute velocity of the water will be less than that due to a fall from the still-water surface down to the level of the point in the orifice; and besides, at all depths in the plane of the orifice except a single medial one, the direction of the flow will be oblique, not normal, to the plane of the orifice. Hence, further, through these two circumstances, jointly or separately as the case may be, it follows obviously that the ordinates of the true curve will everywhere be less than those of the parabola.

Fig. 4 illustrates in like manner the false theoretical and the true actual conditions of the flow over a level upper edge of a vertical plane face, which may be exemplified by the case of a rectangular notch without end contractions, or of a portion of the flow not extending to either end in a very wide rectangular notch. In this case it is to be observed that the ordinates at and near the top of the issuing water in the vertical plane of the orifice must be only slightly less than those of the parabola—because, at the very top or outside of the stream, atmospheric pressure is maintained throughout the length of any stream-line, and so the velocity will be very exactly that due by gravity to the vertical depth of the flowing particle below the still-water surface-level in the vessel; and because, also, the direction of the motion does not deviate much from perpendicularity to the plane of the orifice. Lower down in the plane of the orifice the direction of the water's motion will approach still more nearly to being perpendicular to that plane; but there the pressure will be considerably in excess of the atmospheric pressure, and so the velocity will be considerably less than that due by gravity to a fall through the vertical distance from the still-water surface-level down to the stream-line in the plane of the orifice. At places still further down in the orifice the flow comes to be obliquely upwards; and this obliquity is so great as to render the normal component very much less than the actual velocity, while the actual velocity itself is less than that due by gravity to the depth of the particle below the still-water surface-level. At this region of the flow then, for both reasons, the ordinates of the true curve are less than those of the parabola. Lastly,

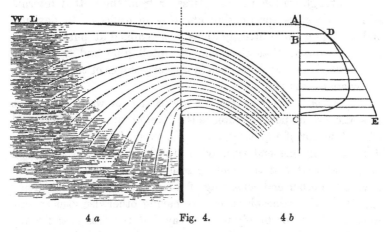

4 a   Fig. 4.   4 b

at the very bottom of the orifice, or immediately over the top of the crest of the notch, the water issues into contact with the atmosphere, and so attains to atmospheric pressure, and must therefore have the velocity due by gravity to its depth below the still-water surface-level. Here, however, its direction of flow is necessarily tangential to the plane face of the vessel from which it is shooting away, and consequently is vertically upwards. Hence the normal component of its motion is zero, and so the ordinate of the true curve at that place is zero in length, instead of the normal component being greater at the bottom of the orifice than at any higher level, and instead of that component being properly represented by the ordinate there of the parabola.

Like explanations to those already given might be offered for other forms of orifices (for circular or triangular orifices or V-notches, and for orifices in general which may be in vertical or horizontal or inclined plane faces, or in faces of other superficial forms than the plane), and it might be shown that in general the ordinary modes of treating the subject are very faulty.

The examples already discussed may suffice to direct attention to the faulty character of the ordinarily advanced theories, and to give some suggestions of directions in which reforms are requisite.

I will now proceed to offer some improved investigations which are applicable to many of the most ordinary and most useful cases in practical hydraulics, in reference to the flow of water through orifices in thin plates, or from the wetted internal surface of vessels terminating abruptly in orifices. In devising and arranging these investigations I have aimed at putting them in such form as that they may be intelligible and completely demonstrative to students even in the early stages of their progress in dynamical studies.

Definition. The *free level* for any particle of water in a mass of statical or of flowing water is the level of the atmospheric end of a column, or of any bar straight or curved, of particles of statical water, having one end situated at the level of the particle, and having at that end the same pressure as the particle has, and having the other end consisting of a level surface of water freely exposed to the atmosphere, or else having otherwise atmospheric pressure there; or briefly we may say that the *free level* for any particle of water is the level of the atmospheric end of its *pressure-column*, or of an equivalent ideal pressure-column.

THEOREM I. *In the case of steady flow from approximate rest of water or any liquid considered as frictionless and incompressible, the velocity of any particle in the stream is equal to the velocity which a body would receive in falling freely from rest through a vertical space equal to the fall of free level which is incurred by the particle in the stream during its flow from rest to its existing position.*

Or, in briefer words sufficiently suggestive, it may be said that, in respect to water or any liquid flowing so as to admit of its being regarded as truly enough frictionless and incompressible, *In steady flow, the velocity generated from rest is that due by gravity to the fall of free level.*

Or if ζ be the fall of free-level sustained by any particle in passing from a statical region of the mass of water to a point in the region of flow, and if v be the velocity of the particle when at that point, then

$$v = \sqrt{2g\zeta}.$$

In fig. 5, let WL be the still-water surface-level, and let $B'BB''$ be a bounding interface separating the region of flow with important energy of motion from the region which may be regarded as statical, or as devoid of important energy of motion. Let $U'UU''$ be another interface crossing the stream-lines at any place in the region of flow.

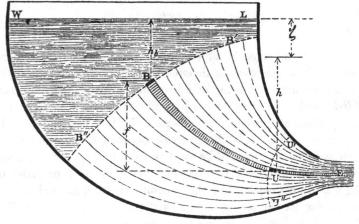

Fig. 5.

Now taking as the unit of volume the cube of the unit of length, taking as the unit of area the square of the unit of length, taking the unit of density as unit of mass per unit of volume, so that the density of a body will be the number of units of mass per unit of volume, taking as the unit of force the force which acting on a unit of mass for a unit of time imparts to it a unit of velocity (that is to say, using the unit of force selected according to the system of Gauss, and which is often called the "absolute" or the "kinetic" unit of force*), and taking water-pressures as being reckoned from the atmospheric pressure as zero, let

* The units of force derivable by the method of Gauss from the various units of length, mass, and time, in common use, though spoken of under general designations such as " *absolute units of force* " or " *kinetic units of force*," have until lately

ρ = density of the water;

v = velocity at U;

h_b = pressure-height at B, or the height of a column of statical
water which would produce the pressure at B;

h = pressure-height at U;

p_b = pressure in units of force per unit of area at B;

p = pressure in units of force per unit of area at U;

f = fall from B to U, measured vertically;

ζ = fall of free level in the flow from the region of statical
water to U;

then $\qquad\qquad\qquad p_b = g\rho h_b,$

and $\qquad\qquad\qquad p = g\rho h.$

Let a small mass, m, of the water, whose volume (or content
voluminally considered) is denoted by c, be introduced into the
stream, its first place being at B just outside of the initial interface
$B'BB''$, and let it flow forward in the stream till it reaches a
second place at U where it is just past the interface $U'UU''$. In
the stream filament BUE the space between the two interfaces
at B and U is traversed alike by both front and rear of the small
mass m; and therefore no excess of energy is given or taken by
the mass in consequence of the pressure on its front and of that
on its rear, for the passage of its front from the interface at B to
that at U, and of its rear over the same space.

been individually anonymous: and this deficiency, notwithstanding the important
scientific and practical uses which these units were capable of serving, has been a
great hindrance and discouragement to their general employment in dynamical
investigations, and even to any satisfactory spread of knowledge of their meaning.
Three years ago, the British-Association Committee on Dynamical and Electrical
Units (*Brit. Assoc. Report*, 1873, Part 1, p. 222), taking the centimetre, the gram,
and the second as units of length, mass, and time, named the force so derived the
Dyne. For the unit of force derived from the foot, the pound, and the second,
the name *Poundal* has been introduced by myself; and it seems likely to come into
use. At this Meeting of the British Association I have proposed the *Crinal* and
the *Funal* as names for the two units of force derived respectively, one from the
decimetre, the kilogram, and the second, and the other from the metre, the tonne,
and the second [see Paper reprinted *infra*, "On Metric Units of Force, Energy,
and Power, larger than those on the Centimetre-Gram-Second System."] The
familiarization of these important units to the minds of students of dynamics
will, in a very important degree, aid the acquisition of clear and true views in
hydrokinetics, as also in dynamics generally.

But work given to it by pressure from behind, while it is passing the initial interface at B, is

$$= p_b . c$$
$$= g\rho h_b . c;$$

or that work is

$$= gmh_b,$$

since $\rho c = m$.

Again, during the emergence of the mass past the interface at U, it gives away to the water in front of it a quantity of work which, in like manner, is

$$= p . c$$
$$= g\rho h . c$$
$$= gmh.$$

Also during the passage of the particle from its first place at B to its place at U it descends a vertical space $= f$; hence during that passage it receives from gravity a quantity of work $= gmf$.

On the whole the mass receives an excess of work beyond what it gives, and that excess of work received is

$$= gmh_b + gmf - gmh$$
$$= gm (h_b + f - h)$$
$$= gm\zeta;$$

and as this is the work taken into store as kinetic energy, we have to put it $= \frac{1}{2}mv^2$. That is,

$$gm\zeta = \frac{1}{2}mv^2,$$

or

$$v = \sqrt{2g\zeta},$$

which is the result that was to be proved in Theorem I.

THEOREM II. ON THE FLOW OF WATER THROUGH ORIFICES SIMILAR IN FORM AND SIMILARLY SITUATED RELATIVELY TO THE STILL-WATER SURFACE-LEVEL. *In the flowing of water, from the condition of approximate rest, through orifices similar in form and similarly situated relatively to the still-water surface-level*, the stream-lines in the different flows are similar in form: also the velocity of the water at homologous places is proportional to the*

---

* Or *free level* of the still water.

square root of any homologous linear dimension in the different flows: and also (pressures being reckoned from the atmospheric pressure as zero) the pressure of the water on homologous small interfaces in the different flows is proportional to the cube of any homologous linear dimension; or, in other words, the fluid pressure (super-atmospheric), per unit of area at homologous places, is proportional to any homologous linear dimension.

Preparatively for the demonstration of this theorem, it is convenient to establish some dynamic principles, which, for present purposes, may be regarded as lemmas or preparatory propositions, and which will be grouped here together under the single heading of Proposition A.

Proposition A. *If there be two or more vessels containing water pent up in an approximately statical condition, and if they have similar orifices similarly situated relatively to the free level of the statical water—and if we imagine the water to be guided in each case to and onward past the orifice by an infinite number of infinitely small frictionless guide-tubes arranged side by side, like the cells of a honeycomb, and having their walls or septums* of no thickness—and if, in the different vessels, these guide-tubes be, one set to another, similar in form, though they may be of quite different forms from the forms which the stream-lines would themselves assume if the flows were unguided—and if, at the homologous terminations of the guide-tubes, fluid pressures be anyhow maintained proportional, per homologous areas, to the cube of any homologous linear dimension, or, what is the same, if pressures be maintained proportional, per unit of area, to the homologous linear dimension,—then the velocity of the water at homologous places will be proportional to the square root of the homologous linear dimension, and the pressure of the water at homologous places on homologous areas will be proportional to the cube of the homologous linear dimension; and the water will press, at homologous places, on homologous areas of the septums, with a force on one side in excess of that on the other, which will be proportional to the cube of the homologous linear dimension.*

Note. For brevity in what follows, pressures at homologous places on homologous areas will be called *homologous pressures*, and pressures per unit of area will be called *unital pressures*;

---

* The English form for the plural of *septum*, when septum is used as an English word, is here purposely preferred to the Latin *septa*.

and any difference of the fluid pressures on the opposite sides of any small portion or element of a septum will be called a *differential pressure.*

The demonstration of the proposition will be aided by first noticing the following relation in respect to two small solid masses in motion. If two similar small solid bodies of masses m and m', having their homologous linear dimensions as 1 to n, are guided to move along similar curves, having likewise their homologous linear dimensions as 1 to n (fig. 6), and if the velocities of the bodies at homologous points in their paths be as 1 to $\sqrt{n}$, then—

First. Their gravities are as 1 to n^3, evidently.

Second. Their "centrifugal forces"[*] applied by them in the plane of curvature and normally to the guide are also as 1 to n^3.

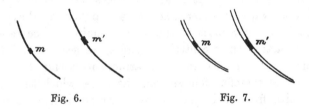

Fig. 6.   Fig. 7.

Let r and r' be put to denote the radii of curvature of the paths at homologous places. Then centrifugal forces are as

$$\frac{mv^2}{r} : \frac{m'v'^2}{r'}.$$

[*] The name "*centrifugal force*" is here adopted in the sense in which it is commonly used. I fully agree with the opinion now sometimes strongly urged to the effect that this name is not a very happily chosen one; for two reasons:—first, because the name *centrifugal* would be better applied to a *motion* of flying from the centre, than to a *force* acting outwards along the radius; and secondly, because the body really receives no outward force, no force in the direction from the centre, but receives a centreward force which, being unbalanced, acts against the inertia of the body, and diverts the body from the straight line of its instantaneous motion. The centreward force actually received by the body, and which is the force acting on it normal to its path, may be called the *deviative force* received by the body. This is equal and opposite to the outward force called "*centrifugal force,*" which is not *received* by the body, but is exerted outwards by it against whatever is compelling it to deviate from the straight line of its instantaneous motion. The name "centrifugal force," however, although objected to, is in too general use throughout the world to allow of its immediate abandonment.

But
$$m' = n^3 m,$$
$$v' = \sqrt{n} \cdot v,$$
$$r' = nr.$$

Hence the centrifugal forces are

$$\text{as } \frac{mv^2}{r} : \frac{n^3 m \cdot nv^2}{nr},$$

or   as $1 : n^3$.

This being understood, it readily becomes evident that if, instead of small solid masses sliding along guides, we have two small homologous masses of water m and m', fig. 7, flowing in similar slender guide-tubes, and if homologous pressures be applied to the two masses in front and behind, which are as 1 to n^3, and if at homologous situations in their two paths their velocities be as 1 to $\sqrt{n}$, then, in respect to all the forces received by the two masses from without, other than those applied by the guide-tubes, and also in respect to the forces required to be received for counteracting their centrifugal forces, we see that all these constitute force systems similar in arrangement and of amounts as 1 to n^3. It therefore follows that the forces which the masses must receive from their guide-tubes must be similarly arranged and of amounts, on homologous small areas, as 1 to n^3.

This being settled, we may now pass to the demonstration of Proposition A, at present in question.

Suppose No. 1 and No. 2 in fig. 8 to represent two similar vessels with similarly guided flows, in all respects as described in the enunciation of this proposition. Let WL and $W'L'$ be the still-water surface-levels, or the free levels of the still water in the two cases. Let BCD, $B'C'D'$ be two similar bounding interfaces, each separating the region of flow with important energy of motion from the region which may be regarded as statical, or as devoid of important energy of motion. Let BUE in No. 1 and $B'U'E'$ in No. 2 be two homologous guide-tubes, and let them for the present be understood as terminating at two homologous cross interfaces E and E', which may conveniently be understood as being each situated at a moderate distance outside of the orifice —for instance, at some such place as that which is usually spoken of as being the "*vena contracta*," or where the water has attained a pressure not differing much from that of the atmosphere, or it may in some cases even be that the atmospheric pressure is there attained; but the exact places at which to suppose the homologous terminations E and E' of the two guide-tubes as being taken are not at all essential to the demonstration.

Let homologous linear dimensions in No. 1 and No. 2 be as 1 to n.

Let the velocity at any variable point U in the guide-tube BUE be denoted by v.

Let the pressure at U, expressed in units of pressure-height, be denoted by h; as shown by the vertical line UT in No. 1, where T is the top of the pressure-column for the point U.

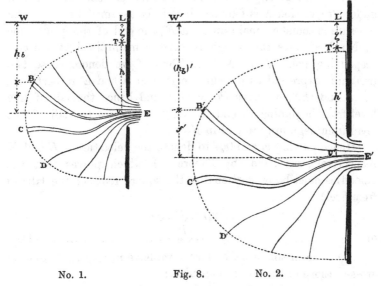

No. 1.   Fig. 8.   No. 2.

Let the pressure at B, the beginning of the tube, on the initial interface, outside of which the water may be regarded as statical, or as having no important energy of motion, be denoted by h_b; or, what comes to the same thing, let the depth from the still-water surface-level down to the beginning of the tube at B be denoted by h_b, as is marked in the figure. It is thus to be noticed that the fall of free level incurred by a particle in flowing along the guide-tube from B to U is the vertical distance from the still-water surface-level, WL, down to T, the top of the pressure-column for the flowing water at U. This fall of free level may be denoted (in conformity with the notation in Theorem I) by ζ.

Let the vertical descent from B to U be denoted by f; so that f is the fall of a particle in passing from B to U. In case of an ascent in any guide-tube, from its beginning to any point U in its course, we shall have the fall f negative.

Let the abatement of pressure-height from B to U be denoted by k, or let $h_b - h = k$. Thus in case of an increase of pressure-height in any guide-tube, from its beginning to any point U in its course, k will be negative.

For No. 2, let the same letters of reference to the diagram, and the same notation, be used as for No. 1, with the modification for No. 2 merely of the attachment of an accent to each letter.

Now as a part of the data on which the present investigation under Proposition A is founded, it is to be assumed that a unital pressure is somehow maintained at E', the end of the guide-tube in No. 2, n times that which is anyhow maintained at the corresponding point E in No. 1. Thus, if we denote these two pressures expressed as pressure-heights, at E and E' respectively, by h_e and $(h_e)'$, we have $(h_e)' = nh_e$; and hence the fall of free level from beginning to end in No. 2 is n times the fall of free level from beginning to end in No. 1.

Hence putting v_e and $(v_e)'$ to denote the velocities at E and E' respectively, we have (by Theorem I, which proves that the velocities must be proportional to the square roots of the falls of free level)

$$v_e : (v_e)' :: \sqrt{1} : \sqrt{n},$$

or
$$(v_e)' = v_e \sqrt{n} \quad \dots\dots\dots\dots\dots\dots\dots\dots(1).$$

Again, from similarity of forms, we have in respect to areas of cross-sections of the two guide-tubes:—

$$\frac{\text{area at } E}{\text{area at } U} = \frac{\text{area at } E'}{\text{area at } U'};$$

or since reciprocals of equals are equal:—

$$\frac{\text{velocity at } E}{\text{velocity at } U} = \frac{\text{velocity at } E'}{\text{velocity at } U'},$$

or
$$\frac{v_e}{v} = \frac{(v_e)'}{v'},$$

or by (1)
$$\frac{v_e}{v} = \frac{v_e \sqrt{n}}{v'},$$

or
$$v' = v \sqrt{n} \quad \dots\dots\dots\dots\dots\dots\dots(2).$$

This applies to any or all homologous points in the two regions of flow.

Now by referring to the figure or otherwise, it will readily be seen that ζ, or the fall of free level from B to U, is $= h_b + f - h$, while $k = h_b - h$; and that therefore $\zeta = f + k$. Hence, by Theorem I, we have

$$v = \sqrt{2g\,(f + k)}.$$

In like manner in No. 2:—

$$v' = \sqrt{2g\,(f' + k')};$$

but by (2) $v' = \sqrt{n}\,.\,v.$

Hence $\sqrt{2g\,(f' + k')} = \sqrt{n}\,.\,\sqrt{2g\,(f + k)},$

or $f' + k' = nf + nk.$

But by similarity of forms

$$f' = nf.$$

Hence, subtracting equals from equals, we have

$$k' = nk \quad \dots\dots\dots\dots\dots\dots\dots(3);$$

but by similarity of forms

$$(h_b)' = nh_b \quad \dots\dots\dots\dots\dots\dots(4).$$

Also, since the pressure at any point in a stream-line, or guide-tube, is its initial pressure minus the relief of pressure, we have

$$h_b - k = h \quad \dots\dots\dots\dots\dots\dots\dots(5)$$

and $(h_b)' - k' = h' \quad \dots\dots\dots\dots\dots\dots(6).$

From this last by (4) and (3) we get

$$nh_b - nk = h', \text{ or } n\,(h_b - k) = h';$$

whence by (5)

$$h' = nh \quad \dots\dots\dots\dots\dots\dots\dots(7).$$

From this, if we put P and P' to denote total pressures on homologous small areas at U and U', it follows that

$$P' = n^3 P \quad \dots\dots\dots\dots\dots\dots(8).$$

This holds good for any homologous places in any homologous guide-tubes, and so it holds for immediately adjacent places in any two contiguous guide-tubes. Hence, in respect to any small element of the septum between two adjacent guided stream-filaments in Flow No. 1, considered comparatively with a homologous element of a septum in Flow No. 2, the homologous differential pressures in No. 1 and No. 2 will be as 1 to n^3.

Thus the demonstration is now completed of all that is included in Proposition A; and we are ready to go forward to the demonstration of Theorem II, for which Proposition A was meant to be preparative. For this we have to observe that the conclusions arrived at in Proposition A hold good, no matter what may be the forms of the guide-tubes, provided that they be similar in both flows; and no matter what may be the distribution of pressures throughout a terminal interface crossing the assemblage of guide-tubes in No. 1, provided that the homologous pressures throughout a homologous terminal interface in No. 2 be anyhow maintained severally n^3 times those in No. 1. Hence, if in Flow No. 1 the guide-tubes be formed so that the water shall flow along exactly the same paths as if it were left unguided, and were left free to shoot away, past the interface at E, to a distance from the orifice great in proportion to the thickness of the issuing stream, without meeting any obstruction— and if the guide-tubes in No. 2 be similar to them—and if in No. 1 the system of pressures distributed throughout the terminal interface at E be made exactly the same as if the water were flowing freely for a great distance past that terminal interface— and if in No. 2 the system of homologous distributed pressures throughout a homologous terminal interface at E' be anyhow maintained severally n^3 times those in No. 1,—it follows that the differential pressure on the two sides of any element of a septum in Flow No. 1 will be zero, as the guide-tubes have there no duty to perform. Then, on the homologous septum element in No. 2, the differential pressure, being n^3 times that in No. 1, will be zero also. Hence in No. 2 the guide-tubes have no duty to perform, and the water flows in them exactly as if it were left unguided, but had still throughout its terminal interface the stated system of distributed pressures somehow applied.

Now, for completing the demonstration of Theorem II, nothing remains needed except to show that this stated system of distributed pressures requisite to be applied throughout the terminal interface at E' will very exactly be applied on that interface backwards by the water in front of it, which constitutes, for the time being, the continuation of the stream past that interface.

For proof of this, conceive any cross interface *FF* (fig. 9), further forward in No. 1 than *EE* is, and conceive a similarly situated cross interface *F'F'* in No. 2. By exactly the same mode of reasoning as before (making use of the like supposed introduction and subsequent removal of guide-tubes), that reasoning being now applied to the two flows commencing at the initial interfaces *BCD* and *B'C'D'*, and continued to the terminal interfaces *FF* and *F'F'*, it results that if the jet in No. 1 be allowed to flow freely to and far past the interface *FF*, the jet in No. 2 terminating at *F'F'* can be left to flow unguided, with stream-lines similar to those in No. 1, and with the same relations of pressures and

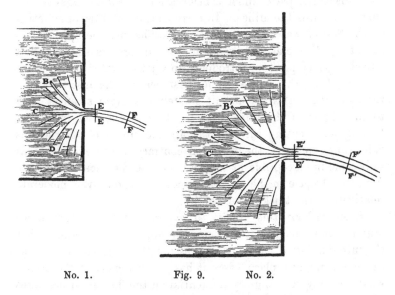

No. 1. Fig. 9. No. 2.

velocities at its various places to the pressures and velocities in No. 1 as have been already proved for the flow terminating at *E'E'*, provided that homologous pressures n^3 times those at *FF* be anyhow maintained at *F'F'*. Thus, then, we see that if adjusted or requisite pressure systems, such as have been already fully explained, be maintained at *FF* and *F'F'*, the two streams, one extending backward from *FF* to *EE*, and the other from *F'F'* to *E'E'*, will transfer backward just such pressures to successive places in retrograde order in their courses as that they will of themselves apply, at the interfaces *EE* and *E'E'*, exactly the already specified requisite pressure systems. Thus we can depart as far as we please from the orifices forward along the two streams to the places where, for purposes of reasoning, certain definite pressures are to be supposed to be applied in two homologous cross interfaces. Now it may be taken as evident that, by going far enough away from the orifice to the terminal cross interface, we can make, for any disturbances or departures from the specified pressure relations, the effects propagated backwards to the water in and near the orifices as small as we please; or that, even if we were to apply not exactly the strictly requisite pressure systems at those terminal places, still the effects of this departure from perfect exactitude would fade away rapidly in either stream as we transfer the place under consideration backwards against the current towards the orifice. In corroboration of this, observation on the flow of water spouting from an orifice may be appealed to as setting this matter beyond doubt, through its showing that any changes of pressure introduced in a jet of water at any place far away from the orifice (as, for instance, by the insertion of a rigid obstruction) will transmit scarcely the slightest effect back to the region of the orifice; or, in other words, that in a free-flowing jet spouting through the air, the effects of obstructions fade away rapidly in the direction contrary to the current, so as to become imperceptible at a very moderate distance taken back from the obstacle in the direction against the flow—very moderate relatively to the thickness of the jet.

Even without this appeal to experimental observation, we might almost intuitively perceive, or might readily admit, that the introduction of more or less pressure than any stated amount in the stream, at a place where it has got well clear of the orifice, would be only very slightly influential on the flow as to pressures and as to velocities and directions of motion within the vessel and near the orifice and contracting vein. A reason for this is, that while an obstruction in a free jet will require a great change in mode of flow of the jet close in front of it, yet the jet approaching to that region need have its outer filaments turned aside only very slightly indeed to allow of all parts moving forward without any of their stream-lines, whether medial or at or near the surface, being subjected to almost any increase of pressure, and consequently without the velocities of any of them being almost at all retarded. This will readily be clearly understood by reference to fig. 10, where the water is shown as spouting against a stone without being made to thicken its stream sensibly in consequence of the obstruction, except for a very short distance at G in front of the stone—that is to say, in the back-stream direction from the stone. If we were to suppose that the stone would have a tendency to produce, at such a place as K, any considerable increase of pressure in the internal or central stream-filaments of the jet, we would have to notice that the external stream-filaments next the atmosphere would fail to resist this augmented pressure; and, instead, they would, with only a very slight change in their own velocities or pressures, yield a little outwards, and so would not exert on the internal filaments the confining action that would be requisite for the maintaining of more than an extremely slight augmentation of pressure in those internal filaments. Then it is obvious that if the pressure is

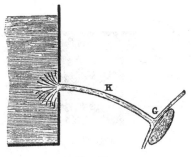

Fig. 10.

very little augmented, the velocity must be very little abated; and so, for this reason, the stream will not tend to thicken itself except very slightly, because any considerable increase of cross-sectional area of the stream would require an important abate-ment of velocity, which, as said before, would require a great increase of pressure in the internal filaments, while the external filaments would fail to exert that necessary confining pressure. These external filaments could, with very little change in their own velocities, allow even of a great augmentation of the cross-sectional area of the jet if the internal filaments, by abated velocity, were requiring to become considerably thicker than before, in virtue of the introduction of the obstruction. It is only the rapid change of direction of motion of the particles of water in the outer filaments in the neighbourhood of G, close to the obstruction, that enables them, by what may be called their centrifugal force, to maintain a greatly increased internal pressure very close to the obstruction, and so to allow of the water in the internal stream-filaments abating its velocity, and of those filaments themselves swelling in their transverse dimensions.

These considerations complete all that is necessary for the demonstration of Theorem II, and it may now be regarded as proved.

FORMULA FOR THE FLOW OF WATER IN THE V-NOTCH.

From the foregoing principle we can find intuitively the formula for the quantity of water which will flow through a V-notch in a vertical plane surface, as in fig. 11. We can see it at once by considering any stream-filament in the flow in one

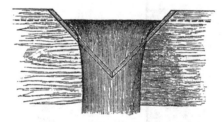

Fig. 11.

notch, and the homologous stream-filament in the similar flow in another notch similarly formed, but having its vertex at a different depth below the still-water surface-level. Let the ratio of the depth of the vertex of the one notch below the still-water surface-level to the depth of the vertex of the other be as 1 to n, so that all homologous linear dimensions in the two flows will be likewise as 1 to n. Then, in passing from any cross section of one of the two homologous filaments to the homologous cross section of the other, we have the cross-sectional area $\propto n^2$, and the velocity of flow $\propto \sqrt{n}$; and the volume of water flowing per unit of time, being as the cross-sectional area and the velocity conjointly, will vary as we pass from the one to the other of the pair of homologous filaments, so as to be $\propto n^2 \sqrt{n}$. Then, as this holds for every pair of homologous stream-filaments throughout the two flows, if we put Q to denote the quantity, reckoned voluminally, flowing per unit of time in each of the two entire flows, we have

$$Q \propto n^{\frac{5}{2}}.$$

Now, as well as considering two separate notches with different streams flowing in them at the same time, we may, when it suits our purpose, consider one single notch with streams of different depths flowing at different times; and if in various cases, either of the same V-notch or of different but similar V-notches, we denote the height of the still-water surface-level above the level of the vertex of the notch by h, we have

$$Q = ch^{\frac{5}{2}} \dots\dots\dots\dots\dots\dots\dots\dots(9),$$

where c is a constant coefficient, which cannot be determined by theory, but can be very satisfactorily determined by experiment for any desired ratio of horizontal width to vertical depth to be adopted for the form of the notch. Experiments determining the values of c for certain forms and arrangements of V-notches, suited for practical convenience and utility, have already been made by myself, and have been reported on to the British Association; and the Reports on them are printed in the British Association volume for Leeds Meeting, 1858*, and in that for Manchester Meeting, 1861†.

INVESTIGATION OF A FORMULA FOR THE FLOW OF WATER IN A RECTANGULAR NOTCH WITH LEVEL CREST IN A VERTICAL PLANE FACE.

It is to be premised that the long-known and generally used formulas for the flow of water in rectangular notches, brought out by the so-called "theories" which I have dissented from in the earlier part of the present paper, have been mainly of the form

$$Q = cgLh^{\frac{3}{2}},$$

where Q denotes the volume per unit of time,

L denotes the horizontal length of the notch,

h the vertical height from the crest of the notch to the still-water surface-level, and

g the coefficient for gravity,

and where c has either been taken as a constant numerical coefficient for want of accurate experiments to determine its values

* [*Supra*, p. 36.]   † [*Supra*, p. 42.]

for different values of L and h, or has been treated as a variable. Poncelet and Lesbros have taken this latter course, and have deduced by experiments extensive tables of its values for different depths of water in notches of the width on which they experimented—a width, namely, of 20 centimetres*. As, however, the coefficient for terrestrial gravity varies but little for different parts of the world, it has most frequently been left out of account, a single coefficient c' being used instead of cg; so that if, for instance, when the foot and second are used as units of length and time, we take 32·2 as a correct enough statement of the value of g for any part of the world, we have $c' = 32\cdot2c$.

A new formula, involving an important improvement in its form and adjusted so as to be in due accordance with numerous elaborate experiments, was developed within or about the time from 1846 to 1855, in America, by Mr Boyden and Mr Francis, both of Massachusetts. It is

$$Q = 3\cdot33 \left(L - \tfrac{1}{10}nh\right) h^{\frac{3}{2}},$$

where Q is the quantity of water in cubic feet per second,

L is the length of the notch in feet,

h is the height from the level of the crest to the still-water surface-level in feet, and

n is the number of end contractions, and must be either 0, 1, or 2.

This formula was offered by Mr James B. Francis, in his work entitled *Lowell Hydraulic Experiments*, and published at Boston in 1855, not as one founded on any complete theoretical views, but as one depending on several assumptions probably not perfectly correct, and yet as one which, through numerous trials and by adjustments introduced tentatively in fitting it to experimental results, had been brought out so as to agree very closely with experiments.

In § 120, at page 72 of his work, Mr Francis says:—"No correct formula for the discharge of water over weirs, founded upon natural laws, and including the secondary effects of these laws, being known, we must rely entirely upon experiments, taking due care in the application of any formula deduced from thence not to depart too far from the limits of the experiments

* *Mémoires de l'Académie des Sciences : Sciences Mathématiques et Physiques,* Tome III. 1829.

on which it is founded." And in §§ 123, 124, at page 74, in respect to the conception of the formula, he further gives the following very clear explanations :—" The contraction which takes place at the ends of a weir diminishes the discharge. When the weir is of considerable length in proportion to the depth of the water flowing over, this diminution is evidently a constant quantity, whatever may be the length, provided the depth is the same ; we may, therefore, assume that the end contraction effectively diminishes the length of such weirs, by a quantity depending only upon the depth upon the weir. It is evident that the amount of this diminution must increase with the depth ; we are unable, however, in the present state of science, to discover the law of its variation ; but experiment has proved that it is very nearly in direct proportion to the depth. As it is of great importance, in practical applications, to have the formula as simple as possible, it is assumed in this work (Mr Francis's book) that the quantity to be subtracted from the absolute length of a weir having complete contraction, to give its effective length, is directly proportional to the depth. It is also assumed that the quantity discharged by weirs of equal effective lengths varies according to a constant power of the depth. There is no reason to think that either of these assumptions is perfectly correct ; it will be seen, however, that they lead to results agreeing very closely with experiment.

"The formula proposed for weirs of considerable length in proportion to the depth upon them, and having complete contraction, is

$$Q = C\,(L - bnh)\,h^a\,;$$

in which Q = the quantity discharged in cubic feet per second ;

C = a constant coefficient ;

L = the total length of the weir in feet ;

b = a constant coefficient ;

n = the number of end contractions. In a single weir having complete contraction, n always equals 2 ; and when the length of the weir is equal to the width of the canal leading to it, $n = 0$;

h = the depth of water flowing over the weir taken far enough upstream from the weir to be unaffected by the curvature in the surface caused by the discharge ;

a = a constant power."

This formula, Mr Francis states, was first suggested to him by Mr Boyden in 1846.

The important novel feature in this formula consists in the subtraction which it makes, from the length L of the notch, of a length for each end contraction directly proportional to the height of the still-water surface-level above the crest in order to find what may be treated in the formula as the *effective length*.

The formula in its general form here last noted expressed only in symbols, as also in its subsequently developed form here previously stated with numerical coefficients arrived at by tentative application of numerous experiments, is thus to be regarded as an ingeniously arranged and valuable empirical formula, but not as one founded on any trustworthy hydrokinetic theory. It is founded partly on the old ordinary false "theoretical" views, and partly on good conjectural assumptions, and is adjusted and approximately verified by elaborate experiments conducted on a scale unusually large, and with unusually good means for attainment of exact results. Mr Francis, it is to be noticed, explains that, in the formula as finally brought out, the index for the power of the height of the water is taken as an exact fraction, $\frac{3}{2}$, in preference to some unascertained fractional expression, different in no great degree from $\frac{3}{2}$, merely for the attainment of facility in calculations in the practical applications of the formula, and not for any theoretic reason. Also it is to be noticed, in respect to the value $\frac{1}{10}$ which he assigned for the symbol b, that the symbol itself was first assumed as a *constant* rather than some unknown *variable* dependent on h, and was afterwards fixed at the particular value $\frac{1}{10}$ for the sake, in both cases, of attaining a convenient degree of simplicity which by trials was found to be attainable, consistently with good accordance between the representations afforded by the formula and the results shown by experiments. He supposed, however, that "many other values of a and b (probably an unlimited number) might be found that would accord somewhat nearer with the experiments [*]."

Many years ago, after my having become acquainted with the empirical formula thus made out by Mr Boyden and Mr Francis, it occurred to me as desirable to attempt to investigate by hydro-kinetic principles, without special experiments, a true formula for

---

[*] *Lowell Hydraulic Experiments*, § 156, p. 118; § 153, p. 116; and the passage quoted above from § 123, p. 74.

the flow of water in rectangular notches in vertical thin plates, or vertical plane faces, on the hypothesis of the water being a perfect or frictionless fluid, and by using in the formula symbols for constant coefficients, which, after the finding of the formula, might be determined by a small number of accurate experiments, and might further be tested as to their trustworthiness, or might be amended so as to become more exact, by a large number of varied experiments. It will be interesting to notice that the formula which had previously been arrived at in America by Mr Boyden and Mr Francis in the way already described is in perfect agreement with the formula which, by my own investigation, is brought out by strict scientific principles as a highly exact formula for water considered as a perfect fluid, and as being a very satisfactory representation of the truth for real water.

It is to be noticed at the outset that obviously a notch may be made so long relatively to the depth of its crest from the still-water surface-level, that, for any additional length, the increase of the flow will be proportional to the additional length. Let mh, in which m is a constant multiplier, be such a length as that, for additional length, the additional flow will be proportional to the addition made to the length. In fig. 12 let AB be the crest of the notch, and let CD be the level of the still-water surface of the pent-up water. Let AE and BF be each equal to $\frac{1}{2}mh$, so that, over the part EF of the crest there will flow a quantity of water exactly proportional to the length of EF if the width of the notch be varied while the depth h of the water remains unchanged. Let the length EF be denoted by l; then

$$l = L - mh.$$

Now, out of the entire flow, conceive the middle portion which flows over EF, and may be regarded as bounded laterally by two vertical planes perpendicular to the plane of the orifice, one passing through ER and the other through FS, to be taken away; and suppose the two remaining parts which flow over AE and BF, with the necessary lateral parts of the notch-plate to be brought together as shown in fig. 13, so as to form one notch having mh for its width and h for the height from crest to still-water level, and in which, therefore, the width of the notch shall bear a constant ratio to the height of the water, when the height varies, the width being always m times the height.

Then, by exactly the same mode of procedure as that already used for finding a formula to show how the quantity of water flowing in a V-notch varies with the depth of the vertex or with any other linear dimension of the flowing stream, we can readily see that if we put Q' to denote the volume of water flowing per unit of time in the case represented in fig. 13, we shall have

$$Q' = \alpha h^2 \sqrt{h} \quad \ldots\ldots\ldots\ldots\ldots\ldots\ldots(10),$$

where α is a constant coefficient.

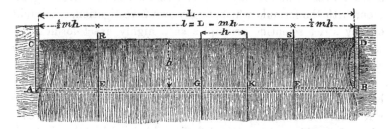

Fig. 12.

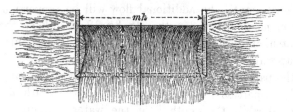

Fig. 13.

Next to find an expression for the quantity (voluminally reckoned) flowing over the middle part EF of the crest, we may consider, first, of that middle part, a portion GK taken always of a length bearing a constant ratio to h; and for simplicity we may take it of length equal to h *. Now in this stream, since the width has in general a constant ratio to the depth, or, in the case more particularly considered, since the width is equal to the depth, the quantity flowing per unit of time will, as in the

* Or, to meet the case in which there might not be, between E and F, a length so great as h, we might as well consider, in another notch having great width and having a height of flow equal to h, a portion of the flow not near either lateral extremity of the notch, and occupying a length of the crest equal to h.

preceding case, be proportional to the $\frac{5}{2}$ power of the depth; or we have

Flow over $GK = \beta h^2 \sqrt{h}$, where β is constant.

Hence if q_1 be the flow, in units of volume per unit of time, over a unit of length in EF, we have

$$q_1 = \beta h \sqrt{h}.$$

By multiplying this by l we get the quantity flowing over the entire middle part EF per unit of time; and so, denoting that quantity by Q'', we have

$$Q'' = \beta l h \sqrt{h}$$

or $$\left. Q'' = \beta (L - mh) h \sqrt{h} \right\} \quad\dots\dots\dots\dots\dots(11).$$

Adding the expressions for Q' and Q'' together, we get for the total flow in the whole notch, which we may denote by Q,

$$Q = \beta (L - mh) h \sqrt{h} + \alpha h^2 \sqrt{h},$$

or $$Q = \beta L h \sqrt{h} - (\beta m - \alpha) h^2 \sqrt{h},$$

or $$Q = \beta \left(L - \frac{\beta m - \alpha}{\beta} h \right) h .$$

But $\dfrac{\beta m - \alpha}{\beta}$ is a constant; and let it be denoted by $2b$; and instead of the constant β we may, in order now to use English letters, put a. Then

$$Q = a (L - 2bh) h^{\frac{3}{2}} \dots\dots\dots\dots\dots(12),$$

which is the desired formula for the flow of water in a rectangular notch with two end contractions.

This formula admits of easy modification to give a formula suitable for a notch with only one end contraction *, thus:—

Let the width of the notch with only one end contraction be denoted by L (as in fig. 14). Then conceive a notch twice as wide

---

* It is to be understood that contraction may be prevented at either end of a notch by there being a vertical plane side face for the channel of approach to the notch, that side face being perpendicular to the plane of the notch, and extending up-stream from the notch so as to reach beyond the region of incipient rapid flow to the notch, and extending for a little way down-stream past the notch, so as to afford the necessary guidance to the issuing stream-filaments. In like manner, by two parallel vertical side walls or side faces to the channel, when the crest of the notch extends quite across from the one wall-face to the other, contraction may be prevented at both ends.

with two end contractions as shown in fig. 15. The flow in this double space will, by the formula last obtained (12), be seen to be $= a\left(2L - 2bh\right)h^{\frac{3}{2}}$; and so if we put now Q to denote the flow in the notch under consideration (shown in fig. 14), which will be half the flow in fig. 15, we have for the notch with only one end contraction

$$Q = a\left(L - bh\right)h^{\frac{3}{2}} \quad \ldots\ldots\ldots\ldots\ldots\ldots(13).$$

Also from (11), by changing, as done before, the letter β into the English letter a, we see that for a notch with no end contraction

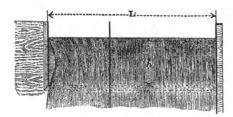

Fig. 14.

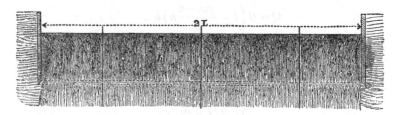

Fig. 15.

(contractions being prevented at both ends by vertical guiding side faces perpendicular to the plane of the notch) we would have

$$Q = aLh^{\frac{3}{2}} \quad \ldots\ldots\ldots\ldots\ldots\ldots\ldots(14).$$

Now the three formulas (12), (13), and (14) may be combined so as to be expressed together, thus:—

$$Q = a\left(L - nbh\right)h^{\frac{3}{2}} \quad \ldots\ldots\ldots\ldots\ldots\ldots(15),$$

where n is the number of end contractions, and must be either 2, 1, or 0.

To determine the constants a and b, all that would be necessary would be to have two very accurate experiments on the flow of water in one notch at different depths, or in two notches of the same kind with the ratio of the width to the depth not the same in both. Then, putting into the formula the measured values of L, h, and Q for the one experiment, and then again those for the other, we would have two equations with two unknown symbols, and so we could find the numerical values of those symbols. It would, of course, be desirable, for experimental verification of the theory on which the formula is founded, as also for mutual verification or testing of the experimental results themselves, to have numerous experiments on the flow for various depths in various notches of different widths, so as to find whether the formula would fit satisfactorily to them all, or to all of them that, after comparison, would be found trustworthy—provided that the width of the notch be not too small in proportion to the depth of the flow, or that in all cases the width be sufficient to allow of there being at least some small part in the middle where the rate of flow per unit of time would be proportional to the length of the part of the crest to which that flow would belong.

Mr Francis's experiments and his reductions of the results carried out in his own way give the formula complete, with its numerical coefficients, as follows :—

$$Q = 3 \cdot 33 \left(L - n \tfrac{1}{10}h\right) h^{\frac{3}{2}},$$

where $Q =$ the discharge in cubic feet per second;

$\quad L =$ the length of the notch in feet;

$\quad n =$ the number of end contractions;

$\quad h =$ the height from the crest to the still-water surface-level in feet.

Mr Francis also states that this formula is not applicable to cases in which the height h from the crest to the still-water surface-level exceeds one-third of the length, nor to very small depths. In the experiments from which it was determined the depths varied from 7 inches to 19 inches; and he remarks that there seems no reason why it should not be applied with safety to any depths between 6 inches and 24 inches.

15. On the "Vena Contracta *."

Extract from a letter of WILLIAM FROUDE, *Esq., C.E., F.R.S., to* Sir WM
THOMSON, *dated Chelston Cross, Torquay, 20th December,* 1875.

[Read before the Philosophical Society of Glasgow, February 23, 1876.]

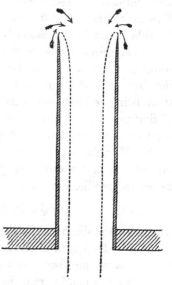

* * * "ONE result I have tried came out well :—the discharge
through an introverted cylinder
with keen edge. Here by theory
the area of the section of the jet
ought to be exactly half that of
the aperture. For the conservation of stream line energy obliges
the velocity to be that due to the
head, while the conservation of
momentum requires that the
pressure on the aperture (which
here is the sole operative pressure
acting in the ultimate direction
of the velocity generated) is only
sufficient to create as much momentum, say, per second, as will
be resident in the length delivered
per second, of a column of discharge of half the sectional area
of the aperture, if its velocity is that due to the head.

"The cylinder was quite smooth outside, and the edge quite
keen. The area ratio came out 0·503, 0·502, &c., instead of
0·500, and the little excess was obliterated, if the head was
counted, to about ¼ the diameter of the aperture below the edge;
as indeed it ought to be (I won't swear to the exact figure ¼),
because till the motion of the particles is purely parallel to the
axis, there must be some acceleration to be effected in the direction
of the axis, and this demands the employment of *some* vertical
pressure.

*   *   *   *   *   *   *

* [The theory here given is due originally to Borda who verified it by experiment. It was rediscovered by Hockin; see *Proc. Lond. Math. Soc.* Vol. III. 1869,
p. 4, with a note added by Maxwell. Cf. Lamb, *Hydrodynamics,* 1906, footnote to
§ 24.]

"In the *vena contracta* experiment with the thin plates and open air between the plates, the  fluid was welcome, if it pleased, to start tangentially from the plane of the aperture as here indicated, and as it appears to do if closely studied. So also with the introverted cylinder; though it was not possible to *see* what happened, I have no doubt that the

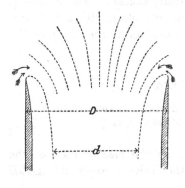

motion of the particles *next* the edge was vertical upwards, the curvature being only such as the pressure in the contiguous stream could satisfy. If the experiment was not adroitly initiated, the water seized the inner surfaces of the cylinder and ran out in an eddied condition, filling the discharge pipe. When, however, it was properly started, the contracted column below issued with beautiful smoothness and symmetry."

After communicating the preceding extract from a letter which he had received from Mr Froude, Sir W. Thomson said that on receiving the letter he had been greatly struck with this passage, as containing the first rigorous investigation of the *vena contracta*, otherwise than by experiment, which had hitherto been made; and that he had therefore asked for and obtained permission from Mr Froude to communicate it to the Philosophical Society. He referred to a letter from Sir Isaac Newton to Cotes, in which the area of the contracted vein in the case of water issuing from a circular aperture in a thin plate was explained by the convergence of the stream lines of the liquid flowing towards the aperture; and an experiment was described in which he (Newton) had actually measured the diameter of the contracted vein, and found it to be $\frac{21}{25}$ of the diameter of the orifice. This makes the area of the contracted vein $(\frac{21}{25})^2$, or ·7058 of the area of the orifice. Froude's dynamical reasoning, which shows the area of the contracted vein to be exactly half the area of the orifice in the case of the introverted tube, shows that it must be more than half the area of the orifice in the case in which water flows from an aperture in a thin plate, which is remarkably in agreement with Newton's experiment.

Sir W. Thomson promised to append to the present communication an extract from Newton's letter, which is contained in a volume of *Correspondence of Sir Isaac Newton and Professor Cotes, including Letters from other Eminent Men*, published in 1850 by Mr J. Edleston, Fellow of Trinity College, Cambridge, from originals in the Library of Trinity College. The following, accordingly, is the extract from Newton's letter, with a foot-note by Mr Edleston, which is interesting as showing the origin of the term "vena contracta":—

NEWTON TO COTES.

"St Martin's Street, by Leicester Fields,
March 24th, 171♀.

  "Sir,

         * * * "That you may have the clearer Idea of the experiments in the beginning of the inclosed paper, let *ABCD*

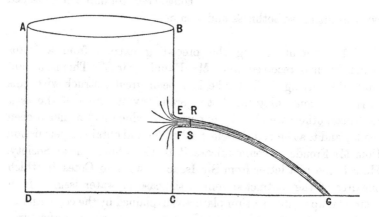

represent a vessel full of water, perforated in the side with a small hole *EF* made in a very thin plate of sheet tin. And conceive that the water converges towards the hole from all parts of the vessel and passes through the hole with a converging motion, and thereby grows into a smaller stream after it is past the hole than it was in the hole. In my trial the hole *EF* was ⅝ of an inch in diameter, and about half an inch from the hole the diameter of the stream $RS*$ was but $\frac{21}{40}$ of an inch. And therefore the streame had the same velocity as if it had flowed directly out of a hole but $\frac{21}{40}$ of an inch wide. And so in Marriott's experim^t the stream had the same velocity as if it had flowed directly out of a hole but $\frac{21}{100}$ of an inch wide. In computing the velocity of the water wch flows out, we are not to take the diameter of the hole for the diameter of the streame, but to measure the diameter of the streame after it is come out of the hole and has formed itself into an eaven and uniform streame. And the velocity thus found will be what a body would get in falling from y^e top of the water; as is manifest also by the distance, CG, to which the stream will shoot itself, and also by the streams ascending as high as the top of y^e water stagnating in the vessel, if the motion be turned upwards.

"I am

"Your most humble & most obliged servant,

"IS. NEWTON.

"For the R$^{erd.}$ Mr Roger Cotes, Professor of Astronomy,
   at his Chambers in Trinity College, in the University
   of Cambridge."

(The following communication from Professor James Thomson, LL.D., C.E., relating to Sir William Thomson's Paper, and to the Discussion upon it, was submitted to the Committee on Papers on 31st March, 1876, and is inserted by authority of the Committee.)

PROPOSITION.

It is proposed to prove that in a jet of water issuing from a circular orifice either in a thin plate (as in fig. 1), or at the extremity of a round nozzle protruding (as in fig. 2), or re-entrant (as in fig. 3), but convergent in every case towards the orifice, the cross-sectional area of the "*vena contracta*," where the water begins to have sensibly parallel stream lines, *is more than half*

---

* "RS is the diameter of the 'sectio venæ contractæ' (a term first used by Jurin, *Phil. Trans.* Sept.—Oct. 1722, p. 185; and afterwards by Dan. Bernouilli, *Hydrodynam.* p. 65. Jurin also uses 'vena contracta' to denote the same thing, and the expression is still retained in works on Hydrostatics, though differently defined by different writers, most of them describing it as that part of the issuing fluid between the orifice and the section whose diameter is RS)."

the area of the orifice; and that this condition only ceases for a re-entrant nozzle when that (as in fig. 4) ceases to be convergent, and becomes the re-entrant tube treated of in Mr Froude's

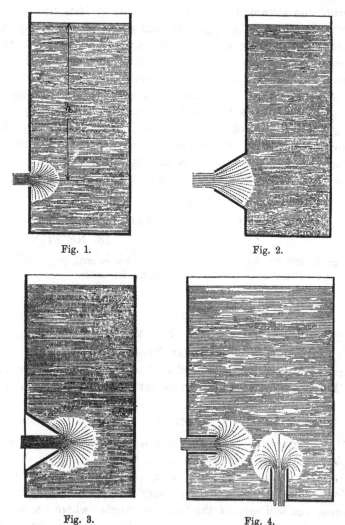

Fig. 1.   Fig. 2.

Fig. 3.   Fig. 4.

hydraulic theorem, which has been laid before the Society in Sir William Thomson's communication.

This proposition is to be understood as being stated only for orifices very small in comparison to their depth below the free surface level of the statical water; and it may be noticed that it is only this case that has usually been contemplated in statements as to the form of the contracting vein, from the time of Sir Isaac Newton's investigations on this subject down to present times. The demonstration is also dependent on the supposition that in cases of this kind the influence of fluid friction or viscosity may without important error be neglected. It is also to be observed that the scope of the proposition is limited to jets from circular orifices; because, for jets from orifices of other forms than the circular, there is no place that can properly be called the contracted vein with flow along stream lines sensibly parallel.

Let a denote the area of cross-section of the contracted vein where the stream lines are parallel.

Let h be the depth from the still-water surface-level in the cistern to the centre of the orifice.

Let v be the velocity at the contracted vein, a velocity which we know in this case must be $= \sqrt{2gh}$.

Let ρ denote the density of the water, expressed as the mass per unit of volume.

Thus we have

$$v = \sqrt{2gh},$$

$$\text{Volume per second} = a \sqrt{2gh},$$

$$\text{Mass per second} = \rho a \sqrt{2gh},$$

and therefore,

$$\text{Momentum per second in direction of the flow} = \rho a \cdot 2gh.$$

But the unbalanced force given by the vessel to the water in a horizontal direction in the case shown in fig. 1, when the orifice is in a vertical plate, or when the flow at the contracted vein is horizontal, must be equal numerically to the momentum produced per second[*]. This is founded on a well-known dynamic principle, and is easily proved, and need not be demonstrated in detail here. Now the reaction force received by the vessel from the water is a force the same in amount but opposite in direction. Let this be denoted in amount by R. Thus we have

$$R = \rho a \cdot 2gh;$$

[*] It is to be understood that the forces referred to in the present paper are expressed numerically in kinetic units of force, according to the method of Gauss; or that the *unit of force is the force which, acting on a unit of mass for a unit of time, will impart to it a unit of velocity.*

and now we see that this, being double of $g\rho ah$, is double the gravity of a column having a, the cross-section of the contracted vein, for cross-section, and h for height.

Suppose now the orifice to be at first stopped with a closely fitting but frictionless cylindrical plug, like a piston, as in fig. 5, and that the plug is afterwards removed to allow the water to flow. While the close fitting but frictionless piston or plug is in the orifice, and is held there against the water-pressure by an

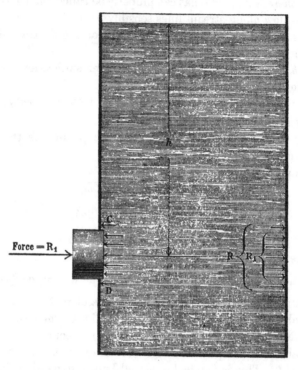

Fig. 5.

externally applied force, the water's reaction against the cistern, which may be called R_1, will be equal to the water-pressure on the plug—that is $= g\rho Ah$, where A is put to denote the area of the orifice. On removal of the piston the water-pressure on the remote face of the cistern, where the water has no important velocity, remains unaltered, but some pressure round the margin of the orifice, represented by the arrows at C and D, is removed. Let this abatement of pressure or of force applied to the vessel

in a direction parallel to the motion of the water in the contracted vein be denoted by P, and let R be put to denote the
total reaction force on the vessel when the jet is flowing; then
we have

$$R = R_1 + P,$$

or,

$$R = g\rho A h + P.$$

But we saw before that

$$R = 2g\rho a h.$$

Hence,

$$g\rho A h + P = 2g\rho a h,$$

or,

$$a = \tfrac{1}{2}A + \frac{P}{2g\rho h},$$

which is what was to be proved, as it shows that a is more
than $\tfrac{1}{2}A$.

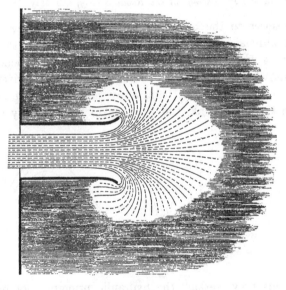

Fig. 6.

The same mode of proof applies obviously to show that the like
result holds good for all cases of the contracted vein issuing from
a circular orifice at the extremity of a round nozzle convergent
towards the orifice, and that the condition only ceases to hold
when the case of the re-entrant tube is reached, which has been
investigated by Froude, and treated of in the preceding paper of

Sir William Thomson. Also, further, it will become obvious in like manner that when that case of the re-entrant cylindric tube is surpassed, as in fig. 6, so that the re-entrant nozzle is divergent towards the orifice, but still such as to permit the water to spout off from the edge of the orifice and to contract, without having further contact with the nozzle, the cross-sectional area of the jet at the place called the "vena contracta" *is less than half the area of the orifice.*

16. ON THE ORIGIN OF WINDINGS OF RIVERS IN ALLUVIAL PLAINS, WITH REMARKS ON THE FLOW OF WATER ROUND BENDS IN PIPES.

[*From the Proceedings of the Royal Society, 4th May, 1876.*]

IN respect to the origin of the windings of rivers flowing through alluvial plains, people have usually taken the rough notion that when there is a bend in any way commenced, the water just rushes out against the outer bank of the river at the bend, and so washes that bank away, and allows deposition to occur on the inner bank, and thus makes the sinuosity increase.

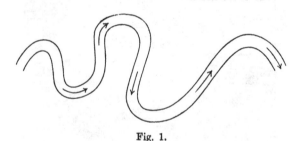

Fig. 1.

But in this they overlook the hydraulic principle, not generally known, that a stream flowing along a straight channel and thence into a curve must flow with a diminished velocity along the outer bank, and an increased velocity along the inner bank, if we regard the flow as that of a perfect fluid. In view of this principle, the question arose to me some years ago:—*Why does not the inner bank wear away more than the outer one?* We know by general experience and observation that in fact the outer one does wear away, and that deposits are often made along the inner one. *How does this arise?*

The explanation occurred to me in the year 1872, mainly as follows:—For any lines of particles taken across the stream at different places, as A_1B_1, A_2B_2, etc. in fig. 2, and which may be designated in general as AB, if the line be level, the water-pressure must be increasing from A to B, on account of the centrifugal force of the particles composing that line or bar of water; or, what comes to the same thing, the water-surface of the river will have a transverse inclination rising from A to B. The water in any stream-line CDE* at or near the surface, or in any case not close to the bottom, and flowing nearly along the inner bank, will not accelerate itself in entering on the bend, except in consequence of its having a *fall of free level* in passing along that stream-line†.

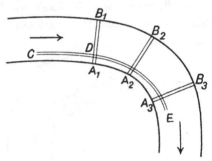

Fig. 2.

But the layer of water along the bottom, being by friction much retarded, has much less centrifugal force in any bar of its particles extending across the river; and consequently it will flow sidewise along the bottom towards the inner bank, and will, part

---

* This, although here conveniently spoken of as a stream-line, is not to be supposed as having really a steady flow. It may be conceived of as an average stream-line in a place where the flow is disturbed with eddies or by the surrounding water commingling with it.

† It must be here explained that by the *free level* for any particle is to be understood the level of the atmospheric end of a column, or of any bar, straight or curved, of particles of statical water, having one end situated at the level of the particle, and having at that end the same pressure as the particle has, and having the other end consisting of a level surface of water freely exposed to the atmosphere, or else having otherwise atmospheric pressure there; or, briefly, we may say that the *free level* for any particle of water is the level of the atmospheric end of its *pressure-column*, or of an equivalent ideal pressure-column.

of it at least, rise up between the stream-line and the inner bank, and will protect the bank from the rapid scour of that stream-line and of other adjacent parts of the rapidly flowing current; and as the sand and mud in motion at the bottom are carried in that bottom layer, they will be in some degree brought in to that inner bank, and may have a tendency to be deposited there.

On the other hand, along the outer bank there will be a general tendency to descent of surface-water which will have a high velocity, not having been much impeded by friction; and this will wear away the bank and carry the worn substance in a great degree down to the bottom, where, as explained before, there will be a general prevailing tendency towards the inner bank.

Now, further, it seems that even from the very beginning of the curve forward there will thus be a considerable protection to the inner bank. Because a surface stream-line CD, or one not close to the bottom, flowing along the bank which in the bend becomes the inner bank, will tend to depart from the inner bank

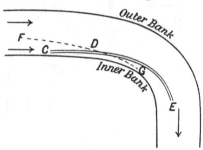

Fig. 3.

at D, the commencement of the bend, and to go forward along DE, or by some such course, leaving the space G between it and the bank to be supplied by slower-moving water which has been moving along the bottom of the river perhaps by some such oblique path as the dotted line FG.

It is further to be observed that ordinarily or very frequently there will be detritus travelling down stream along the bottom and seeking for resting-places, because the cases here specially under consideration are only such as occur in alluvial plains; and in regions of that kind there is ordinarily*, on the average, more

* That is to say, except when by geological changes the causes which have been producing the alluvial plain have become extinct, and erosion by the river has come to predominate over deposition.

deposition than erosion. This consideration explains that we need not have to seek for the material for deposition on the inner bank in the material worn away from the outer bank of the same bend of the river. The material worn from the outer bank may have to travel a long distance down stream before finding an inner bank of a bend on which to deposit itself. And now it seems very clear that in the gravel, sand, and mud carried down stream along the bottom of the river to the place where the bend commences, there is an ample supply of detritus for deposition on the inner bank of the river even at the earliest points in the curve which will offer any resting-place. It is especially worthy of notice that the oblique flow along the bottom towards the inner bank begins even up stream from the bend, as already explained, and as shown by the dotted line *FG* in fig. 3. The transverse movement comprised in this oblique flow is instigated by the abatement of pressure, or lowering of free level, in the water along the inner bank produced by centrifugal force in the way already explained.

It may now be remarked that the considerations which have in the present paper been adduced in respect to the mode of flow of water round a bend of a river, by bringing under notice, conjointly, the lowering of free level of the water at and near the inner bank, and the raising of free level of the water at and near the outer bank relatively to the free level of the water at middle of the stream, and the effect of retardation of velocity in the layer flowing along the bed of the channel in diminishing the centrifugal force in the layer retarded, and so causing that retarded water, and also frictionally retarded water, even in a straight channel of approach to the bend, to flow obliquely towards the inner bank, tend very materially to elucidate the subject of the mode of flow of water round bends in pipes, and the manner in which bends cause augmentation of frictional resistance in pipes, a subject in regard to which I believe no good exposition has hitherto been published in any printed books or papers; but about which various views, mostly crude and misleading, have been published from time to time, and are now often repeated, but which, almost entirely, ought to be at once rejected.

ORIGIN OF THE WINDINGS OF RIVERS IN ALLUVIAL PLAINS.

[From the *Glasgow Herald* of Sept. 8, 1876.   Report of the British
Association Meeting at Glasgow.]

Professor James Thomson submitted an experimental illustration of the origin of windings of rivers in alluvial plains.   The
Professor referred to a communication which he made to the Royal
Society in the month of May last, and which was subsequently
published in *Nature**.   In that communication he had given a
new theory of the motion of water in round bends in rivers and
in pipes, and had explained the reason why in alluvial plains the
bends of rivers go on increasing by the wearing away of the outer
bank and the deposition of mud, sand, and gravel on the inner
bank.   The theoretical view which he had then formed he now for
the first time had verified by practical experiment, and this
experiment he showed in the meeting.   The chief point of the
new view now experimentally proved was that the water in
turning the bend has centrifugal force, but a thin bed of the water
at the bottom is retarded by friction with the bottom, and so has
less centrifugal force than the great body of the water flowing
over it.   Consequently the bottom layer flows onward obliquely
across the channel towards the inner bank, and rises up in its
retarded condition between the inner bank and the rapidly flowing
water, and protects the inner bank from the scour, and brings
with it sand and other detritus from the bottom, which it deposits
along the inner bank.   A working apparatus, he further stated,
had been placed in the Kelvingrove Museum.

17. EXPERIMENTAL DEMONSTRATION IN RESPECT TO THE ORIGIN
OF WINDINGS OF RIVERS IN ALLUVIAL PLAINS, AND TO THE
MODE OF FLOW OF WATER ROUND BENDS OF PIPES.

[From the *Proceedings of the Royal Society*, No. 182, 1877.]

In a paper which I had the honour of submitting to the
Royal Society rather more than a year ago, and which is printed
in the *Proceedings* for May 4, 1876, I proposed, on hydrokinetic
principles, a theoretical view of the mode of flow of water round

* [No. 16 *supra*; reprinted in *Nature*, Vol. XIV. p. 122.   June 1, 1876.]

bends of rivers and of pipes, and offered under that view
explanations of the origin of the windings of rivers flowing
through alluvial plains. Wishing to bring under the test of
experiment the views then put forward, and to render very clearly
perceptible the phenomena anticipated, I constructed, in the
summer of 1876, a small artificial river, about eight inches wide
and an inch or two deep, having a bend turning about a half-round,
or 180°, so that the course of the river might be likened to the
capital letter U. The water flowing in this river showed very
completely, and very remarkably, the phenomena which had been
anticipated, and which are to be found described in the paper
referred to. The courses of the water's flow at the various parts
of the river, along the bed, and at the upper surface, and at places
anywhere within the body of the current, were made to show
themselves in several ways. One way was by means of threads of
suitable length (about an inch or two long), some of which were
anchored at bottom, while others were attached at various depths
in the river to pins or slender wires standing upright like thin
posts in the river. These threads, by the lines of direction which
they assumed, showed very well the directions of the flow at
bottom and at various depths. Another way, and one which
proved very satisfactory for showing the bottom currents, was by
dropping into the river granules of various kinds, such as sand,
and peas selected of good round form, and other small round seeds,
such as clover-seed and poppy-seed. Granules such as these
showed very clearly numerous phenomena, not only of the flow
of the water, but also of the transmission of material like detritus
forward along the bottom in straight parts, and very obliquely
across the bottom in the bend; and gave imitations on a small
scale, easy for observation, of the processes of accumulation of
detritus along the inner banks of the bends of rivers, and presented also interesting suggestions and considerations as to some
of the details or secondary actions involved in the processes*.

* The experiments here described were shown in the Mathematical and Physical
Section of the British Association at the meeting held at Glasgow, in September
1876, and further in the temporary collection prepared in the Kelvingrove Museum
at Glasgow, for that meeting of the Association. As they were arranged expressly
for testing and illustrating the theoretical views contained in a paper previously
submitted to the Royal Society, the present brief account of them is offered here to
the Society as a sequel to that previous paper.

18.   On the Flow of Water round River Bends.

[From the *Proceedings of the Institution of Mechanical Engineers.*   Read at a meeting in Glasgow, 6th August, 1879.]

Professor James Thomson, F.R.S., gave a short description of an apparatus, figs. 1 and 2 [p. 103], illustrative of the mode of the flow of water round bends of rivers, and also round bends of pipes.   On this subject he had long felt dissatisfied with all such views as had been put forward; and had considered that the various formulæ which had been published in regard to the resistance offered by bends in pipes to the flow of water, and which had long been in use, more or less, among hydraulic engineers, were certainly not founded on any reasonable theory whatever of the behaviour of the water.   Lately a new view had occurred to him, which he had brought before the Royal Society; and he had afterwards constructed an experimental apparatus which gave confirmation to the theory, by exhibiting very distinctly the phenomena which the theory had led him to anticipate.   This apparatus, it was thought, would be of interest to members of the Institution of Mechanical Engineers; and arrangements were made for showing it in action in an adjoining room.

In the windings of rivers in alluvial plains, such as were manifested very strikingly in the Mississippi and Ohio rivers, and indeed all over the world in rivers whether large or small, there were very interesting phenomena and considerations involved, both as to hydraulic questions, and as to questions of physical geography: both in reference to the behaviour of the water, and in reference to the modes of growth or modes of change of the river bends.   It was well known that, in rivers flowing in alluvial plains, windings, already existing, tended to go on increasing by the scouring away of material from the outer bank, and the deposition of detritus along the inner bank.   The sinuosities often went on increasing till a loop was formed, with only a narrow isthmus of land left between two encroaching banks of the river; and the process continued till a " cut off " occurred, a short passage for the water being opened through the isthmus, and the loop being left separated from the river course and remaining as a

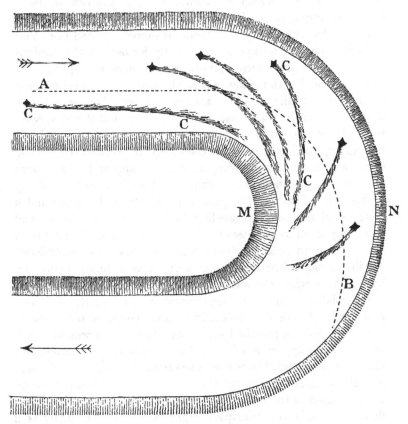

Fig. 1.

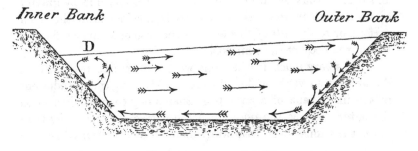

Inner Bank　　　　　　　　*Outer Bank*

Section at M N.

Fig. 2.

horse-shoe shaped lagoon or swamp, often, in the case of large rivers, stretching as much as five or ten miles away from the new course of the river channel. The usual supposition had been that the water, tending always to go directly forward in the straight line of its existing course, simply rushed outwards against the outer bank and wore it away, and at the same time caused deposits at the inner bank; and that thus sinuosities when once begun tended to increase. That view was not wrong, but it was very far from the whole truth. There was an important principle in hydraulics, which was not very commonly known, and the bearing of which on the present subject did not appear to have been noticed in any of the theories which had been advanced previously to his own. This was the principle that water flowing round a bend would necessarily, according to true dynamical theory and according to actually observed facts, get to run more quickly in feet per second near the inner bank than near the outer bank. An illustration of that might be seen in an ordinary wash-hand basin, which, when the water was whirled round a little and the plug withdrawn, presented approximately the "whirlpool of free mobility*." As the water flowed in towards the centre, the character of the whirlpool approached very closely to that of a true whirlpool of free mobility, in which the linear velocity was throughout inversely proportional to the distance from the centre: so that not only the angular velocity, but the linear velocity in feet per second was increased as the water flowed towards the centre. The water flowing inwards in a whirlpool might be regarded virtually as if flowing downhill, or more properly it was to be regarded as actually flowing from a region of higher to a region of lower *free level*†; and it must thus acquire the accession of velocity due by gravity to its *descent*, in the case of a particle flowing down the open surface, or to the *fall of free level*, in the case of a particle flowing within the body of the whirlpool.

It was very much in the same way, and for like reasons, that the water in a river bend flowed more quickly along courses adjacent to the inner bank of a river bend than along courses adjacent to the outer. Hence the question had occurred to him :—why then did not the inner bank wear away more ? As a matter of fact we

* [*Supra*, p. 1.]

† [A footnote repeating the definition of *free level* given in the note on p. 97 *supra* is omitted.]

knew that it did not; but why was this so, if the water scoured more quickly along the inner bank than along the outer? For a long time he could not explain this; but afterwards he had found the true explanation. The water of a river, in flowing round a bend, pressed outwards in virtue of centrifugal force, and so there was a transverse slope of the water surface rising from the inner bank towards the outer, as shown in the cross section, fig. 2 [p. 103]. Thus at any one level in the stream there was for this reason a greater pressure near the outer bank than near the inner. The case might be better stated thus. The free level for any particle of the water near the outer bank was higher than the free level for any particle near the inner bank at the same level and in the same cross section; and the water tended to flow from the place of higher free level to the place of lower free level; but the centrifugal action kept the water outwards against that inward tendency.

Further the most special point of the explanation was the following. At the resisting bottom there was a thin lamina of water at all times prevented by fluid friction, or viscosity, from having so high a velocity as the general average velocity of the body of the stream above it; and consequently it had not enough centrifugal force to keep it out against the inward tendency given by the higher free level at the outside than at the inside. Thus the water at the bottom must have a prevailing tendency from the outer bank towards the inner, and must tend to carry inwards with it gravel, sand, mud, and other detritus, and leave them deposited at the inner bank. Conjointly with this inward flow at the bottom, there must be in the middle and upper body of the stream a prevailing flow towards the outer bank; but the outward flow would be more gentle than the inward, being participated in by a much larger body of water.

That theory he had submitted to the Royal Society in a paper* read 4th May 1876, and printed in the *Proceedings* of the Society, 1876, p. 5. Afterwards he had proceeded to test it experimentally, and the experiments turned out completely to verify the theory†. An artificial channel, on a small scale, including a bend, was formed in imitation of a river, and upright pins were stuck into the bottom, like posts in a channel; to these pins little pieces of thread were attached, some at the bottom, some at the water surface, and some

* [*Supra*, p. 96.] † [*Supra*, p. 100.]

at intermediate levels. These, like flags flying, showed the different directions of the currents at different depths. The courses of the bottom currents, and their action in carrying detritus across the channel in the bend towards the inner bank, so as to form accumulations at or adjacent to the inner bank, were brought into notice by throwing in sand and small seeds in imitation of detritus.

In the apparatus, as fitted up in an adjoining room for this meeting of the Institution of Mechanical Engineers, the phenomena would be shown by the use of seeds serving as detritus, and small objects floating on the surface; and also in a new way* by little specks of aniline dye introduced so as to adhere to the bed of the channel at various places on the bottom and banks under water. From each speck of the adhering dye, the flowing water carried forward colouring matter, which showed that the course of the thin film of the stream, close to the bed, was along lines such as CC in the plan, fig. 1, the stream there flowing strongly towards the inner bank; while a floating object placed on the water surface would move along a line such as AB, in fig. 1, going towards the outer bank. The whole of the upper part of the river flowed out-wards slowly, while the stream along the bottom flowed inwards very quickly. There was a little eddy at D, fig. 2, whose existence was manifested very clearly by the aniline dye. The theoretical considerations connected with this eddy were very interesting; but, for brevity, he would not enter upon its discussion on that occasion.

19. ON THE FLOW OF WATER IN UNIFORM *RÉGIME* IN RIVERS AND OTHER OPEN CHANNELS.

[From the *Proceedings of the Royal Society*, No. 191, 1878.]

In respect to the mode of flow of water in rivers, a supposition which has been very perplexing in attempts to form a rational theory for its explanation, has during many years past, during at least a great part of the present century, been put forward as a result from experimental observations on the flow of water in

* [Suggested by Mr, now Professor, Archibald Barr.]

various rivers, and in artificially constructed channels. It was, I presume, put forward in the earlier times only as a vague and doubtful supposition; but, in later times it has, in virtue of more numerous and more elaborately conducted experimental observations, advanced to the rank of a confirmed supposition, or even of an experimentally established fact. This experimentally derived and gradually growing supposition was perplexing, because it was in conflict with a very generally adopted theory of the flow of water in rivers which appeared to be well founded and well reasoned out.

That commonly received theory, which for brevity we may call the *laminar theory*, was one in which the frictional resistance applied by the bottom or bed of the river against the forward motion of the water was recognized as the main or the only important drag hindering the water, in its downhill course under the influence of gravity, from advancing with a continually increasing velocity; and in which it was assumed that if the entire current is imagined as divided into numerous layers approximately horizontal across the stream, or else trough-shaped so as to have a general conformity with the bed of the river, each of these layers should be imagined as flowing forward quicker than the one next below it, with such a differential motion as would generate through fluid friction or viscosity, or perhaps jointly with that, also through some slight commingling of the waters of contiguous layers, the tangential drag which would just suffice to prevent further acceleration of any layer relatively to the one next below it. Under this prevailing view it came to be supposed that for points at various depths along any vertical line imagined as extending from the surface of a river to the bottom, the velocity of the water passing that line would diminish for every portion of the descent from the surface to the bottom.

The experimentally derived and perplexing supposition for which no tenable theory appears to have been proposed, though the want of such a theory has been extensively felt as leaving the science of the flow of water in rivers in a state of general bewilderment, is, that inconsistently with the imagination of the water's motion conceived under the laminar theory, *the forward velocity of the water in rivers is, in actual fact, sometimes or usually not greatest at the surface with gradual abatement from the surface to the bottom;* but that when the different forward velocities are compared which are met·with at successive points along a vertical line traversing the water from the surface to the bottom, it may often be found that the velocity increases with descent from the surface downwards through some part of the whole depth, until a place of maximum velocity is reached, beyond which the velocity diminishes with further descent towards the resisting bottom.

That the superficial stratum of water flowing downhill under the influence of the earth's attraction should not have its forward velocity continually accelerated until, by its moving quicker than the bed of water on which it lies, a frictional drag would be communicated to it from below, by that supporting bed of water, sufficient to hold it back against further acceleration, has appeared very paradoxical. In various cases, during a long period of time, the alleged result appeared so incredible that the experimental evidence was doubted, or was dismissed as untrustworthy. In some cases the phenomenon was admitted as a fact, but was attributed to a frictional drag or resistance applied to the surface of the water by the superincumbent air, even in case of the air being at rest with the water flowing below, or more strongly so when the wind might be blowing contrary to the motion of the river.

Omitting to touch on the experimental results, and the opinions of various investigators in the older times, as I have not had sufficient opportunity to scrutinise them in detail, I have to refer to the investigations conducted at about the year 1850 by Ellet on the Mississippi and Ohio Rivers*. He was led to the conclusion† from his own experiments on the Mississippi, that the mean velocity of that river (or at least the mean velocity of the great body of its current, as the part near the bottom or bed of the river had not been definitely included in his researches) instead of being less, is in fact greater than the mean surface velocity. He attributed this phenomenon, which he regarded as indubitably proved, and which if true must certainly be very remarkable, to a frictional drag or resistance, against the forward motion, applied to the surface of the water by the atmosphere in

---

* Ellet on the *Mississippi and Ohio Rivers*. Philadelphia: 1853. This is a republished edition of a Report to the American War Department by Ellet on his investigations, which were made under authority of an Act of Congress.

† Pages 37 and 38 of the book referred to in the preceding note.

contact with the surface. Like suppositions had previously been made by some observers and theoretical investigators in Europe, as may be gathered from D'Aubuisson, *Traité d'Hydraulique*, 2nd edition, 1840, p. 176, and from other sources of information.

Other experimental researches on the flow of the Mississippi River, much more elaborate than those of Ellet, were made in the period between 1850 and 1861 by Captain Humphreys and Lieutenant Abbot, with others acting under authority from the American Government, and an account of them was published as a Report by Humphreys and Abbot in 1861*. These experiments and the investigations exhibited in the report, where the observed results are combined in various ways so as to bring out average results and more or less probable conclusions for various circumstances, lead very clearly and very convincingly to the conclusion that ordinarily the maximum velocity is not at the surface but at some depth below it, usually much nearer to the surface than to the bottom, and often at some such depth from the surface as $\frac{1}{4}$ or $\frac{1}{3}$ of the whole depth of the water. These investigators (Humphreys and Abbot) show further (at pages 285, 288, and 289 of their Report) that this phenomenon is not wholly nor even mainly due to any frictional resistance applied by the super-incumbent atmosphere to the forward flow of the surface of the water; because they found that even when the wind is blowing in the direction of the river current, and advancing at the same velocity as that current, so that the air lies on the surface of the water without relative motion, the phenomenon manifests itself almost in as great a degree as when the air is lying at rest relatively to the land; and found yet further that the phenomenon still manifests itself even when the wind is blowing in the direction of the flow of the river much faster than the current, so that it blows the water surface forward instead of applying a resisting drag or backward force to the surface.

At about the middle of the present century very important experiments on flowing water were made in France by Boileau, and by Darcy and Bazin; and elaborate accounts of these researches were published†.

---

* Report on the *Physics and Hydraulics of the Mississippi River*. By Captain A. A. Humphreys and Lieutenant H. L. Abbot. Philadelphia: 1861.

† Boileau: *Traité de la mesure des eaux courantes*. Paris: 1854. Darcy: *Recherches expérimentales relatives au mouvement de l'eau dans les tuyaux*. Paris: 1857.

The experiments comprised among the researches of Boileau and of Darcy and Bazin, to which I have to refer as bearing on the special subject of the present paper, relate to the flow of water in long channels and conduits constructed artificially, some in wood and some in masonry and other materials. The channels or conduits in different cases were of widths comprised between half a metre and two metres. In some of the more important experiments the channels were constructed in wood, and were open above, and had a flat bottom and vertical sides, so that the current was rectangular in cross section. Channels of various other forms were also used, and the mode of flow of the water in them was scrutinized. The results arrived at by these experimenters tend very much towards establishing the supposition which forms the subject of the present paper—the supposition, namely, of the prevalence or frequent occurrence of a distribution of velocities having the maximum velocity not at the surface but at some moderate depth below. Boileau, by his experiments, was led to announce as one of his conclusions (page 308), that in the medial longitudinal vertical section of a rectangular canal with uniform *régime*, the maximum of velocity is situated not at the surface, but at a depth which is a fraction more or less considerable of the total depth of the current. He also announced, as a conclusion, that the decrease of velocity, from the place of maximum velocity up to the surface, must be attributed to some new cause different from that which produces the diminution of velocity from the place of maximum down to the bottom. This new cause, he says, cannot be solely the resistance of the bed of air in contact with the liquid surface acting like the face of a pipe or conduit; and he assigns, in proof of this, the reason that the mobility of this bed of air does not permit of our attributing to it a retarding influence so great as that which is implied in the rapid abatement of velocities in approach towards the surface in the upper part of the current. He recounts his own special experiments, made in 1845, on the influence of wind on the velocities in currents,— a subject which he says had up to that time been very little

Darcy et Bazin: *Recherches Hydrauliques*. Paris: 1865. This last book constitutes a memoir by Bazin on researches commenced by Darcy, and continued for some time by him with the aid of Bazin; and, after the death of Darcy in 1858, continued by Bazin, and by him completed and worked out in the discussion of their results.

investigated by hydraulicians. He deduces from his experiments conclusions (page 313) to the effect that in spite of varied disturbances produced by wind blowing over the water with varied intensity, yet there is manifested a very sensible tendency to a decrease of velocities of the water for approach towards the liquid surface; and that the maximum velocity is yet below the surface, even when the wind blows forward with the current, and has a velocity greater than that of the current. Judging, then, that resistance of the air cannot be the cause of the phenomenon, he says that it is then principally in the mutual actions which bind among one another the liquid particles, and in the oblique and rotatory movements which result, under the influence of these forces, from the difference of velocities of neighbouring particles, that it is necessary to seek for the explanation of the phenomena of the decrease of velocities in the approach towards the surface of currents. He goes on to say that we have to conceive, in fine, that these oblique movements, producing transverse living forces (*forces vives*), diminish, according to certain general laws, the living forces of forward motion which the hydrometric instruments are adapted to indicate.

I have cited this passage from Boileau very fully, because it seems to me to contain the nearest approach towards an explanation of the phenomenon in question of any that have been attempted, so far as any such attempts have come under my notice. It involves, I think, at least a glimmering towards a true explanation; but I regard it as being in great part erroneous, and importantly so in principle, and as being besides altogether incomplete. I do not think it has been offered by the very able investigator himself, who has proposed it, as being at all sufficient; but I think it has been offered only as tending to throw some light over the region for further search, and some indication towards courses in which speculation and research might well advance.

Bazin's experiments, of the general character already mentioned, were very extensive in their scope, and were carried out in great detail, and with some remarkable refinements of method. The velocities were measured mainly or wholly by a modification devised by Darcy of the well-known instrument called Pitot's tube. Bazin, in the case of canals not very wide relatively to the depth of the current, found very clearly and decisively the pheno-menon in question of the maximum velocity being below the surface. But, in the case of rectangular channels of more considerable width, channels having the width of the current so much as four or five times the depth or more, Bazin by his scrutiny and consideration of his experimental results, was led to conclude that the diminution of velocity for approach towards the surface in the upper part of the current is to be found only in the side parts of the current—the parts flowing along the two side walls. He judged that throughout the whole of the current, except two side parts, each having some moderate width, which might be equal to about twice the depth of the current, the maximum of the velocities for all points, situated in a vertical line, is to be found at the surface; and that the rate of diminution of velocity for descent from the surface would begin as nothing at the surface, and would go on increasing with descent to the bottom. His experiments, according to his own careful analysis and combination of them, appeared to be in agreement with this assumption, or to bring this supposition out as a result.

I do not, however, regard this conclusion as being trustworthy. His experiments for the case of great width relatively to depth had not, in any instance, a depth of water exceeding ·38 of a metre, or $1\frac{1}{4}$ foot, and thus the depths were so small absolutely as not to admit of a fine enough discrimination of minute changes of velocity for minute changes of depth of the point where the velocity was observed, nor of measuring velocities close enough to the surface. So far as experimental researches go, some doubt I presume must still remain over this part of the subject. Indeed, the Indian experiments, next to be mentioned, show results in disagreement with this conclusion offered by Darcy.

Quite recently, in 1874–75, experiments were conducted in India on the Ganges Canal, close to Roorkee, by Captain Allan Cunningham, R.E.* These experiments bring out among their results, very remarkably, the frequently alleged phenomenon of the maximum velocity of the water being not at the surface, but at some moderate depth below. And further, it is deserving of special notice that those of his experiments, which have chiefly to be referred to as throwing light on this subject, were made in an aqueduct about 85 feet wide, and with an approximately level

* "Hydraulic Experiments at Roorkee, 1874–75," by Captain Allan Cunningham, R.E., published in *Professional Papers on Indian Engineering*. Thomason, College Press, Roorkee, 1875: also Spon and Co., London, &c.

bottom; and that the depths of the water in different experiments ranged from about 6 feet to about $9\frac{1}{2}$ feet, so that the width was on different occasions from about nine times to about fourteen times the depth, and yet the maximum of the velocities at mid-channel (or the maximum velocity in the longitudinal medial vertical section) came out by averages of numerous results, and, by varied modes of experimenting, to be very decidedly below the surface.

Experiments carried out lately on a very large scale on the Irawaddy river by Robert Gordon, Executive Engineer, British Burmah, Public Works Department, go to confirm the truth of the same phenomenon. These experiments of Mr Gordon, how-ever, although valuable in many respects, appear to be subject to some doubt as to whether, through the mode of experimenting, the level of supposed maximum velocity has not been brought out too low, that is to say, too near the bottom. On this point Mr Gordon (in his Introductory Note, § vii, p. ii of date 16th June, 1875) intimates his intention to make further experiments with other instruments, but still asserts his confidence in his previous methods and results.

Until about two years ago I had not happened to become acquainted with any of the evidence for the phenomenon in question except the unsatisfying experimental results given by Ellet; but about two years ago I met with accounts of some of the more recent and more convincing experimental investigations. It then appeared to me that if the asserted phenomenon must really be accepted as a truth, there ought to be some mode possible of accounting for it: and a theory occurred to me which I now propose to submit.

The mode of thought which near the beginning of the present paper I have described as constituting the laminar theory, I must premise, has long appeared to me to be an erroneous and a very misleading view. It was a very prevalent mode of thought, and was usually too influential on people's minds even when they did entertain decidedly, though often not clearly enough, the con-sideration of eddies and transverse movements or commingling currents with different velocities. The great distinction between the mode of flow of a very viscid fluid, such as treacle or tar, and the mode of flow of water in ordinary circumstances in pipes and in open channels, has not been enough generally and enough consistently attended to. The laminar theory constitutes a very good representation of the viscid mode of motion; but it offers a very fallacious view of the motion in the flow of water in ordinary cases in which the inertia of the various parts of the fluid is not subordinated to the restraints of viscosity.

In the flow of water in an open channel in ordinary circumstances the earth's attraction is perpetually tending to accelerate the forward motion of the water throughout the whole body of the current in consequence of the surface declivity; or we may say, with more complete expression, in consequence of the fall of *free level** which, in virtue of the surface declivity, occurs to all particles in the current as they advance in their down-stream course. The tendency to increase of velocity, if we neglect the backward or forward force, usually very small, or it may be nothing, applied by the air to the water surface, we may say is counteracted solely by a backward resisting force-system applied by the wetted face of the channel to the water momentarily in contact with it. The wetted channel face, it must be observed, is ordinarily more or less rough with gravel, mud, weeds, or other asperities. It is not a true view to imagine a smooth channel face washed by a thin lamina of water, which imagined lamina of water receives a backward or resisting force-system applied tangentially by the so imagined channel face, and transmits tangential backward force to another lamina of water lying next to itself on the side remote from the channel face. It is not the case that from any layer of water whatever, thick or thin, spread over the channel face, resisting forces are transmitted to the interior of the body of the current in any great degree by mere viscid resistance to change of form in the intervening fluid, as would be the case if it were like treacle or tar. But, very differently, indefinite increase of velocity of the water situated in the interior of the current is prevented by continual transverse flows thereto, and commingling therewith, of portions of water already retarded through their

---

* The *free level* for any particle of water, in a mass of statical or of flowing water, is the level of the atmospheric end of a column, or of any bar of statical water, straight or curved, having one end situated at the level of the particle, and having at that end the same pressure as the particle has, and having the other end consisting of a level surface of water freely exposed to the atmosphere, or else having otherwise atmospheric pressure there. Or, briefly, we may say that the *free level* for any particle of water is the level of the atmospheric end of its pressure-column, or of an equivalent ideal pressure-column.

having been lately in close proximity to the resisting channel face; and, jointly with that, by the condition that portions of the fluid which have been flowing forward temporarily in the interior of the current, and have been gaining forward acceleration there, are gradually expelled, or do gradually flow from that region, and come themselves into close proximity to the resisting channel face; and so, in their turn, do receive very directly backward forces from the face, because in proximity to it processes of fluid distortion subject to viscid resistance are going on with great activity and intensity.

The transverse motions have their origin primarily in the rush of the water along the wetted channel face. When that face is rough or irregular with lumps and hollows or other asperities, reasons for the origination of transverse currents may be sufficiently obvious. But even if the channel face is extremely smooth, so as to present no sensible asperities, still there is good reason to assert that transverse flows will come to be instituted in consequence of the rapid flow of the main body of the current along a lamina, very thin it may be, of water greatly deadened as to forward motion by viscid cohesion with the channel face, and throughout and across which, if regarded as only very thin, in virtue of its thinness, the backward force applied by the face can be transmitted by mere viscosity. The thin lamina of deadened water will tend by the scour of the quicker going water always moving subject to variations both of velocity and of direction of motion to be driven into irregularly distributed masses; and these, acted on by the quicker moving water scouring past them, will force that water sidewise, and will be entangled with it and will pass away with some transverse motion to commingle with other parts of the current*.

---

* This principle I noticed myself in the connexion in which it is here adduced; and the idea has since been confirmed to me and rendered more definite through additional considerations mentioned to me lately by my brother, Sir William Thomson, which have originated with him in some of his theoretical investigations in quite another branch of hydraulic science, and which relate to finite slip in a frictionless fluid. He pointed out that if, for water theoretically regarded as frictionless, or devoid of viscosity, we imagine a long smoothly formed straight trough or channel with a thin vertical longitudinal plane septum dividing it into two parts each uniform in cross-section throughout its length, and if we imagine the space on one side of the septum to be occupied by still water, and a current to be flowing along on the other side; and if, while this is in progress, we imagine the vertical partition to be withdrawn so as to leave the current flowing along a

If we watch the surfaces of flowing rivers, or of tidal currents flowing in narrows or *kyles*, we may often have opportunity to observe very prevalent indications of rushes of water coming up to the surface and spreading out there. These rushes often may be seen to keep rising in quick succession in numerous neighbouring parts of the water surface, and they may be seen presenting appearances of spreading out till they meet one another and give indication of momentary downward sinking at their places of meeting.

From whence do these transverse currents come to the surface? It seems to me they must have had their origin in the deadened water scouring along the bottom, or along the wetted side-faces of the channel, in such ways as have just now been briefly sketched out. Thus it seems that there are tendencies bringing about the result that the superficial stratum of the river receives perpetually renewals of its substance by water currents arriving to it, and spreading out there, which have very recently departed from the bottom before coming up to enter into that superficial stratum. But their substance, having come in great part from the bottom, must be largely made up of the deadened or slow-going bottom-water. It is to be understood that this deadened water, in rising through the current towards the surface, is partly urged forward in the down-stream direction by the surrounding quicker-going water, but that it arrives at the surface without having attained fully to the down-stream velocity of that intermediate stream.

It may readily be perceived that it is from the washed face of the channel alone, or from that and the retarded layer of water in proximity to it, that any strong transverse impulses can be applied to any parts of the current. No rapid transverse current will originate in the middle of the body of the river; for there is no cause for the origination of transverse currents there, unless perhaps we were to regard as such any slight transverse motions which may be produced through the gliding forward of parts of the water there relatively to others near them going with different

plane face of still water, the motion with the finite slip thus instituted will be essentially unstable. Reasons for this, when once it is brought under notice, are very obvious from consideration of the centrifugal forces, or centrifugal actions, which would be introduced on the slightest beginning being made of any protuberance or hollow in the originally plane interface between the still water and the current. [Cf. Sir William Thomson, *Math. and Phys. Papers*, Vol. IV. pp. 330 *et seq.*]

velocities, and unless we were to regard as such any transverse disturbances that may be imparted to forward-flowing water there by the intrusion and commingling of partially deadened water from the channel-face.

We may now have great confidence, I think, in taking as a well-established truth, or at least as a very probable view, the supposition already laid down to the effect that very commonly the superficial stratum of a river receives perpetually renewals of its substance by water currents arriving to it and spreading out there, which have very recently departed from the bottom or sides of the channel before coming up to enter into that superficial stratum; and that the substance thus perpetually renewing the surface stratum is largely composed of deadened or slow-going bottom-water, or of water going slower forward than the water through which it traverses in ascending to the surface. It is further to be noticed that the water which at any moment constitutes the superficial stratum is, in its turn, very soon overflowed by later arrivals from the bottom. So it gradually descends from the surface into the interior of the body of the river. But during this action it is always flowing downhill, or we may better say it is experiencing *a fall of free level*, in consequence of the surface declivity. It is thus receiving forward acceleration in the downhill direction, and its velocity goes on increasing until at some depth from the surface it reaches a maximum, from whence, during further lapse of time and further descent of this water towards the bottom, the retarding influences imparted to it from the bottom are predominant over the downhill accelerating influence of gravity. These retarding influences, chiefly acting through transverse rushes of water from the bottom commingling more numerously and more briskly with the descending water under consideration the more it gets into the neighbourhood of the bottom, bring about the result that the water goes forward with less and less velocity as it approaches nearer and nearer to the bottom.

I have now to offer, by consideration of an imaginable case different from that of an ordinary river, an illustration which will aid in the forming of clear ideas on what I have been presenting as a true theory of the real behaviour of the water in rivers.

Let us imagine a flowing river composed mostly of water, but with a layer of oil floating on the top, the oil being of some such depth as a tenth or a twentieth part of the whole depth of the river. Let us suppose the width of the river to be so very great relatively to the depth as that in considering the flow in a middle portion of the river, we may regard it as experiencing no sensible retarding influences, either through the water or the oil, from the sides of the river; and let the flow to be kept under consideration be only that middle portion without the lateral portions which would be sensibly affected by retarding influences from the sides. Here we have a case differing from that of an ordinary river of water in this important respect, that, while in the ordinary river the superficial stratum of fluid is perpetually changing its substance, and is, as I suppose, perpetually receiving new supplies of deadened water from the bottom, in the imagined case now adduced the substance of the superficial layer being of oil floating at top, does not undergo any such change. The oil then, it seems very certain, would really rush down what we may call the inclined plane of water on which it lies, and would go on accelerating its motion until, by advancing very much faster than the water, it would introduce a frictional drag between itself and the water sufficient to hinder its further acceleration*; or rather until, without

* *Postscript note, 1st November,* 1878.—An observed phenomenon, which, if duly taken into consideration, must doubtless be found to be closely allied in its nature to the supposed behaviour of the imagined layer of oil on a flowing river of water above adduced, and which is certainly of much interest, both for its own sake and in reference to theoretical views which have been held as to its origin and its indications, has come under my notice since the time when the present paper in manuscript was presented to the Royal Society. The book by Bazin, which may be briefly named as Darcy et Bazin, *Recherches Hydrauliques*, Paris, 1865 (*see* a previous footnote in this paper), contains prefixed to it a report, dated 1863, of a Committee of the Academy of Sciences on the memoir of M. Bazin, "Sur le Mouvement de l'Eau dans les Canaux decouverts." In that report the committee remark (as confirmatory of the view which they accept, to the effect that in deep rivers, especially when not very wide relatively to their depth, the place of maximum velocity is at a considerable depth below the surface) as follows:—"Il y a longtemps que les bateliers du Rhin et nos pontonniers savent qu'un bateau chargé et ayant un fort tirant d'eau, marche, en descendant, plus vite que l'eau qui le soutient ou que les corps flottants à la surface." This obviously conveys the opinion that a heavily loaded boat, sinking deep into the water, and thereby having its deeper part immersed in water which is flowing quicker than the surface water, is dragged forwards by that deeper and quicker moving water, and so is made to advance quicker than the surface water does. The idea seems to be that the boat has some average velocity less than that of the water at its bottom, and greater than that of the surface water. The view which thus appears to be held in respect to the observed phenomenon seems to me to be inadequate and erroneous. On the principle put forward above in the present paper in reference to the imagined case

attaining to that stage of great relative velocity, it would at an earlier stage ruffle up the mutual face of meeting of itself and the water into protuberances and hollows, somewhat like waves, on the principle referred to already in a foot-note as having been proposed by Sir William Thomson, and would carry this action on to the extent of causing commotion and commingling of the water and oil. The contrast between this case and that of an ordinary river of water is so remarkable as to aid the forming of a clear comprehension of the very different mode of action which I have been attributing to the water in ordinary rivers and other open channels.

It is further worthy of notice that if, from any local cause, the water flowing forward in some part of the width of a river has in its motion a component downward from the surface towards the bottom, and is free from intrusion of upward currents or rushes of deadened water from the bottom, or of water retarded by the influence of the river-bed, we ought to expect the forward velocity to increase from the surface to very nearly the bottom. The accelerative influence of gravity due to the surface inclination, and more particularly due to the fall of free level experienced, as an accompaniment of that inclination, by the water throughout the body of the current in its onward flow would generate in every portion or particle of this water increase of velocity for advance along its course; because, in the absence of rushes of deadened

of a river with an upper layer of oil, I would suppose that a large and heavy boat, even if flat-bottomed and of shallow draught of water, would run down the river-course quicker than the water in which it swims; for the reason that while all the water surrounding it makes occasional visits to the bottom of the river, and meets with great retardation there, the boat does not dive to the bottom, and is free from any such retardation, and so is only held back by the surrounding water against taking from gravity a perpetually increasing velocity. Thus it must go faster than the surrounding water which has to hold it back. The boat of deeper draught referred to by the committee I would suppose would advance quicker than the surface water, for the same reason, and not merely because of its bottom being situated in water moving quicker than that at the surface. The principles I have assigned would afford ample reason for our supposing that the boat of deep draught might swim forward much quicker not only than the surface water, but also than the water at its bottom, or indeed than any part of the water of the river surrounding the boat. Very small floating objects, such as sticks or leaves, would present, in proportion to their small masses, so much resistance to motion through the surrounding water that they would be constrained in fact to move sensibly at the same velocity as that of the water surrounding them. The phenomenon would thus be presented of the boat swimming forward past the small floating objects around it.   J. T.

water from the bed, such as it appears do commonly intrude into the body of the current, there would be no retardative influence to counteract the gravitational accelerative influence; since the mere viscosity of the water unaided by transverse commingling is, I consider, insignificantly small and quite ineffectual as a resisting influence or means of transmitting resistance from the bed to any part of the water in the body of the current out of close proximity to the bed. But as this forward moving water is also descending towards the bottom while it is gaining forward velocity, it follows that, in the circumstances of flow supposed, we ought to expect the forward velocity to increase with descent from the surface to very nearly the bottom. It is to be understood that the freedom supposed from upward rushes or intrusions of deadened water will not be maintained in the water when it arrives into proximity to the bottom. In approaching very near to the bottom the water must begin to receive important resisting forces communicated to it from the bottom through commingling of deadened water, and by intense distortional actions with viscosity.

It is also to be noticed in connexion with the case under consideration that if, in one part of the width of the river, there is a prevailing descent towards the bottom, there will be upward flows to compensate for this in other parts of the width. Then obviously the whole character of the action of the water will be very different in the regions where ascent prevails from that in the regions where there is a prevailing descent; and the distribution of forward velocities throughout any vertical line in the one region will be quite different from the distribution of forward velocities throughout any vertical line in the other region. Local circumstances casually affecting the flow in the way here described I think may perhaps account for some of the apparent anomalies in respect to the distribution of velocities through different parts of the depth from surface to bottom which have been met with by various experimenters, and have been included among the recognised causes of the perplexity and bewilderment with which this branch of hydraulic science is pervaded.

I wish next to draw attention to one of the results of observation and experiment announced by Captain Cunningham in his book already referred to (*Hydraulic Experiments at Roorkee*). In his discussion of his experimental results on the flow of water in each of two artificially-formed channels on the Ganges Canal, one of them, 168 feet wide, and the other 85 feet wide, and each having the water often about from 6 feet to 9 feet deep, he states (p. 46, article 35): "There is a constant surface motion (deviation) from the edges towards the centre, most intense at the edges and rapidly decreasing with distance from the edges."

This experimental conclusion, on the supposition of its being decidedly trustworthy, as Mr Cunningham asserts with confidence that it is, I think may probably be satisfactorily explicable through considerations intimately connected with those which I have already given for an amended theory of the flow of water in rivers.

I wish, however, not to prolong the present paper by entering on any detailed discussion of this branch of the subject, and besides I prefer to reserve this for some further consideration before venturing to put forward the views in reference to it which at present appear to me likely to be tenable. It may be noticed, however, that Captain Cunningham's experimental result, if decidedly correct, throws additional light on the subject of the abatement of surface velocity comparatively to the velocity at some depth below the surface being found in Bazin's experiments to occur in a much greater degree near the sides of rectangular and various other channels than at middle. Bazin thought indeed from his own experiments (as I have already had occasion to mention) that the relative retardation or slowness of the surface occurred not in the middle of wide channels (that is to say, of channels wide relatively to the depth of the water) but only near the sides; but this supposition I have referred to as appearing not to be trustworthy. With these brief suggestions I will now leave for further consideration the subject of the special phenomena of the influence of the sides.

Historical Note.

Subsequently to my having formed, in all its primary or more essential features, the new view now explained of the flow of water in rivers, and before I had met with the book of Humphreys and Abbot, I happened to see in the writings of another author (paper of Mr Gordon already referred to) the following remark in reference to their views as to the velocity at the surface being less than at some depth below. "Humphreys and Abbot attribute the fact to transmitted motion from the irregularities of the bottom; but confess themselves dissatisfied with their own explanation."

These words seemed to me to indicate a probability of Humphreys and Abbot having anticipated me in some part at least of the theory which I had been forming. On obtaining their book, however, and reading the passage referred to, not by itself alone, but with its context, it appeared to me that it involved no real anticipation, although one clause of a sentence in it, read by itself, might be supposed to do so. The passage is to be found in their work at p. 286. They begin by saying, that their experimental observations detailed in their previous pages " prove that even in a perfectly calm day there is a strong resistance to the motion of the water at the surface as well as at the bottom," and that this resistance at the surface " is not wholly or even mainly caused by friction against the air." They go on to say :—" One important cause of this resistance is believed to be the loss of living force, arising from upward currents or transmitted motion occasioned by irregularities at the bottom. This loss is greater at the surface than near it. The experiment of transmitted motion through a series of ivory balls illustrates this effect. It is likewise illustrated on a large scale by the collision of two trains of cars on a railway, in which case it has been observed that the cars at the head of the train are the most injured and thrown the farthest from the track ; those at the end of the train are next in order of injury and disturbance ; while those in the middle of the train are but little injured or disturbed. Other causes may and probably do exist, but their investigation has, fortunately, more of scientific interest than practical value. For all general purposes it may be assumed that there is a resistance at the surface, of the same order or nature as that which exists at the bottom."

Now although this passage does contain the words " *arising from upward currents or transmitted motion occasioned by irregularities at the bottom,*" yet the illustrations, by means of the series of ivory balls, and of the collision of railway trains, show that the authors attribute to those words no clear and correct meaning, but, on the contrary, I would say they put forward quite a false view of the actions going on. Besides I myself do not admit that, except from the air, there is a resistance at the surface. According to my supposition the already resisted and retarded bottom water comes to the surface and spreads out there, but receives no new resistance there, and on the contrary receives acceleration from gravity in running down hill.

20. ON FLUX AND REFLUX OF WATER IN OPEN CHANNELS OR IN PIPES OR OTHER DUCTS.

[From the *Philosophical Magazine* for October 1888.   Having been read at the British Association, Section A, Bath, 1888.]

IN the autumn of 1872 I was staying at a place named Castlerock, on the north coast of Ireland, between the mouth of the Bann River and the entrance to Lough Foyle.   There was an extensive sandy beach there, lashed by the great waves of the Atlantic Ocean.   At a part of that beach a small river or stream flowed to the sea; but the sandy beach had been thrown up as a bank, at about high-tide level, obstructing what might have been the direct outfall course for the stream into the sea, and causing the stream to turn to its right and to flow eastward for some distance along the back of that sandy bank before finding an opening for flow out to the sea.   Thus, at the back of the bank, a little estuary was formed, along which, when the tide was down, the stream would have for a considerable length a nearly level bed; and into which, when the tide was up, the sea-water entered so as to fill it up to various depths according to the height of the sea-surface.

I happened to be watching with interest the motions of the water in the little estuary at a time when a considerable depth of water (such as a few feet depth along its mid-channel) was maintained in it by the height of the sea-water outside, and when the slow rising and sinking of the ocean-waves was producing in the estuary a flux and reflux on a small scale like that of the tidal flow in large estuaries*.   The motions of the water being indicated by numerous little pieces of sea-weed carried in suspension, I noticed that the water at or very close to the channel-bed reversed its landward or seaward flow always much earlier than did the main body of the water in the channel, less affected by contiguity to the bed.   The phenomenon

---

* The period of these oscillations may be about from 10 to 20 seconds; as I have been informed that Prof. Stokes has found, by observations on that coast, that the period from one wave to the next, in the large Atlantic waves there, is at most about 17 seconds. [See *Scientific Correspondence of Sir George Stokes*, Vol. II. p. 142 *seq.*]

being noticed, the reason at once became apparent. The lamina contiguous to the bed, or channel-face, would be always hindered by the frictional resistance of that face from getting into so great a velocity, either seaward or landward, as that which would be attained to by the main body of the water. Then, when the water at the sea-end of the estuary was raised in level, by the arrival of an ocean-wave, so as to give a gravitational propulsive influence tending to cause the water to flow landward along the estuary, the main body of the water, in virtue of its inertia with seaward momentum, would continue to flow for some time seaward, flowing as it were uphill*; while the frictionally restrained lamina at the channel-face, being nearly devoid of inertial tendency seaward, would readily yield to the landward gravitational propulsive influence due to the landward surface declivity of the water in the estuary.

Exactly a like explanation, *mutatis mutandis*, is applicable to the case of reversal of the flow from having been landward to its becoming seaward. The channel-face lamina makes its reversal of flow, just as in the other case, *earlier* than does the main body of the water, and for like reason.

It may now further be noticed that precisely corresponding phenomena would present themselves in the flux and reflux of water in a pipe; if, for instance, the pipe were connecting two cisterns, and a plunger were kept oscillating upwards and downwards in one of them so as to cause the alternating flow through the pipe. The phenomena might be very interestingly manifested in an open trough connecting two cisterns, arrangements being made, by a plunger or otherwise, for causing flux and reflux along the trough, and the motions of the water being indicated by small visible particles in suspension in the water, or by the dropping in of granules of aniline dye.

It may now be worthy of remark that the hydraulic principle brought into notice in the present paper, in respect to Flux and Reflux along Channels, is closely allied to, and is in some respects identical with, the leading principle set forth in previous papers by myself on the Flow of Water round Bends in Rivers, &c. In that case the frictionally resisted and retarded lamina in contiguity with the channel-face, or bed, flows transversely (or

* Or, in more precise terms, flowing from a place of lower to a place of higher free level.

rather obliquely) across the channel towards the inner bank of
the bend, impelled inwards by gravitational propulsive influence
(that is, downhill as it were), while the main body of the stream,
flowing quicker in the bend, exerts centrifugal force outwards, or
tends inertially out towards the outer bank. The papers here
referred to on Flow of Water round Bends in Rivers, &c. are to be
found in the *Proceedings of the Royal Society* for May 1876*; in
the British Association Report for the Glasgow Meeting, 1876,
Section A, page 31 of Transactions of the Sections; in the
Proceedings of the Royal Society, 1877, No. 182, page 356†;
and in the *Proceedings of the Institution of Mechanical Engineers*,
August 1879, p. 456‡. Also some other important cases in which
like principles of fluid motion come into play (in Whirlwinds &c.)
are adduced in a paper by myself in the British Association
Report for the Montreal Meeting, 1884, Section A, page 641§.

---

21. ON CERTAIN CURIOUS MOTIONS OBSERVABLE AT THE SURFACES
OF WINE AND OTHER ALCOHOLIC LIQUORS.

[From the *Philosophical Magazine* for November 1855. Read at the British
Association, Section A, Glasgow, 1855, pp. 16, 17.]

THE phenomena of capillary attraction in liquids are accounted
for, according to the generally received theory of Dr Young, by
the existence of forces equivalent to a tension of the surface of the
liquid, uniform in all directions, and independent of the form of
the surface. The tensile force is not the same in different liquids.
Thus it is found to be much less in alcohol than in water. This
fact affords an explanation of several very curious motions observable, under various circumstances, at the surfaces of alcoholic
liquors. One part of these phenomena is, that if, in the middle
of the surface of a glass of water, a small quantity of alcohol or
strong spirituous liquor be gently introduced, a rapid rushing of
the surface is found to occur outwards from the place where the
spirit is introduced. It is made more apparent if fine powder be
dusted on the surface of the water. Another part of the phenomena

* [*Supra*, No. 16.]   † [*Supra*, No. 17.]
‡ [*Supra*, No. 18.]   § [*Infra*, No. 26.]

is, that if the sides of the vessel be wet with water above the general level of the surface of the water, and if the spirit be introduced in sufficient quantity in the middle of the vessel, or if it be introduced near the side, the fluid is even seen to ascend the inside of the glass until it accumulates in some places to such an extent, that its weight preponderates and it falls down again.   The manner in which I explain these two parts of the phenomena is, that the more watery portions of the entire surface, having more tension than those which are more alcoholic, drag the latter briskly away, sometimes even so as to form a horizontal ring of liquid high up round the interior of the vessel, and thicker than that by which the interior of the vessel was wet.   Then the tendency is for the various parts of this ring or line to run together to those parts which happen to be most watery, and so there is no stable equilibrium, for the parts to which the various portions of the liquid aggregate themselves soon become too heavy to be sustained, and so they fall down.

The same mode of explanation, when carried a step further, shows the reason of the curious motions commonly observed in the film of wine adhering to the inside of a wine-glass, when the glass, having been partially filled with wine, has been shaken so as to wet the inside above the general level of the surface of the liquid ; for, to explain these motions, it is only necessary further to bring under consideration that the thin film adhering to the inside of the glass must very quickly become more watery than the rest, on account of the evaporation of the alcohol contained in it being more rapid than the evaporation of the water.

That this part of the explanation is correct, or that these motions of the film in the wine-glass are really due to evaporation, may be shown by a very decisive experiment.   If a vial be partly filled with wine and shaken, and then allowed to rest, no motion of the kind described will be found to occur in the thin film wetting the inside, provided that the vial be kept corked.   On the cork being removed, however, and the air contained in the vial, and saturated with the vapour of wine, being withdrawn by a tube, so as to be replaced by fresh air capable of producing evaporation, a liquid film is instantly to be seen creeping up the interior of the vial with thick or viscid-looking pendent streams descending from it like a fringe from a curtain.   These appearances are quite of the same kind as those met with in the open wine-glass.

Another experiment may be made to show, in a very striking way, the phenomenon of the more watery portion of the surface of a mixed liquid drawing itself away from the more alcoholic portion as follows:—If water be poured to the depth of about a tenth of an inch or less on a flat silver tray or marble slab, previously cleaned from any film which could hinder the water from thoroughly wetting its surface; and if then a little alcohol or wine be laid on the middle of that water, immediately the water will rush away from the middle, leaving a deep hollow there, and indeed leaving the tray bare of all liquid except an exceedingly thin film of the spirit, which continues always thinnest close to the margin of the water, because the water draws out to itself every portion of the spirit which approaches close to its margin.

The experiment alluded to near the commencement of the present paper, in which spirit was to be introduced into the middle of a surface of water previously dusted over with fine powder, may be well conducted as follows:—A tube for supplying the spirit should be provided*, which may be three or four inches long, half an inch or three-quarters in diameter, and terminating at bottom in a small open point, which, if found too wide, may be partially stopped by the insertion of a piece of thick soft thread, such as a strand from the wick of a spirit-lamp. A knot on the thread inside of the tube will serve as a valve to curtail or stop the flow of the spirit when required. The surface of the water should be clean and free from any kind of pellicle, such as is often met with, and is sometimes not easily avoided. It should then be lightly dusted over with some fine powder not apt to be quickly wet: Lycopodium powder will serve the purpose. Then the tube filled with spirit is to be dipped with its open point into the surface of the water, and instantly a nearly circular patch round the point of the tube will be seen occupied with liquid rushing outwards and completely divested of the covering of powder, while on the part outside of that patch there will be seen, by the motions of the powder, one, two, three, or many radial streams flowing outwards from the middle, and other return streams or eddies flowing backwards to the margin of the patch, on arriving at which each particle seems suddenly as if driven outwards with a rapid impulse. The margin of the central patch is usually to be seen formed like as of

---

* The tube of a small glass syringe as sold by apothecaries will serve the purpose well.

leaves of a plant growing out all round, and some superimposed on others, and all in rapid motion. The nature and causes of these forms of the margin, and of the eddies outside of the margin, I have not as yet been able satisfactorily to explain.

Another experiment may be made which is quite in accordance with the explanations already given, and which, being due to condensation of alcohol on a surface of water, is interesting when viewed in comparison with that in which the motions were shown to be produced by evaporation:—If a silver spoon, perfectly wetted with water so that a thin film adheres to it, be held over an open cup or vessel containing strong alcohol, the surface of the liquid will become greatly agitated with numerous motions, which are to be attributed to the unequal and varying condensation of the vapour of the alcohol at different parts of the surface of the film, according as the vapour is wafted about in fumes by the air.

While engaged in the investigation of the phenomena which I have now described, my attention has been turned to some other very interesting phenomena previously observed by Mr Cornelius Varley, and described by him in the fiftieth volume of the Transactions of the Society of Arts. He observed, with the aid of the microscope, numerous motions of extremely curious and wonderful characters in fluids undergoing evaporation. Although I have not yet had it in my power to examine into all the phenomena he has discovered relative to these motions, yet I think that many of them, or all, are to be explained according to the principles I have now proposed.

I have not had access to the Transactions of the Society of Arts to read Mr Varley's paper in full, but I quote the following abstract of his results from Queckett's *Treatise on the Microscope*, 1st ed. p. 413:—"The plan recommended is as follows: take an animalcule-cage of moderate size, and upon the tablet place a drop of turpentine or spirits of wine, &c., then slide over it the thin glass cover, but do not compress the fluid very much; the microscope being placed in the vertical position, and provided with a magnifying power from 40 to 100 diameters, the contents of the cage are to be examined in the same way as if animalcules were contained in it. As the evaporation of either of these fluids takes place, numerous currents and vortices will be seen, especially if a small quantity of finely-powdered coal be ground into them; the particles of coal being very light, are held in suspension whilst the evaporation is going on, and are whirled about by the currents in different directions." The following fluids Mr Varley has given as the best for the illustration of the currents :—

" 1. A drop of spirits of wine, or of naphtha, exhibits two, three, or four vortices or centres of circulation, according to the size of the drop; and if these vortices are viewed laterally, the lines of particles will be seen forming oblique curves from top to bottom of the drop.

" 2. Oil of turpentine shows a rapid circulation in two continuous spirals, one to the right, the other to the left, around the drop. These meet in the opposite diameter, from which the particles are carried slowly across the diameter to the place of starting, and this continues while there is fluid enough to let it be seen.

" 3. If, however, the drop does not exceed one-tenth of an inch in diameter, it presents the appearance of particles continually rising up in the middle, and radiating in gentle curves to the circumference.

" 4. If the liquid be put into a very small vial, similar motions are perceived, the particles when they have reached the side of the vial going down to rise up afterwards in the centre or axis.

" 5. If a bubble of air be enclosed in the liquid, motions similar to those described in No. 2 are observed in the part immediately in contact with the bubble.

" 6. In a flat drop of new wine laid on the tablet or disc of the aquatic live-box, but not compressed by the cover, the motion was a regular uniform circulation, the particles rising from below at one end of the drop, then passing straight across on the surface, and descending at the other end *."

* [See Clerk Maxwell, *Theory of Heat*, 4th edition, 1875, p. 293 : " This phenomenon, known as the tears of strong wine, was first explained on these principles by Professor James Thomson. It is probable that it is referred to in Prov. xxiii. 31, as an indication of the strength of the wine."]

22. ON SMOKY FOGS.

[From *Northern Whig* of Feb. 17, 1868. Abstract prepared by the author.]

A PRIVATE meeting of the Belfast Natural History and Philosophical Society took place at the Museum, on the evening of Wednesday, the 12th inst. The President (Mr Joseph John Murphy) occupied the chair.

Professor James Thomson read a paper on Smoky Fogs. The question which he proposed to discuss was—How are we to account for the well-known and frequently-noticed fact of the smokiness of the air in towns often being excessively aggravated in times of fog; or, at least, of the smokiness becoming often in such times more perceptible and more disagreeable than usual? He said it cannot be that the chimneys give out more smoke then than at other times, nor can it be merely that the air is then calmer than usual, for we often have in towns cold, calm weather, with tolerably clear or bright, fresh air. Also, it cannot be merely that the smoke, with the fog super-added, appears more dark and thick than usual, for when this foggy, smoky air comes into our well-warmed houses, the fog is dissolved or dissipated by the heat; but the smoke remains distinctly perceptible to an unusual degree even within the limits of our warmed apartments. Neither could he admit, what is frequently offered as an explanation, that the smokiness in question is to be attributed to the particles of smoke becoming, in foggy weather, loaded with moisture, so as to be prevented from rising freely up and away from the streets and houses. Last winter, he had noticed smoky fog accompanying thaw after severe frost; and this not infrequent occurrence had suggested to him an explanation, which he thought likely to prove valid, for many, or perhaps all, cases of the phenomenon in question. The rapid change often met with from frost to thaw, and often even occurring in the night time, and without wind at or near the ground, must be attributed to the arrival of a vast mass or stratum of air from warm regions of the earth into the space high above the ground, and above a bed of cold air previously on the ground and still remaining there. The mingling of the two masses of air at different temperatures tends, according to

well-known laws, to produce cloud or mist at their junction. At
this stage, with a warm, cloudy sky, the thaw on the ground may
begin and advance rapidly. The moving current of warm air
above may continue till it has carried away nearly the whole of
the cold bed of air next the earth, so as to leave at the earth
only a stratum of the cloudy mixture of the previously warm
with the previously cold air. If, now, the forward motion of the
warm air gradually abates, and a calm condition of the atmosphere
ensues, we have at the ground, and among the streets of a town,
a bed of air which is cold and consequently heavy, and which is
also foggy; while the atmosphere immediately above is warmer
and lighter. The coldest and heaviest air, being at the bottom,
tends to stagnate down about the houses; all upward and downward currents, such as might often at other times occur, being
then arrested. Thus, in the stagnating lower stratum, the smoke
emitted as usual from chimneys accumulates excessively, accompanied by the fog due to the cooling of the lower part of the
newly-arrived, warm moist air.

[The following MS. note by the author is dated Nov. 28, 1870:]

I find in Herschel's meteorology, published separately from
Encyclopædia Britannica, 1862, page 98, that Herschel attributes
the dense yellow suffocating fogs which infest London, especially
in the winter months, to the sooty particles acting as nuclei of
radiation and causing the fog.

I do not think the explanation offered by Herschel can be
tenable. Without seeing the phenomenon he alludes to I could
not attempt decidedly to explain it. He says:—"A very peculiar
phenomenon exhibited by the London atmosphere has been described to us by the Astronomer Royal and appears to be referable
to the same cause. On calm evenings after sunset, as seen from
the Royal Observatory on Greenwich Hill, the vast irregular mass
of smoke hovering over London appears to subside. Its heaped
and turbulent outline becomes flat and sinks rapidly into a low
level cloud bank, with a very definite outline, and fair sky above.
It would seem that each particle of soot, acting as an insulated
radiant, collects dew on itself, and sinks down rapidly as a heavy
body."

Now I presume the turbulent outline is really the result of
many ascending and descending currents due on the calm day to
the heat of the sun warming the town and warming the smoky

air; but that after sunset this cause of upward and downward currents soon ceases to act:—and I suppose it very likely that radiation from the air and soot etc. may cool the low stratum after sunset and cause it to lay itself down flat:—but this flattening I think must be due much more to a subsidence of a heavy cool stratum of air with included smoke than to the falling of heavy dew-bemoistened particles of soot out of the air where they were to a lower mass of air so as to clarify by precipitation of soot the sooty air. I think also it may be well to keep it an open question whether the "very peculiar phenomenon" referred to by Herschel may not sometimes or in some degree be due to the supervention of a flow of warmer air above the smoky atmosphere: though I think this scarcely likely to account for the phenomenon described. At present it seems to me most likely that the sooty particles in the air would radiate heat out to the cold sky, and that they, by becoming cooled, would cool the air surrounding them, and that the *air thus becoming heavier* would sink, and that its tendency to sink would be increased by the weight of the soot and of the dew attached to it if such there be; but I think the *main levelling action* must be that of a cooler and denser fluid spreading itself out flat below a warmer and lighter one (very like the self levelling of syrup at the bottom of a vessel of water) and I distinctly think the precipitation of sooty particles from air containing them *cannot possibly* account for the phenomenon described. I think they could not go down so as to leave clear and transparent the air which contained them. The turbulent roughnesses would be, I am sure, 50, 100, 500 or more feet high; and it is inconceivable that any such masses of air could clarify themselves, by precipitating their soot down through themselves and out of themselves. The different sized pieces would fall at different rates and the finest particles of smoke, I suppose, would not descend in the whole course of an evening so as to clarify a stratum 10 ft. deep. The smoke of a smoky fire in a dwelling room will last a long time without settling by precipitation. It is usually got rid of out of the air mainly by ventilation though partly by precipitation. Now over London it is quite conceivable that sometimes after sunset the smoky air may become cooler than before, by radiation; and that through various causes (the passage of a barometric wave for instance perhaps, etc. etc.) a stratum of warmer and lighter and purer air above may be slowly depressed, while the smoky turbulent mass flattens itself down while gently gliding out sidewise from London. Some such action as that, I think, would account for the sinking of the turbulent smoky canopy to a smoothed and lowered sheet and for vision being free at a level which was smoky before. Probably too there would be a gentle wind, not a perfect calm in the air high up, and thus new pure air would replace some lofty air slightly vitiated with smoke.

23. ON SMOKY FOGS. (Second Paper.)

[From *Northern Whig* of Dec. 6, 1870.]

THE third joint meeting of the [Belfast] Natural History and Philosophical Society and the Naturalists' Field Club, for the session 1870–71, was held in the Museum, on Wednesday evening, the 30th November. The chair was occupied by the President of the Natural History and Philosophical Society, Robert Patterson, F.R.S. Dr James Thomson offered some remarks on Smoky Fog.

Many persons in Belfast will remember having noticed the air in their houses on Saturday evening last, the 26th of November, as being in an excessively smoky condition. People in many cases at first thought the smoke pervading their apartments and staircases was coming from some of the fires within their own house. But this idea was dispelled when they found their chimneys drawing just as well as usual; and the truth became apparent that Belfast was enveloped in a densely smoky fog. It cannot be supposed that the fog itself could continue to exist in the air after entrance into the well-warmed apartments in houses, as it would instantly dissolve away under the influence of the warmth; but certainly the smoke remained most disagreeably evident, apparent to the eyes, and perceived almost suffocatingly in the throat. At that time on a Saturday evening the mill chimneys must have ceased to emit their usual smoke, and the dwelling-houses would be only giving off about their usual quantity. What, then, could be the reason why the air of the town should at that particular time be ten-fold, fifty-fold, or, perhaps, a hundred-fold as smoky as usual? Mere calmness of the atmosphere would not suffice to account for such excessive smokiness, because often we have in towns calm weather with very tolerably clear air, fresh and pleasant. He thought the explanation of the phenomena in question was afforded by a theory which he had submitted to the Natural History and Philosophical Society in February 1868. The view he then gave, and which he thinks meets with confirmation from the atmospheric phenomena of Saturday and Sunday last, was to the effect that the usual or general cause of excessively smoky fogs is the arrival in the sky over a town of a vast influx of warm air, much warmer than the air lying calmly on the ground and immediately over the houses. The mingling of the two masses of air at different temperatures must tend, according to well-known laws, to produce cloud or mist at their junction, especially if both, although clear, be nearly saturated with moisture in the gaseous state. If the substratum of cold air be of considerable depth, the first indications of the arrival of the warm air may be a clouding of the sky, and a rise of temperature at the ground due to radiation of heat downwards from the newly arrived air, the warmth at the ground often in such cases showing itself by an incipient thaw if there has been frost before. Then as time advances, the cold substratum may be gradually thinned away by the scouring off of its upper part by the current of warm air flowing above it, till the foggy junction where the intermixture is proceeding extends down to the houses and the ground. In this way there may be brought about a condition in which the lowest air will be the coldest, and in which there will be a gradual increase of temperature in ascending for a considerable height from the earth upwards. If now the newly arrived warm air maintains a brisk flow, there certainly will be no smoky fog, because the smoke will be carried away as quickly as it is evolved from the chimneys. But if, with the distribution of temperature already supposed to have been attained, a calm supervenes, we have a set of conditions present together which can scarcely fail to produce a smoky fog, because the coldest and heaviest air, being at bottom, must tend to stagnate there, all upward and downward currents, such as often at other times occur, being then arrested. Thus in the stagnating lower stratum the smoke emitted as usual from the chimneys must accumulate excessively; and if the newly arrived warm air be moist enough to produce fog by its mixing with the cooler air below, as must often be the case, the accumulating smoke will be

accompanied by fog, and so the smoky fog will be produced. If
the newly arrived air over the substratum of colder air be warm
and not very moist, an excessive accumulation of smoke may occur
in the atmosphere of a town in calm weather without there being
necessarily any fog, the main conditions tending to an extraordinary accumulation of smoke being a calm atmosphere, with a
cool stratum at bottom, and warmer air above, and with the cool
stratum extending high enough to cover the chimneys of a town,
and receive their smoke. On the other hand, if the upper air be
cool, having recently come from some cold region of the earth,
and the lower air be warm, as it often may be when the sun is
shining on the ground and on houses and streets, the air of the
town may be kept clear and fresh without any perceptible wind
by a constant gentle interchange of air upwards and downwards,
the warm air of the town floating upwards and being replaced by
the cool air from above. Last Saturday was a sunny, frosty day
in Belfast and the neighbouring country. The frost was not
severe, as some parts of the roads were wet, while others were
hard frozen. In gardens in the town sheltered from the sunshine
there was hoar frost on the leaves, and water exposed had a crust
of ice. During the day time the air was more or less foggy in
various parts of the town, and some appearances of unusual
smokiness were observed also. The frost continued till after
sunset. In the evening, at about 6, 7, 8, and 9 o'clock, the town
was enveloped in an intensely smoky fog; and the smokiness of
the air was very unpleasantly perceptible even in well-warmed
rooms in which the fog itself would be completely dissipated.
This condition of things, then, when considered under the theory
already explained, appeared to prove that a vast mass of warm
moist air had arrived aloft, while a cold stratum of air was still
left stagnating below and enveloping the town. The surmise
that such must be the case was made on the Saturday evening,
and was abundantly confirmed on the following morning; for,
on the Sunday morning, the frost was all gone, the smoky fog
was gone, and the air was remarkably mild, and moist, though
clear, and there was sunshine during much of the day, and the
stars were visible in the evening. The occurrence of the thaw
and warmth in the night time, when no sun was shining,
affords strong proof of the influx of warm air having really taken
place. In fact, all the previous cold air in which the smoke had

accumulated was carried away during the night time, and the town was enveloped with newly arrived air from some warmer region of the earth. The warmth and moisture of the air were clearly shown by copious deposition of condensed vapour which was to be seen trickling down the surfaces of painted stone columns and other large non-absorbent objects which had been exposed to the cold of previous days.

24. ON A CHANGING TESSELATED STRUCTURE IN CERTAIN LIQUIDS [DUE TO CONVECTIVE CIRCULATION*].

[Abstract of Paper read to the Philosophical Society of Glasgow,
Proceedings, 15th February, 1882.]

IN June, 1870, the author, when on an excursion of the Belfast Naturalists' Field Club, happened to notice in a tub of opaque-looking water, in the yard of a roadside inn, a structure showing itself, which appeared remarkable, and excited a good deal of curiosity among members of the Club then present, as to what might be its nature and mode of origin. The water, viewed from above, presented a tesselated appearance, considerably resembling such patterns as are exhibited by vertical basaltic columns when exposed to view in a horizontal section—such, for instance, as may be seen at the Giants' Causeway or at Staffa. On the water being disturbed the tesselated structure was destroyed, and eddying streams of pearly lustre showed themselves in its stead; but the same kind of tesselated structure as before, speedily reappeared, and, on being watched for some time, it was found to be perpetually, though slowly, changing. Inquiries made as to how the water had been previously used did not elicit any definite information. A sample of the water was then brought away by the author for further examination. It presented, when shaken in a clear glass bottle, a remarkable satin-like, streaky appearance, with the pearly lustre before mentioned.

Like phenomena were afterwards noticed occasionally in soapy water, and so it became probable that the water originally noticed

* [This subject has been investigated independently and with extensive developments by Prof. Bénard of Bordeaux in a Thesis published in the *Annales de Chemie*, 1901 (cf. also a note by him in the forthcoming part, Dec. 1911, relating to the present paper): see also Bénard, *Revue Générale des Sciences*, 1900; and also in connexion with Solar phenomena, Deslandres, *Annales de Meudon*, Vol. IV. 1910.]

might have been soapy, and that soapyness had probably been associated with the production of the phenomena.

Various experiments and observations on soapy water have since been made, and results which seem worthy of notice have· been arrived at. The more obvious features of the phenomena must of course have been seen thousands of times before by washerwomen and others; but perhaps the conditions may have been left unheeded, and without receiving any scientific attention or scrutiny.

The occurrence of the phenomena seems to be associated essentially with cooling of the liquid at its surface where exposed to the air, when the main body of the liquid is at a temperature somewhat above that assumed by a thin superficial film. A very slight excess of temperature in the body of the fluid above that of the surrounding air is sufficient to institute the tesselated changing structure, provided that the soapy water is in other respects in good condition for taking the actions to which the phenomena are due.

An easy method of procedure for making suitable experiments and observations, and which was shown to the Philosophical Society in illustration of the communication, will now be explained. A glass pan was filled to a depth of four or five inches with soapy water of proper consistency, arrived at by previous trials. Then, to warm it slightly, one or two small tin cans of hot water were plunged into the soapy water, so that heat would be given to the soapy water by conduction through the sheet metal. The warm water cans were then removed, and the soapy water was stirred to bring the whole to about a uniform temperature. Then the pan and its contents were left undisturbed, and the subsequent behaviour of the liquid was watched. Swirling flows, conspicuous by their pearly streaks, showed themselves for some time, but with gradually abating speed, and in the course of five or ten minutes the eddying motions, accompanied by spirally-formed moving streaks, were gradually dying out, and the moving patterns on the surface were, by slow degrees, assuming a tesselated character. After the cessation of all such swirling motions, and disappearance of spirally-moving streams, the tesselated configuration might be found to continue, perpetually changing in all its individual details from minute to minute, but retaining the same general character for many hours together.

Fig. 1 represents the general appearance that the surface of the soapy water in the pan would usually exhibit when left standing in the manner described. By continuous watching, it may be noticed that the smaller enclosed patches are generally diminishing in size, being encroached on by the larger ones until

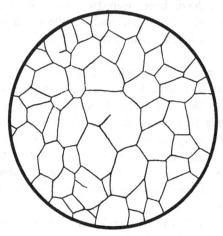

Fig. 1.

they collapse and cease to exist. At the same time the large ones show tendencies to sever themselves into two or more new ones, which in their turn either increase and split again, or diminish and go out of existence. To describe the numerous and varied

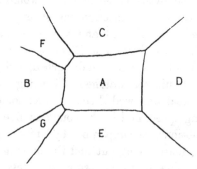

Fig. 2.

features of transition in words would not be easy, but figures 2, 3, and 4 show three successive conditions which have been observed as occurring.

In fig. 2 the patch *A* has larger neighbours—*B, C, D,*

and *E*—contiguous with it, and it has also two narrow patches
—*F* and *G*—contiguous with small portions of its boundary.
The patch *A* is comparatively narrow in the direction between *C*
and *E*, and is rapidly encroached on by those two large patches *C*
and *E*, and becomes narrower than before, without necessarily being
shortened in length. Also, during the same time, the narrow
patches *F* and *G* collapse, each at its place of meeting with the
boundary of *A* ; and in each case where the collapse takes place a
bounding line is left between the two, which come together, as is
shown by *mn* and *pq* in fig. 3, the patches *F* and *G* of fig. 2
having retired to their new positions, shown as *F* and *G* in Fig. 3.
A little later, and the patch *A* is observed to have disappeared
entirely, by collapsing into the line *st* in fig. 4.

Multitudinous varieties of changes, partaking more or less of
the general character of those here described, may be noticed in

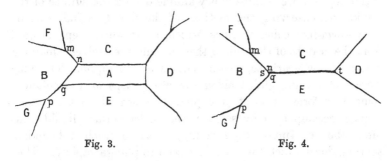

Fig. 3. Fig. 4.

various watchings of the behaviour of the soapy liquid from time
to time. The patches, with their boundaries, when viewed with
favourably applied light, show appearances as if each patch were
formed as the top of a column of slowly rising fluid, having a
mother-of-pearl-like lustre, seemingly at a little distance below the
surface of the liquid, and with a thin layer of more translucent
liquid lying on the tops of these pearly patches. It seems to be
that there is on the whole a very slow ascending motion in these
columnar spaces, and that above, there is a thin superficial layer of
cooler and seemingly more translucent liquid, receiving perpetually
new supplies from the rising substance of those columns, and
flowing outwards over each column top, and that the two sheets,
spreading out on each of two contiguous columns, plunge downwards, where they meet in the mutual bounding line of the two
spaces or column tops, and so form a descending current or septum

between the two columnar spaces. Where three of the bounding lines meet at the junction of three column tops, the downward current seems to be more active than at the other parts of the septums. The various ascending flows here spoken of as columns, or columnar spaces, for want of any better nomenclature, may probably not exist like separate columns with septums between them of descending liquid except near the surface. It seems likely that the down-flowing liquid of the so-called septums may tend to gather into thicker streams descending from the corners of the surface patches where three of such spaces meet; but the internal motions, being concealed from view, remain as yet obscure, and in a great degree unknown.

Scrutiny of such motions as have here been briefly mentioned and speculated upon has been carefully made in many ways, but the one which gave the most remarkable indication was by sprinkling minute crumbs of dry aniline dye on the surface of the liquid, and observing the motions of the liquid, as indicated by long worm-like coloured lines, which, as it were, crept forward from the crumbs of the aniline, these crumbs themselves remaining stationary, or nearly so, on the surface of the liquid. The blue worms of aniline glided along the pearly tops of the columns, sometimes forming a pit or depression, down which they would plunge, accompanied by some of the cooler surface liquid; but often they continued their course, so as to reach a bounding septum, down which they could be seen to plunge rapidly. They gave very clear indication of there being a down-flow of the cooler surface liquid in every septum.

As to the physical and optical conditions to which the pearly appearance may be due, the author offered some suggestions, but stated that much must remain still to be found out by microscopical or other researches, which he inclined to recommend to any whose opportunities and inclinations might lead them to take interest in the prosecution of such researches. He thought it probable, judging from numerous appearances, that the soapy liquid might contain in suspension multitudinous flattened or elongated particles, like scales or crystals, or comminuted glass hairs, which, in the motions and distortions of the liquid, would tend to place their flat faces or their elongated sides in approximately-parallel directions, and so would give special optical effects by their reflexion of light. He offered any such suggestions, however, with great diffidence, wishing to leave the subject entirely open for better researches than any he had had opportunity to attempt.

A like tesselated formation, but without the pearly lustre, is often to be noticed in a plate or bowl of beef-tea or other clear soup containing minute flocculent particles in suspension and left cooling. The flocculent particles are capable of showing the tesselated structure, and of indicating clearly the existence of motions such as those described as operative in the formation of the tesselated structure in soapy water, but such particles are not capable of reflecting the light so as to produce the pearly lustre.

The motions now described and explained as occurring in the soapy water and other liquids when showing the tesselated structure he thus believes constitute one special case of what is often called convective circulation. He points out, however, that this case, with a thin film cooled at the surface while the great body of the liquid is at a somewhat higher temperature, presents essential distinctions from the case in which convective circulation is caused by heat applied at the bottom of the vessel. The surface phenomena and the actual motions throughout the body of the liquid are very different in the two cases, as may easily be seen by a little consideration, both theoretical and observational.

Professor Thomson followed up the communication described in the above abstract, by giving to the Society an account of views which, a good many years ago, had been developed, partly by his brother Sir William Thomson, and partly by himself, on a somewhat kindred subject—Calm Lines on a Rippled Sea. Those new views he had submitted to the Belfast Natural History and Philosophical Society in a paper read on 7th May, 1862, of which an abstract is to be found in the *Philosophical Magazine* for September 1862 *. (Fourth Series, Vol. XXIV. page 247.)

The phenomena in the two cases—those of the Tesselated Changing Structure, and of the Calm Lines on a Rippled Sea, present certain features of close resemblance, and other features of wide distinction; and to facilitate the consideration of the two subjects together, an abstract of the paper on the Calm Lines is here subjoined.

* [*Infra*, p. 142.]

25. On the Cause of the Calm Lines often seen on a Rippled Sea.

[From the *Philosophical Magazine*, Fourth Series, Vol. XXIV. (1862), pp. 247, 248. Read before the Belfast Natural History and Philosophical Society, May 7, 1862, and also printed in its *Proceedings*.]

In this paper the object of the author was to offer a new explanation of the origin of lines of glassy-calm water, usually long and sinuous, which are often to be seen extending over the surface of a sea darkened elsewhere by a ripple. He adverted to the commonly received supposition that these lines are due to some kind of oily film on the surface of the water, and to the prevalent idea that the oil is somehow given off from shoals of fishes. These suppositions, he thought, although having some slight foundation in the facts of the case, did not form the true explanation of the phenomenon. His brother, Professor William Thomson, had observed, and had pointed out to him, that the water at the calm lines always contains considerable quantities of small floating objects, such as little detached pieces of seaweed, leaves of trees, or the like, and had accounted for the smoothness of the water by the friction induced among the little undulations by the presence of those solid objects. The question still remained, however, as to what might be the cause of the leaves and seaweed being arranged in such long and sinuous lines. One of these calm lines was noticed last autumn in Brodick Bay by Professor William Thomson and the author; and on rowing into it, they found leaves and seaweed abundantly diffused through the water there, while the rippled water on both sides of it was comparatively free from such objects. The line of calm water evidently sprang from a point on the shore where a small river entered the sea, and there could be no doubt that the leaves were supplied by that river. Still it appeared an untenable supposition that the river water could extend so very far out to sea, and wind about over the surface of the sea as the calm line did, sometimes even narrowing instead of spreading out as a broad sheet, which the light fresh water flowing over the heavy salt water should be expected to do. It occurred to the author that the water of the river would actually spread out as a broad sheet over the surface of the sea; and that in its outward lateral flow it would constantly be carrying with it the leaves with which it was originally charged, and all such small pieces of seaweed as it might meet with in the sea-water, and that these would accumulate in a boundary line of the region of dispersion, which might be determined by some slight flow of the surface of the adjacent sea-water meeting this outspreading fresher water, and causing a downward or sinking motion, however slight, of the two meeting currents. The author does not mean to attribute the calm lines in general to the spreading of a sheet of fresh water over the salt water of the sea, but he thinks the general explanation is readily developed from the observations and explanation in the foregoing particular case. He supposes in general that in estuaries and channels of the sea, and in lakes and rivers, the water must often be affected by various causes, such as tides, breezes, currents, and circulation due to differences of heat, so as to be made to rise occasionally at some places and to sink at others. Now along the line of meeting and sinking of two opposing surface currents, all floating objects carried by those currents will be collected together; and there they will act as dampers, or floating breakwaters, for the small ripple undulations. The slightest possible inequalities of the forces of the two opposing currents will suffice to account for the great sinuosities which the calm lines show. If there should happen to be any oily scum thinly diffused on the surface of the water, this will be brought together, along with seaweeds, etc., from a wide area, to the line of meeting and descent of the two opposing surface currents; and to an oily film having been frequently noticed on the calm places, though really brought to them in the way now suggested, the author inclines to attribute the supposition that it is their cause; but he agrees that very possibly, too, an oily scum thus sometimes brought together may cooperate with the seaweed and other floating objects in resisting the propagation of the ripple.

In the discussion which followed the reading of this paper, it was remarked by members of the Society that, in drawing nets through the sea near its surface for collecting small marine animals, they had often found that these creatures were present in vastly greater abundance in the calm lines than in the rippled sea; and it was suggested that, as the bodies of many of these contain oil, this, on their dying, might be given off from their bodies, and might float on the surface, and might cause the smoothness there observed; and it was also stated that shoals of herrings are much more abundantly met with under the calm lines than at other parts of the sea.

The author of the paper, in replying, pointed out that, if marine animals giving off oil from their bodies were really the cause of the calm lines, it would still remain as an unsolved question, and one which would then be of great interest, Why should those animals be found to congregate in long sinuous lines, extending often continuously for miles over the surface of the sea? He thought, however, that the true state of the case would probably be, that the small animals are brought together by the same currents which, according to his supposition, collect the seaweed, leaves, and other floating objects into lines, and that the animals, not wishing to descend with the meeting currents into deep water, remain near the surface, and that the herrings or other fishes congregate to the same lines in order to feed on the smaller animals.

26. On the Grand Currents of Atmospheric Circulation.

[From MS. notes of a paper read to the British Association, Dublin, 1857.]

THE prime mover of atmospheric currents is the heat of the sun, and the general or grand currents constituting the circulation of the atmosphere originate in the difference of temperature of the equatorial and polar regions. The currents resulting from this source of motion are, however, greatly influenced by the motion of the earth on its axis; or, in other words, they are very different from what they would be if the earth were at rest and if the sun were merely a source of heat revolving round the earth according to its present apparent motion.

To the combined agency of the equatorial heat and the rotation of the earth the origin of the Trade Winds has been long ascribed, and in this respect the ordinarily received theories of the winds appear to me perfectly satisfactory. The same agencies have also been taken into consideration with more or less correctness to explain the prevailing motion from the west in temperate and polar regions. I am not aware, however, that

any satisfactory theory has as yet been promulgated showing the
nature of the atmospheric circulation except in, or about, the
Torrid Zone; and to supply what appears to be wanting in this
respect is the object of the present paper.

In proceeding to explain my theory of atmospheric circulation,
I assume first as a well ascertained fact, that round the earth,
at or near the equator, there is a zone of air ascending in virtue
of its lightness produced by expansion by heat. This air, on
rising to the upper regions of the atmosphere must divide into
two great currents floating away from the equatorial zone towards
the poles. For the sake of simplicity of expression, I shall now
confine attention to the great currents of the northern hemisphere.
It is then to be observed that the air floating away in the upper
regions of the atmosphere from the equatorial ascending zone is
possessed of the rapid motion of rotation from west to east which
pertains to the equatorial regions of the earth in virtue of their
great distance from the earth's axis. The velocity of this motion
of rotation amounting as it does to a movement once round the
equatorial circumference of the earth in 24 hours may be stated
in round numbers at 1000 miles per hour. Endued with this
velocity at the outset, the great current floating away towards
the polar regions, comes successively into latitudes each having a
less velocity of rotation of the earth's surface than the preceding
one. The air then, tending to retain its original equatorial
motion from west to east, which, at the equatorial zone, was the
absence of motion in reference to the earth below it, comes by
degrees to have a rapid motion from west to east relatively to the
earth below it, in virtue of the velocities of the earth's surface
becoming less and less as the latitude increases. The air, of
course, is not to be understood as retaining its equatorial motion
of 1000 miles per hour undiminished in its course northward, for
it meets with retardation communicated upwards to it from the
slower moving regions of the earth's surface, through the medium
of the air below. For the present purpose it is necessary to refer
to this retardation merely with a view to show that there will
not be maintained such a powerful vortex or whirling movement
as would, by centrifugal force, hinder the motion towards the
higher latitudes. This motion towards the pole, accompanied by
a motion from west to east, quicker than that of the earth in
each latitude, in my theory I suppose to go on throughout the

entire of the upper stratum of the air extending from the equatorial rising zone to the cold regions of the north; and the grand current from equatorial to polar regions I suppose to occur in the upper regions alone.

The cold of the polar regions increasing the density of this air gives it a tendency to descend, and it does descend, and flows as a grand under current towards the equator to take the place of other air ascending in the equatorial rising zone, and in its turn to rise in the same regions and repeat the circulation which has now been briefly described. Throughout the regions extending from about the 30th parallel of latitude northward, winds from the west are observed to prevail, and those are to be ascribed to the equatorial motion still remaining in part in the air when it descends to the earth, and only to be removed by friction and impacts against the earth's surface.

It is obvious that, as great a torsional force or couple must be applied in these regions of west winds in retarding the equatorial motion, as was applied in the equatorial regions in generating that motion, and hence may be derived some conception of the necessarily great extension and prevalence of these west winds.

At latitude 30° the atmosphere on the surface of the earth is found to have no average motion from west to east nor from east to west: and it follows that, in passing from that latitude to the equatorial rising zone, the air must fall behind the earth in respect to rotatory motion; or, in other words, the air will revolve from west to east slower than the earth does, or will have a motion from east to west relatively to the earth. This constitutes the north east trade wind.

There is, however, a very important motion of the air, distinct from the primary circulation which I have now described. It is an observed prevailing motion of the winds on the earth's surface, from south to north in the region occupied by the west winds. This motion has generally, I believe, been considered as rather paradoxical, and no explanation that appears to me to be tenable has hitherto been proposed; but, on the contrary, in Maury's *Physical Geography of the Sea*, a recent and highly valuable work, a theory of atmospheric currents is put forth, comprising this motion as one of those that must be explained, and in that view assuming a grand action in the upper regions of the atmosphere, which I look on as the contrary of what must really occur, and as seeming so paradoxical that I can scarcely conceive of its being satisfactory to any mind.

(Here show Maury's diagram*: give explanation as in Maury, pp. 72, 73.)

Instead of this explanation I suppose the primary circulation I have already described to occur in reality. Thus the atmosphere over the temperate and frigid zone must be considered as in the condition of a great vortex or whirl, the air revolving quicker than the earth. Thus a partial vacuum or a diminished pressure in the polar region must arise from the centrifugal force of the whirling air. The film or stratum of this vortex however, which is in contact with and in immediate proximity to the earth, must have its rotatory motion retarded by friction and impacts on the earth's surface.

Thus its centrifugal force comes to be less than that of the great mass of atmosphere above it. It is therefore not aided by centrifugal force so much as the parts above are in keeping out from the polar region of diminished pressure, and consequently it is actually drawn in by the vacuum towards the centre of the vortex or in other words it receives a motion towards the pole of the earth. Thus then does the friction of the earth's surface serve to explain clearly and simply the motion towards the pole which otherwise would appear paradoxical and which has led to the supposition of such complicated and inexplicable currents as have been supposed by Maury to occur.

[The foregoing is printed from MS. in the Author's hand. The abstract of the paper in the Report of the *British Association* proceeds as follows :—]

This is the substance of Mr Thomson's theory; and he gives, as an illustration, the following simple experiment :—If a shallow circular vessel with flat bottom, be filled to a moderate depth with water, and if a few small objects, very little heavier than water, and suitable for indicating to the eye the motions of the water in the bottom†, be put in, and if the water be set to revolve by being stirred round, then, on the process of stirring being terminated, and the water being left to itself, the small particles in the bottom will be seen to collect in the centre. They

---

* [For Diagrams see No. 28 *infra.*]
† A few tea-leaves taken from a teapot will suit the purpose well.

are evidently carried there by a current determined towards the centre along the bottom in consequence of the centrifugal force of the lowest stratum of the water being diminished in reference to the strata above through a diminution of velocity of rotation in the lowest stratum by friction on the bottom. The particles being heavier than the water, must, in respect of their density, have more centrifugal force than the water immediately in contact with them; and must therefore in this respect have a tendency to fly outwards from the centre, but the flow of water towards the centre overcomes this tendency and carries them inwards; and thus is the flow of water towards the centre in the stratum in contact with the bottom palpably manifested.

27. WHIRLWINDS AND WATERSPOUTS.

[From *Nature*, Oct. 30, 1884, Vol. xxx. p. 648. Read at the British Association, Montreal, 1884. Printed in *Transactions*, p. 641.]

WHIRLWINDS, whether on sea or on land, have their characters in great part alike. For simplicity it will be convenient to begin by taking up only the case of whirlwinds on sea, as thus the necessity for alternative expressions to suit both cases, that of sea and that of land, will be avoided.

It may be accepted as a fact sufficiently established, both by dynamic theory and by barometric observations, that at the sea-level the pressure of the air is less in the neighbourhood of the axis of whirl than it is at places farther out from the axis, though within the region of the whirl. The apocentric force (centrifugal force) of the rapidly-revolving air resists the inward pulsive tendency of the greater outer than inner pressure. But close over the surface of the sea there exists necessarily a lamina of air greatly deadened as to the whirling motion by fluid friction, or resistance, against the surface of the sea; and all the more so because of that surface being ruffled into waves and often broken up into spray. This frictionally-deadened lamina exerts, because of its diminished whirl speed, less apocentric force than the quicker-revolving air above it, and so is incapable of resisting the inward pulsive tendency of the greater outer than inner pressure already mentioned. Hence, while rushing round in its whirl, the air of that lamina must also be flowing in centre-ward.

The influx of air so arriving at the central region cannot remain there continually accumulating; it is not annihilated, and it certainly does not escape downwards through the sea. There is no outlet for it except upwards, and as a rising central core it departs from that place. This is one way of thinking out some of the conditions of the complex set of actions under contemplation; but there is much more yet to be considered.

Hitherto, in the present paper, nothing has been said as to the cause or mode of origin of the diminished barometric pressure which, during the existence of the whirlwind, does actually exist in the central region. Often in writings on this subject the notion has been set forth that the diminished pressure is caused by the rapid gyratory motion of the whirling air; but, were we to accept that view, we would have still to ask, How does the remarkably rapid whirling motion receive its own origin? The reply must be that the view so offered is erroneous; and that, in general, a diminished pressure existing at some particular region is the cause rather than the effect of the rapid whirling motion; though in some respects indeed these two conditions can be regarded as being mutually causes and effects, each being essential to the maintenance of the other, while there are also some further promoting causes or conditions not as yet here mentioned.

It seems indubitably to be the truth that ordinarily for the genesis of a whirlwind the two chief promoting conditions are: firstly, a region of diminished barometric pressure, this diminution of pressure being, it may be presumed, due to rarefaction of the atmosphere over that region by heat, and sometimes, further, by its condition as to included watery vapour; and, secondly, a previously existing revolutional motion, or differential horizontal motion, of the surrounding air, such revolutional or differential motion being not necessarily of high velocity at any part.

The supposed accumulation of air rarefied by heat or otherwise, for producing the abatement of pressure may, the author supposes, in some cases extend upwards throughout the whole depth of the atmosphere; and in some cases may be in the form of a lower warm lamina which somehow may have been overflowed or covered by colder air above, through which, or into which, it will tend to ascend: or the lower lamina may in some cases be warmed in any of several ways, and so may get a tendency to rise up through the colder superincumbent atmosphere.

On this part of the subject the author believes there is much scope for further researches and advancements both observational and considerational;—that is to say, by encouragement of a spirit towards accurate observation; and by collection and scrutiny of observed facts and appearances; and by careful theoretical consideration founded on observational results or suppositions.

To the author it seems probable that the great cyclones may have their region of rarefied air extending up quite to the top of the atmosphere; while often whirlwinds of smaller kinds, many of the little dust whirlwinds, for instance, which are frequently to be seen, may terminate, or gradually die out, at top in a layer or bed of the atmosphere different in its conditions, both as to temperature and as to original motion, from the lower layer in which the whirlwind has been generated. In many such cases the upper air may probably be cooler than the lower air in which the whirlwind originates.

On the subject of the actions going on at the upper parts or upper ends of whirlwind cores, in most cases, the author feels that he is able to offer at present little more than suggestions and speculative conjectures. In very many descriptions of the appearances presented by those whirlwinds with visible revolving cores, which are called waterspouts, it is told that the first appearance of the so-called waterspout consists in the rapid shooting down from a dense cloud of a black cloudy streak, seemingly tortuously revolving and swaying more or less sidewise. This is said rapidly to prolong itself downwards till it meets the surface of the sea; and the water of the sea is often imagined and described as rising up bodily, or as being drawn up, into the partial vacuum or central columnar place of diminished pressure. The frequently entertained notion—a notion which has even made its way into writings by men of science and of authority in meteorology—that the water of the sea is sucked up as a continuous liquid column in the centre of waterspout whirlwinds, is by some writers and thinkers repudiated as being only a popular fallacy, and it is affirmed that it is only the spray from the broken waves that is carried up. In this denial of the supposition of the water being sucked up as a continuous liquid column, the author entirely agrees, and he agrees in the opinion that spray or spindrift from the sea, set into violent commotion by the whirlwind, is carried up in a central ascending columnar core of air.

On the other hand, the commonly-alleged inception of the visible waterspout phenomena, in a descending, tortuously-revolving, and laterally-bending or swaying cloudy spindle protruding from a cloud, the author supposes to be so well accredited by numerous testimonies that it must be seriously taken into account in the development of any true theory and explanation of the physical conditions and actions involved. He ventures to hazard a suggestion at present—perhaps a very crude and rash one. It is that the rising central core may perhaps, in virtue of its whirling motion and centrifugal tendency, afford admission for the cloudy stratum to penetrate down as an inner core within that revolving ascending core now itself become tubular. The cloudy stratum may be supposed not originally to have been endowed with the revolutional motion or differential horizontal motion with which the lower stratum of thermally expanded air has been assumed to be originally endowed. The upper stratum of air from which the cloudy spindle core is here taken to protrude down into the tubular funnel is not to be supposed to be cold enough to tend to sink by mere gravity. Though it were warm enough to allow of its floating freely on the thermally expanded air below, it could still be sucked down into the centre of the revolving ascending core of the whirlwind.

Not to proceed further on this occasion with attempts towards explanation of the difficult subject of the actions at the upper ends of waterspout whirlwinds, the author wishes to have it understood that his main object in proceeding to prepare the present paper was to put forward clearly the theory he has given as to influx at the bottom in consequence of abatement of whirl in the lamina close to the sea-surface by frictional resistance there.

Addendum.—A few brief explanations and references will now be added to assist in the understanding of some of the principles assumed in what has been already said. It is to be clearly understood that, in a whirling fluid, even if the velocity of the whirling motion be very small at great distances from the axis, if the fluid be impelled inwards by forces directed towards the axis, the absolute velocity will greatly increase with diminution of distance from the axis. Thus in the *whirlpool of free mobility*, in which the particles are perfectly free to move outward or inward, the velocities of the particles are inversely proportional to the distances from the axis, the fluid being understood to be inviscid or frictionless. On this subject reference may be made to a paper by the author on "Whirling Fluids," published in the British Association volume for the Belfast Meeting, 1852*. Again, as to the inward flow caused in a frictionally retarded bottom lamina of a whirlwind or whirlpool with vertical axis, by the frictional retardation from the bottom on which the whirling fluid rests, reference may be made to a paper by the author, "On the Grand Currents of Atmospheric Circulation" in the British Association Report, Dublin Meeting, 1857, Part II. p. 38†. On another case of the manifestation of the same principle, reference may be made to a paper by the author in the *Proceedings* of the Royal Society for May 1876‡, in respect to the "Flow of Water round Bends in Rivers, &c.," with reference to the effects of frictional resistance from the channel in the bends; and to another paper by him, on the same subject, in the *Proceedings* of the Institution of Mechanical Engineers (August 1879, p. 456§), where the inward flow is explained as experimentally exhibited.

Postscript of date August 16.—Prof. James Thomson wishes now to offer in continuation of his paper on "Whirlwinds and Waterspouts," despatched two days ago for Montreal the following postscript, which will extend the considerations there already put forward, and will tend to modify or amend some of them; but will leave unchanged the theory as to influx of the bottom lamina of the whirlwind towards the central region in consequence of the frictional resistance offered by the surface of the sea to the air whirling in close contiguity upon that surface.

He wishes to put forward the question as to whether it may not be possible, in some cases of whirlwinds, for the barometric pressure in the central or axial region to become abated through the combined influences of rarefaction by heat (increased, perhaps, by conditions as to included moisture) on the one hand, and the whirling motion on the other hand, very much beyond the abatement that could be due to heat, or heat and moisture, alone, without the whirling motion. He thinks it very likely that in great whirlwinds, including those which produce the remarkable phenomena called waterspouts, it may be impossible for the

* [No. 1, *supra*, p. 1.]   † [No. 26, *supra*, p. 144.]   ‡ [No. 16, *supra*, p. 96.]
§ [No. 18, *supra*, p. 102.]

whirling action to be confined to the lower region of the atmosphere; but that, even if commenced there, it would speedily be propagated to the top. It seems also not unlikely, and in some trains of thought it comes to appear very probable, that the whirling fluid, ascending by its levity, would drive outwards from above it all other air endowed with less whirling energy, and would be continually clearing away upwards and outwards the less energetic axial core which enters from below, and any, if such there be, that has entered from above. He is unable at present to offer much in further elucidation (possibly it might only prove to be in further involvement) of this very difficult subject. He thinks the question should at least be kept open as to whether the whirling and scouring action may not go forward growing more and more intense, promoted always by energies from the thermal sources which have produced differences of temperature and moisture in different parts of the atmosphere, and that thus a much nearer approach to vacuum in the centre may be caused than would be due merely to the levity of the superincumbent air if not whirling.

He also wishes to suggest that the dark and often frightful cloud usually seen in the early stages of whirlwinds and waterspouts, and the dark columnar revolving core often seen apparently protruding downwards from the cloud, may be due to precipitation of moisture into the condition of fog or cloud, on account of abatement of pressure by ascension in level, and environment with whirling air, which by its centrifugal tendency acts in protecting the axial region from the pressure inwards of the surrounding atmosphere.

28. BAKERIAN LECTURE.—ON THE GRAND CURRENTS OF ATMOSPHERIC CIRCULATION.

[From the *Philosophical Transactions of the Royal Society*. Received March 10, Read March 10, 1892.]

[Professor Thomson died on May 8, 1892, before this Lecture was printed.]

IN the early times of the Royal Society (a little more than 200 years ago) a spirit of inquiry and of speculation as to the causes of the Trade Winds arose among its members. The papers which we may presume to have first brought the subject into special notice in the Society, and which were published in the *Transactions*, offered views which, in the light of subsequent knowledge and theory, show themselves as being untenable, and in part even grotesque. But those papers were soon followed by, and probably had an effect in leading to, a much more important paper by the eminent astronomer Edmund Halley; and this was followed 49 years later by one, more important still, by George Hadley, in which we may with confidence judge that a substantially true theory of a large part of the system of Atmospheric Circulation in its grandest and most dominant conditions was for the first time offered to the world through the pages of the *Philosophical Transactions*.

Further speculations on the subject and advances in our knowledge of it have been made in later times and have been brought into notice in various ways. I believe that I have myself arrived at some improved considerations which are to a large extent trustworthy and go far towards completing the true theory of the grand currents of atmospheric circulation, and I entertain the ambition to have my views placed on record by this Society—the Society in which the subject had its most important beginnings.

With this in view it appears indispensable that some historical recital should be adduced of the progress made by others previously: but still, for those who may at any time wish to direct their attention specially to the physical conditions irrespective of the history of the progress of thought or of discovery on the subject, it appears desirable that an exposition of the resultant theory which I have devised should be presented without being itself encumbered by historical details of the courses through which it has been ultimately arrived at. I propose, therefore, to present, in a first section, a historical sketch of all the speculations and theories which, as far as known to me, have conduced in any important way towards the resulting theory that I have to offer as being tenable and trustworthy; and then to set forth that new theory itself divested as far as possible of historical or personal references; and to conclude with some considerations as to the reasons for or against the views put forward by various persons.

The first opening up of considerations and discussions in the Royal Society on the subject of Atmospheric Circulation appears to have been made in a paper submitted to the Society, in 1684, by Dr Martin Lister, Doctor of Physic of the University of Oxford, and published in the *Transactions**. As an illustration of the scanty and crude condition of knowledge and of thought on this great subject at that time—the middle period of the life of Sir Isaac Newton—I may be permitted to cite the views of Dr Lister in his own words as offered briefly in that paper:—

"Among the known *Sea Plants*, the *Sargosse*, or *Lenticula Marina*, is not to be forgot; this grows in vast quantities from 36 to 18 *Degrees Northern Latitude*, and elsewhere, upon the deepest Seas. And I think (to say something by the by of that great *Phenomenon* of the *Winds*) from the daily and constant breath of that *Plant*, the *Trade* or *Tropick Winds* do in great part arise: because the matter of that *Wind*, coming (as we suppose) from the breath of only one *Plant* it must needs make it constant and uniform: Whereas the great variety of *Plants* and *Trees* at Land must needs furnish a confused matter of *Winds*: Again the *Levant Breezes*† are briskest about *Noon*, the *Sun* quickening the *Plant* most then, causing it to *breathe* faster, and more vigorously; and that *Plants* mostly languish in the *night* is evident from many of them which contract themselves and close at that time; also from the effects of our winters upon them, which cause them to cast both fruit and leaves too; whereas they are said (the same *Plants* for kind) universally to flourish all the year alike within the *Tropicks*.

"As for the *direction* of this *Breeze* from *East* to *West*, it may be owing to the *General current* of the *Sea*, for a gentle *Air* will still be led with the *stream* of our *Rivers*, for example. Again every Plant is in some measure an Heliotrope, and bends itself, and moves after the *Sun*, and consequently emits its vapours thitherward, and so its direction is in that respect also owing in some measure to the *Course* of the *Sun*."

(NOTE.—The above is the whole passage given by Dr Lister about Trade Winds. The rest of his paper relates to entirely different subjects, chiefly to salt springs and brines.)

In scrutinizing these utterances of Dr Lister, we may notice that he must have been in possession of some information, more

---

* *Phil. Trans.* No. 156, p. 494. Date February, 1683-84.

† By "Levant Breezes" here Dr Lister obviously means breezes from the east, in fact, the Trade Winds of the tropics.—JAMES THOMSON.

or less vague, to the effect that over extensive regions of the great oceans between the Tropics, or near to them, winds blowing from east towards west are prevalent; and that he has attempted to explain this prevalence by attributing it to the breath of a plant floating on the sea and turning "as an heliotrope" so as to blow its breath westward according to the direction of the Sun's diurnal relative motion through the sky from its rising in the east to its setting in the west. He does not indicate any knowledge of the fact that on the two sides of the Equator in tropical regions there are two Trade-Wind zones, one on each side, in each of which the wind prevails from east to west, with an accompanying motion in each case towards the Equator.

We may, indeed, suppose that such knowledge was only gradually acquired, chiefly by mariners, and was but vaguely and imperfectly intercommunicated among them, and was spread very little among others during a long period of time. I do not suppose that any remarkable step in the discovery and promulgation of knowledge of the prevalent courses of the winds in those seas in and about the Torrid Zone is to be attributed to any one person in particular, nor that there was, indeed, any very important and clear promulgation of the floating knowledge on the subject until the time when the astronomer, Halley, collected and systematized a large amount of valuable information, and presented it to the Royal Society, in his paper in the *Transactions* of 1686, to which I shall make particular reference a little further on.

It may be well at the present stage, before going further into the history of speculations, to draw attention to the chief features of the Trade Winds and other perennially prevalent air currents, as they present themselves very manifestly to the notice of mariners.

The mariners on board a ship at sea, it is to be observed, however, have direct cognizance only of the wind blowing at the spot on the ocean's expanse where for the time being their ship is situated. They can make no observations on the winds blowing at the same moment 100 miles away, and the vault of the sky above them presents to their eyes no adequate indication of the upper currents, or of the places whence these come or whither they are going in their circuits. But even long ago, by the collation among navigators of facts contemporaneously observed by various seamen, important knowledge was acquired gradually as to the general character of contemporaneously existing air currents at the surface of the sea, without the aid of any trustworthy theory as to the continuations of such currents in circulation through the upper regions of the atmosphere. It is further to be noticed that the geographical distribution of sea and land, presenting as it does great regions of ocean, and large continents themselves varied with mountain ranges and low-lying plains, introduces great local variations in the conditions determining the courses of winds, and prevents the institution of any complete uniformity in the character of the air currents all round the Equator, or throughout zones between any parallels of latitude.

But in the Atlantic and Pacific Oceans there are extensive regions within, and adjacent to, the Torrid Zone, in which the winds blow with remarkable constancy from the east while converging also from north and south at the two sides, towards a medial belt of calms and rains which is situated along, or very near to, the Equator.

These remarkably persistent winds blowing in the northern hemisphere from the north-east, and in the southern from the south-east, are called the Trade Winds. The outer limits of the two Trade Winds vary in different seasons of the year, and are affected by casually varying conditions of the atmosphere in other parts of the world, and by the geographical configurations of the surrounding continents affecting them unequally in different parts; but, without minute exactitude, they may be regarded as occupying some such breadth as perhaps 25° or 30° on each side of the Equator.

It was also found by mariners in those early times previous to the development of theories of atmospheric circulation, that in the great oceans, in the higher latitudes, outside of the trade-wind bands, west winds are prevalent in frequency and strength over winds in other directions. It became the practice of traders when going on a voyage from east to west to make their way into the trade-wind region, where they were sure of finding favouring breezes, and on their return voyage to get into higher latitudes, so as to take advantage of the prevailing west winds there.

Until recent years no information was definitely gathered from observations or otherwise as to whether or not there be any prevalent general average tendency in those west winds to blow in their variations more towards the Pole or towards the Equator; and I avoid entering on any statements on the subject at the present historical stage, as that matter will be better associated with the subsequent progress of theories than with the early history.

The explanations just given in words as to the chief features of the trade winds and of the west winds of higher latitudes may be supplemented so as to come more vividly before the imagination by aid of fig. 1.

This figure is sketched without regard to the disturbing influences of continents and mountain ranges. It may be regarded as being suggestive of the most remarkable features which would probably present themselves in the winds if the surface of the world were all ocean, or were ocean mottled very uniformly with small islands.

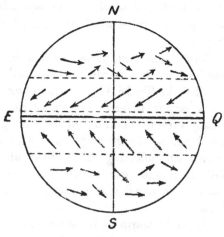

Fig. 1.

Now, to revert to the historical sketch already entered on, of speculations and theories as to perennially prevalent winds, and to variable winds which manifest perennial prevalence in special directions, the next theory to which I have to refer, is that of Dr Garden of Aberdeen, which, about one year after that of Dr Lister, was placed on record in the *Transactions of the Royal Society*. Dr Garden, in his paper*, attributes the east to west

* *Phil. Trans.* Vol. 15, No. 175. September and October, 1685.

motion of the Trade Winds of the Atlantic and Pacific Oceans to the supposed vortices of a supposed ether, or all-pervading atmosphere, which, according to the planetary system proposed by Des Cartes, and at that period still believed in by some, were imagined to be the agents carrying on or sustaining the revolutions of the Planets round the Sun, and of the Moon round the Earth, and of the Earth round its own axis. Dr Garden's paper gives indication of his having some knowledge, not only of prevalence of winds from the east within the tropics, but also of prevalence of winds from the west in higher latitudes outside of the tropical regions. He gives no indication of knowledge of the Trade Winds having, along with their westward motion, also motions towards the Equator from both sides; and is, in this respect, apparently on an equality with Dr Lister. They had both made praise-worthy exertions in collecting and bringing into notice important results from the observations of mariners and other travellers. When, however, Dr Garden offers explanations of his supposed reasons for the blowing from east to west within the Tropics, and from west to east in latitudes higher than those of the tropical regions, his statements, in their meaninglessness, quite transcend the inadequacy of the explanations in the amusing attempt of Dr Lister.

The papers of these two men may probably have had a beneficial effect in instigating Halley to prepare, for the Royal Society, a paper presenting the results of his researches as to the observable facts of the winds, and his speculations to account for the prevalent directions of their motions. In 1686, about one year after Dr Garden's paper, Halley, then at the age of thirty, submitted to the Society an elaborate and very clear account of the information as to the winds in different parts of the world which he could collect from numerous sources, including obser-vations carefully made by himself on voyages and on land between the Tropics. The title of his paper in the *Transactions*, is "An Historical Account of the Trade Winds, and Monsoons observable in the Seas between and near the Tropicks, with an attempt to assign the Physical Causes of the said Winds*." His description of the observed facts and his theoretical considerations on the subject, have constituted an important step in the development of the science of that subject, even though his theory in its most

* *Phil. Trans.* No. 183, p. 153.

important part—that which relates to the east to west motion of the Trade Winds—turns out to be fundamentally untenable. He adduced, no doubt, in his explanation, an important part of the real truth as to causes of the wind, a part which, if not first suggested by him, was clearly either not generally known or not generally adopted at the time.

This true element in his theory consisted in his assigning as the primary motive cause of the winds, the expansion of the air of hot regions, accompanied by its outflow in its upper parts from those regions towards places of less heat and entailing a diminished pressure at the base of the ascending heated current, and consequently entailing an influx at bottom from the lower part of the atmosphere at the colder places where descending currents are generated. In applying this general principle further to the explanation of the observed winds, he rightly explained the influx of the air from both sides towards the Equator or some medial part of the trade-wind region as being due to the more intense heating effects of the Sun in the Equatorial regions. But in the more important, because less obvious, element for explanation of the Trade Winds and of atmospheric circulation generally—that which is requisite for explaining the east to west motions of the Trade Winds, and the prevalence of winds from west to east in higher latitudes—he quite missed the true explanation. He attributed the east to west flow of the Trade Winds to the diurnal revolution round the equatorial zone from east to west of the maximum of accumulation of heating effect from the daily sunshine, which gives an accumulation of heat in the afternoon in each successive locality. Briefly, he said to the effect that as the maximum of accumulated heat runs round the Torrid Zone from east to west, passing each place at a few hours after noon of that place, and as the maximum of heat in travelling round always causes an indraught towards itself, so the atmosphere of the Torrid Zone must be brought into flowing round from east to west likewise. But this conclusion from the submitted premises is really quite inconsequential.

In reference to this speculation, and treating for the present the direction which we will call the forward direction round the Torrid Zone as being that of the Sun's progress from east to west, we may entertain considerations such as the following:—That consequent on the indraught from all sides towards the hot region, where the barometric pressure is most reduced, the backward-tending forces acting on the air in front of the maximum may be acting as much in respect to time and duration backward on the air in front of the maximum as do the forward-tending forces on the air behind that maximum, and that, through this consideration by itself, we might not be entitled to suppose that any resultant tendency to the generation of a current round the Torrid Zone one way or other, east to west, or west to east, would be produced. But when we further consider the unsymmetrical character of the conditions of the two influxes towards the maximum region from before and from behind, and the to us very unknown accompanying frictional conditions between these unsymmetrically conditioned currents of air and the surface of the earth or sea over which they pass, we may be led to think it very unlikely that the forwarding and backwarding influences would exactly counteract one another; and I certainly think they would not do so, and I think some resultant flow from east to west, or from west to east would be produced, but in which way, east to west or west to east, it would occur I am quite unprepared to say*.

The theory or speculation in the terms in which it was set forth by its author makes no reference to the inertial conditions of the atmosphere concerned in its diurnal revolution along with the Earth, to which, as a matter of fact, it clings so as to have at all times and all places almost the same revolutional speed or angular velocity of diurnal rotation as the Earth has. In fact, Halley's theory would be equally applicable to the case of the world being non-rotative and having the Sun, or an equivalent source of heat, revolving round it from east to west.

But, in view of the very powerfully influencing conditions subsequently brought to light in the theory of Hadley which will next be adduced, any such feeble causes as those relied on by Halley must fall practically into insignificance, the indubitable cause shown in Hadley's theory being such as to be dominant.

In 1735 George Hadley (brother of the John Hadley who invented the instrument commonly known as Hadley's Quadrant) submitted

---

* As a matter of curiosity I think it might be interesting in a time of comparative leisure for some person to make experiments with a spirit lamp or other heater kept revolving slowly round in a circular path under a circular tray filled with water, the path being of a little smaller radius than the tray. The question being, would or would not the water be set into revolutional motion, and if so would it revolve in the same direction as the lamp or other source of heat does?

to the Royal Society the paper of which I have made mention
already as supplying for the first time a substantially true theory
of the primarily dominant conditions of atmospheric circulation *.
The paper is entitled "Concerning the Cause of the General
Trade-Winds," and it is right here to notice that Hadley applied
the name General Trade-Winds, not merely to those winds of
equatorial regions to which the name Trade Winds is ordinarily
restricted, but uses it as including also the west-to-east winds
known to be prevalent in higher latitudes, and used in trade by
mariners for ocean passages from west to east. Thus the scope
of his theory must be understood as being much wider than what
would be conveyed in ordinary nomenclature by the name, Theory
of the Trade-Winds.

In his paper, Hadley commences by adopting, as a part of the
whole truth, the view already in his time currently held by others,
that the Sun's heat, intensely applied and greatly accumulated
in the equatorial regions of the Earth, conjointly with the cooler
temperatures of the regions in higher latitudes, is the main and
primary cause of the Trade Winds and other currents of the
atmosphere. In this way he supposes that at the Equator or
near to it there is a belt of air ascending because of its high
temperature and consequent rarefaction, and an influx from both
sides towards a zonal region of diminished pressure at its base;
and that from its upper part currents float away to both sides,
northward and southward, and that these continue in the upper
regions of the atmosphere advancing pole-ward until, by cooling
in the higher latitudes, their substance gradually becoming less
buoyant sinks down gradually and returns towards the equatorial
regions as a lower current along the Earth's surface, thence to
renew the circulation by ascent again in the equatorial region.
While indicating virtually that such atmospheric circulation would
be generated, whether in an irrotative world with a source of
heat revolving round it corresponding to the Sun in its apparent
diurnal revolution, or in a world revolving on its axis as does the
Earth, he shows that in the latter case—the case, namely, of the
revolving Earth—in addition to such circulation as has just been
described, east-to-west and west-to-east motions relative to the
Earth's surface would necessarily come into being for reasons which
may be stated or suggested as follows:—

* *Phil. Trans.* Vol. xxxix. No. 437, for April, May, and June, 1735, p. 58.

If we consider the air in a nearly calm region at the outer limits of the trade-wind zone, and regard the air at that place as being at rest relatively to the Earth's surface, and if we consider it to be drawn over the surface by indraught towards the Equator without application to it of any other force than that of the indraught, except what it may receive by friction from the surface of the Earth, be that land or ocean, this air in arriving at places always lower and lower in latitude (and consequently further and further out from the Earth's axis) is coming to places in succession each moving eastward quicker than the previous one; and thus the air is arriving successively at places each going quicker eastward than the air itself was going when at the previous place; consequently the air in arriving at each new place must obviously have a slower motion eastward than the Earth's surface at that place has.

Thus throughout that course the Earth must be rushing forward under the air eastward quicker than the air goes, and that is the same as to say that the air must be blowing westward over the surface of the Earth.

In connection with this part of his theory he brings into notice that, while the surface of the Earth at the outer edges of the Trade Winds has much less of absolute velocity eastward in diurnal revolution round the Earth's axis than the surface at or near the Equator has, yet the trade-wind air, on arriving at the foot of the equatorial belt of rising air after its course from those outer parts in higher latitudes, has become imbued with eastward velocity little less than that of the equatorial surface of the Earth, the only deficiency in this eastward velocity from that of the equatorial surface being what is manifested as wind blowing westward over the Earth's surface, or having in relation to that surface a moderate westward velocity. He shows, for an example, that the eastward velocity of the Earth at either of the tropic circles is less than that at the Equator by about 87 miles per hour, but yet that the air which comes from calm regions near the tropic circles to the equatorial belt has, on its arrival at that belt, an eastward absolute velocity which is only a few miles per hour in defect of the velocity of the Earth there, the actual defect being manifested in the relative velocity with which the wind at the equatorial parts blows westward over the surface of the land or sea. He explains that this result is brought about by reason that the air, during its course from the outer edge of the trade-wind

11—2

zone to the foot of the equatorial rising belt, is perpetually
being dragged forward eastward by the quicker-moving land or
sea below it, and so its velocity is kept nearly assimilated to that
of the part of the Earth over which, for the time being, it exists,
and is allowed only to be a little less than that velocity.

Such, then, is Hadley's theory, in so far as it relates to the origin
of the Trade Winds of the equatorial regions on both sides of the
Equator. His theory further extends to explain the cause of the
prevalence of winds from west to east in latitudes higher than
those of the winds of equatorial regions, to which, except in the
nomenclature of Hadley himself, the name *Trade Winds* has been
usually restricted; and this part of his theory may be represented
as follows :—

The equatorial surface of the Earth has a velocity of diurnal
revolution from west to east of about 1000 miles per hour. The
air of the land and sea at and near the Equator participates
nearly in the same velocity. The ascending equatorial belt of
heated air retains as it ascends an absolute velocity from west to
east nearly the same as that of the equatorial surface of the
Earth. He supposes, then, in his theory, that the air floating
out from the upper part of the rising belt to north and south
over the equatorial zones of Trade Winds, and thence, still in the
upper parts of the atmosphere, spreading over extensive regions
of land and sea in latitudes higher than those of the Trade
Winds, will, on reaching those regions whose velocities of diurnal
revolution are much slower, be rushing forward from west to east
quicker than do the portions of the Earth's surface over which
it successively arrives in floating poleward; that greater speed
of eastward motion of the air than of the Earth beneath being,
however (as he indicates with a fair approach to clearness), kept
in moderation by influences from the surface of the land or sea
offering resistance to relative motions of the air above it. Further,
he supposes that this upper air, while moving eastward quicker
than does the Earth below it, gradually loses a great part of its
previously acquired heat, and becomes less buoyant, and consequently descends gradually towards the surface of the Earth, the
supply above being always maintained by fresh arrivals from the
equatorial regions; and he supposes that the descending air
brings from aloft perpetually new supplies of west-to-east motion
relative to the surface, and so maintains winds blowing over the

surface from west to east. The air then, after its descent from
the sky towards the surface throughout extensive regions, must,
I think, necessarily, under his theory—although he does not
explicitly mention this—be supposed to flow gradually back in
the lower levels of the atmosphere towards the Equator, while
also blowing prevalently from west to east, till it reaches again
the outer border of the trade-wind region, thence to go forward
repeating such a circulation as has just been described.

Hadley concludes his paper with a short passage which, considered in reference to the crude condition of progressive opinions
prevalent in respect to atmospheric circulation up to the time
of the promulgation of his theory, is to be regarded as suggesting,
though in somewhat vague and not entirely correct expression, a
very notable and important principle.

The passage is as follows:—" That the N.E. and S.E. Winds
within the Tropicks must be compensated by as much N.W. and
S.W. in other Parts, and generally all Winds from any one
Quarter must be compensated by a contrary wind somewhere or
other; otherwise some Change must be produced in the Motion
of the Earth round its Axis."

The really important idea which it appears to me is suggested
in this passage, is that in respect to the Earth's rotation round
its axis the sum of all the forward turning-force-influences
applied by the winds to the surface of the Earth, land and sea
included, must be equal to the sum of all the backward turning-
force-influences likewise applied to the Earth's surface; so that
these force influences may be such as conjointly to produce no acceleration or retardation in the revolution of the Earth round its axis.

In putting forward this idea he was doubtless assuming as a
principle that we are not to attribute to the thermal influence
of the Sun any effects in altering the rotation of the Earth by
producing winds blowing upon the Earth more effectually on the
whole forward than backward, or the reverse. He did not, nor
probably did anyone else till long after his time, notice the now
known principle that the Sun and Moon can, by their attractions,
apply to fluids on the Earth—to the sea or to the atmosphere—
turning forces* which these fluids must communicate to the solid

---

* Any system of forces which can be balanced by what under the nomenclature
of Poinsot is called a *couple*, may be described as a turning-force-influence, and
may now with advantage be called a *torque*.

earth, and which must, in very long periods of time, make changes on the Earth's rotation. Such influences, however, are certainly so very small comparatively to those Hadley had under consideration as occurring in the action of the equatorial Trade Winds from the east, and the winds of higher latitudes from the west, that his not knowing of them is not to be regarded as derogating from the practical or substantial truth and validity of his Theory of the Winds in its main features.

In the account I have given of Hadley's theory of the primarily important perennial features of atmospheric circulation, I have endeavoured faithfully to give a fair and favourable account of the truths which he brought to light. I have not held it as a duty to bring under review every statement or phrase to which objection might be taken by an adverse critic. There is one mistake, however, into which Hadley fell, and which is too important to be passed over without notice. This error, although incorporated by himself along with his true explanations in respect to the causes of the equatorial Trade Winds and of prevalent westerly winds of higher latitudes is quite separable from those true explanations; and its elimination does not make any breakdown in any essential part of his reasoning as to the real conditions of the atmospheric motions. His error pertained not to his suppositions as to the actual motions of the real air, but to supposed motions and behaviour of air in an ideal case which he adduced as a simplified illustration intended to be helpful to the consideration of the more complex conditions of the real case. The two cases—the ideal and the real—are not explicitly and distinctively specified by himself, but they are brought implicitly under consideration in his statements to the following effect :—

Firstly.—That air having been in an approximately calm condition at one of the Tropic Circles, and having moved thence in the Trade Wind to the Equator, will, on arriving at the Equator, retain still the same absolute eastward velocity that it had when at the Tropic, and so will at the Equator have less velocity of absolute eastward motion than the Earth there has, by 2083 miles per day, or 87 miles per hour, and that so it will be moving relatively to the Earth there as a wind blowing at the rate of 2083 miles per day from east to west.

And *Secondly* :—That as an amendment on the previous

statement, it is to be considered that "before the air from the Tropicks can arrive at the Equator, it must have gained some motion eastward from the surface of the earth or sea, whereby its relative motion will be diminished, and in several successive circulations may be supposed to be reduced to the strength it is found to be of."

In reading these two statements conjointly we may with confidence judge that the first of them is not meant to convey the actual truth in respect to the real behaviour of the atmosphere, but that it is only a theoretical utterance as to an ideal case, in which the frictional drag between the surface of the ocean and the atmosphere is left out of account, and that the second is that which is meant to convey the real truth. Now the important error into which he has here fallen, consists in his supposing that in an ideal case, in which the trade-wind air is regarded as frictionless and free from receiving any eastward or westward force-influences from the ocean below it, or as I will add, from the atmosphere immediately above it, it ought to be expected on arriving at the Equator, from a calm at the tropic circle, to retain the same amount of eastward absolute motion which it had when at the tropic. Instead of that, in the ideal case, if fully specified with due limitations, such as we may suppose were tacitly contemplated, without being fully thought out, the true averment would have been that the air on arriving at the Equator would have a velocity of eastward absolute motion less than that at the tropic, in ratio inverse of that of the distances of the two places respectively from the Earth's axis. What I mean here to say, may, perhaps, without elaborate definitions and specifications, be tolerably well suggested in brief words, by saying that, in a vortex of free mobility, with circular motions round an axis, the velocities at different distances from the axis must be inversely as those distances.

But now, in truth, the ideal case which Hadley touched upon, was quite outside of the scope of the real conditions of the atmospheric motions, which he professed to explain better than had been done in the attempts of others before his time. He had amply sufficient reason for his averment, to the effect that the real trade-wind air in its approach from the tropic to the Equator, under the influence of indraught towards the Equator, should be expected at each new place nearer to the Equator than the previous one, to have a less velocity of eastward absolute motion than the surface of the sea has at that new place, and that the frictional drag eastward applied to the air by the sea surface will only act towards assimilation of the eastward velocity of the air to that of the water, while still in principle, as in fact, leaving the air to go slower eastward than does the water—that is, to blow as a westward wind relative to the ocean. If what he professed to do had been to bring into notice a special variety of vortex motion, constituting what we may call a vortex of free mobility in a frictionless fluid, and to offer a dynamic theory of its motions, and if his theory had included such an error as the one in question; then his theory would have been fundamentally erroneous. But such was not at all what he professed to do. He proposed to explain certain large and very remarkable phenomena of the observed winds. This he did well, and in doing so he made a very important advance in development of true theory in respect to atmospheric motions.

I have touched in some detail on these matters, because I think that remarks making inadequate recognition of the importance of Hadley's true discoveries have sometimes been put forward in our own times.

During a period of more than a century from the time of the promulgation of Hadley's theory, in 1735, there was, I consider, little if any remarkable progress made in development of new speculations for better or for worse in respect to the grand or perennial currents of atmospheric circulation. A long time elapsed, in which there seems to have been little or no vigorous spirit of investigation into the significances or the relative merits of the speculations which had been propounded, or of effort to amend the existing theories, or to discover new truths on the subject. In confirmation of this it may be noticed that we find that 58 years after the publication of Hadley's paper, Dalton arrived independently at substantially the same theory as that part of Hadley's which dealt with the equatorial Trade Winds, and in his book entitled *Meteorological Observations and Essays**, which in 1793 he was preparing for publication, he gave an account of his theory, supposing it to be original, but he

---

* *Meteorological Observations and Essays*, by John Dalton, D.C.L., F.R.S., 1793. Of this work there is also a second edition, which is a verbatim reprint issued by Dr Dalton himself, in 1834.

discovered, before the book was issued to the public, that he had been completely anticipated by Hadley's paper, of the existence of which he had not been previously aware. In his preface to that book, after making recognition of Hadley's priority, he goes on to say:—" I cannot help observing here, that the following fact appears to be one of the most remarkable that the history of the progress of natural philosophy could furnish.—Dr Halley published in the *Philosophical Transactions* a theory of the trade-winds which was quite inadequate and immechanical, as will be shown, and yet the same has been almost universally adopted; at least I could name several modern productions of great repute in which it is found and do not know of one that contains any other."..." On the other hand G. Hadley, Esq., published in a subsequent volume of the said *Transactions* a rational and satisfactory explanation of the trade-winds, but where else shall we find it?"

It is right here to remark further that Dalton in his own speculations did not touch at all upon the prevalence of west winds in extra-tropical regions, either as to its explanation or even as to its existence: and that he does not seem to have noticed or appreciated the great importance of Hadley's theory in this respect.

Not only before, but also after this episode of Dalton's speculations and researches so published, the theory of Hadley must certainly have remained but little read in its author's original paper.

Within the first half of the present century writings on the winds, including the Trade Winds and general circulation of the atmosphere, have been very numerous, some of these have appeared in our encyclopædias, and others in works on meteorology and navigation, and have been widely diffused in atlases containing maps and charts on physical geography.

In such ways many sketches have been presented to the public as explanations of the Trade Winds and other currents of the atmosphere related to them, embodying more or less of the fundamental principles of Hadley's theory, but often without reference to his name, and usually without due appreciation of the meaning and importance of his theory. In many of these cases we may suppose that the authors had never seen his own original paper, but had obtained their information indirectly through the writings of others.

On the other hand, within the period just mentioned—the first half of the present century—real progress was made in many ways, in the gaining of new knowledge and the making of a few new discoveries, chiefly in connection with the temporary and local disturbances of the atmosphere, and in the bringing together of information of various kinds to help in the elucidation of the subject of the winds. The influence of moisture in air of any given temperature and pressure in rendering the fluid more buoyant was brought effectually into consideration.

The attainment of information from the practical observations of mariners and travellers, and especially explorers of the polar regions, and also from meteorological observatories, was making gradual but important progress. Considerable progress was made in the collecting and correlating by many persons of observational results as to winds and weather and barometric pressures in various latitudes, and in the presenting for practical use among navigators and others of the generalized conclusions so derived.

In that course of progressive labours there were included various speculations or theories as to great storms, commonly designated as hurricanes, tornadoes, or cyclones. In beginning to touch on this subject I have to mention that from among the many persons who may have taken part in researches and speculations regarding cyclones, those whom I deem the most noteworthy are Capper, Dove, Redfield, Thom, Reid, and Piddington.

Now, within the period which we have at present under consideration—the first half of the present century—by a very gradual course of experience, chiefly maritime, and of speculation based on such experience, it came to be promulgated that violent storms were generally great whirlwinds; and so the old name *tornado*, of Portuguese origin, suggestive of turning, and the new name *cyclone*, used in the sense not merely of circular form, but also of revolving motion, came to be accepted as well suited for the designation.

Also it was found that in the centre of a cyclone there is a region of comparative calm, and that the centre does not remain stationary, but travels at some moderate speed, taking generally a curved course over the surface of sea or land.

The discovery also was established beyond room for doubt that cyclones in the northern hemisphere revolve in the direction opposite to that of the hands of a watch situated in their locality with its face up; while in the southern hemisphere they revolve in the same direction as do the hands of a watch situated in their locality with its face up.

Also it was discovered and promulgated that in the central region of a cyclone the barometric pressure is remarkably diminished as compared with that of the general surrounding atmosphere, and that this condition must necessarily subsist as a concomitant of the centrifugal tendency or "centrifugal force" of the revolving air, but whether the diminished pressure was to be regarded as a result of the centrifugal force of the revolving air, or as one of the primary causes of the institution of the cyclonic revolution, seems commonly to have been left unnoticed or to have been adverted to under erroneously imperfect views.

Dove, for instance, when discussing the tremendously violent whirling motion which is met with in the inner part of a cyclone immediately around the central region of remarkable calm, says, *"the diminution of barometrical pressure is not the cause of the violent disturbance of the air, but rather a secondary effect of it*,"* and through that passage with its context it seems doubtless that, while entertaining the view that the rapid revolving motion of the air somehow instituted maintains by centrifugal tendency the diminished pressure in the central region, he fails to notice the more complete truth, that without the actual occurrence of centripetal motion caused by predominating influence of inward suction the rapid revolving motion would not institute itself at all.

This being said, however, there is yet, of the whole truth, another element which must be brought into notice, and which I here briefly describe, with some perhaps new ideas that have occurred to myself.

It is, that while for a beginning an accumulation of buoyant air at bottom elongates itself upwards into a shape approaching to a columnar form, and so effects an abatement of pressure at its base; and this abatement of pressure (or suction) induces a centripetal flow towards that place from outer regions where some slight, though it may be almost imperceptible, motions having revolutional momentum (or, in other words, moment of momentum) round that place may already exist, and the revolving mass of air through the action of the centreward forces applied to it, takes

* Dove, *Law of Storms*, English translation by Robert H. Scott, M.A., p. 198.

an increasingly rapid revolving motion; and further, this rapid motion reacts on the buoyant central column, *keeping that from scattering through the air around it,* and so institutes a very lofty continuous column of the buoyant air.

To make this clearer we may notice that if a buoyant central column were for a moment existing surrounded by non-rotative air having greater pressure in its lower parts than that in the column at the same level, that column could not continue its existence. The outer air with its greater pressure would press in on the column, and would increase the pressure in its substance instantly, but the weight of the upper portion of the buoyant column would be inadequate to resist the upward thrust so produced in the lower part, and so the lower parts would shoot those above them upwards with violently accelerating motion. Through the rushing upwards so generated a breaking up of the column would supervene, and its substance would scatter itself in rolling masses among the surrounding air; and the two commingling would ascend gradually, and at the same time the pressure of the surrounding air would communicate itself to the region where the base of the column had been.

But now, on the other hand, if the mass of air around the central buoyant column be whirling, it will keep itself out by the centrifugal tendency accompanying its own rapid revolution, and so will not press in upon and break up that central column of air of diminished pressure, and thus the abatement of pressure at the foot of the column will be maintained and will become further intensified.

To Redfield is due much credit for his able and long-sustained labours in collecting and correlating observed facts as to cyclones and the smaller kinds of whirlwinds. He gathered and published* a very interesting collection of accounts of violent columnar whirlwinds which formed themselves over large fires of circular masses of brushwood, the flame and smoke in each case ascending as a lofty rotating column; and this has had part in suggesting to me some elements in the theoretical considerations here briefly sketched out. In his remarks on these whirlwinds he emphatically

* "Some account of Violent Columnar Whirlwinds which appear to have resulted from the action of large Circular Fires," by W. C. Redfield. Read before the Connecticut Academy of Arts and Sciences, Jan. 22, 1839. Printed in the *American Journal of Science and Arts* (Silliman's), 1839, Vol. xxxvi. p. 50.

brought into contrast the distinction between the flames and smoke ascending without whirling motion from hot furnaces and various ordinary fires, on the one hand, and, on the other hand, the revolving columns of flame and smoke often met with in those great fires of brushwood in the open air. By his various researches into the actions and effects of great storms, Redfield contributed more, perhaps, than any other man to the advancement of observationally-derived knowledge of their cyclonic character and features.

Wild and fantastic notions were, however, afloat in those times as to the origin of cyclones. Thus Piddington, in his well-known work entitled the *Sailor's Hornbook*, even in the edition so late as 1860*, in stating his resultant opinions and conclusions, makes such statements as the following :—

That he considers cyclones to be flat circular disks which may be formed at the sides and upper and lower surfaces of clouds, and which, once formed, may either rise higher or descend downwards, and may extend themselves greatly or contract in diameter, and which may be "parallel to the surface of the globe" or "inclined forwards"; he goes on to say: "It appears to me that a simple flattened spiral stream of electric fluid generated above in a broad disk, and descending to the surface of the Earth, may amply, and simply, account for the commencement of a Cyclone†."

After making careful search through numerous writings on the subject of cyclones, I have to say that I have no reason to think that the investigators who took part in the discovery of the directions of turning of cyclones in the northern and southern hemispheres had generally, or that any of them in particular had, any clear dynamic theory explanatory of the connection between these modes of turning and the rotation of the Earth, nor even of the origin of the very rapid whirling motion itself, but I have found strong indications of deficiency of such knowledge. Even Herschel, so late as 1857, in his article on "Meteorology," in the *Encyclopædia Britannica‡*, stated that a complete account of the phenomena of cyclones had been afforded "by Hadley's theory as developed by Dove in his *Law of Rotation*, and applied to this

---

* *Sailor's Hornbook*, third edition.
† *Sailor's Hornbook*, third edition, p. 338, section 410.
‡ *Encyc. Brit.* eighth edition, Vol. xiv. p. 650.

specific class of aërial movements by Professor Taylor," and then
went on to give what we may presume to be that explanation,
but the explanation he gives, although containing enough of
truth to prove the connection between the direction of the
Earth's rotation and that of the mode of turning of cyclones in
each hemisphere, is incomplete, and is vitiated by important
errors of principle.

Mr Wm. Ferrel, of Nashville, Tennessee, in a paper of date
1856 (to be referred to further on in connection with other
matters), adduced dynamic considerations of more advanced
character for explanation of causes of the gyratory motions of
cyclones; but his treatment, although in some respects usefully
suggestive and indicating sufficient reason for the direction of
turning in each hemisphere, I cannot regard as being on the
whole to very good effect.

Also, as a further result of the researches and scrutinies and
efforts towards generalization told of already, it came gradually
into notice and into acceptance as an established truth that in
the latitudes outside the limits of the Trade Winds extending
far towards the poles, sometimes for brevity called the middle
latitudes, the wind, while prevailing from the west as had been
long previously known, prevails also for each hemisphere more
from the Equator towards the Pole than from the Pole towards
the Equator, so that, on the whole, to take for simplicity the case
of the northern hemisphere, the prevalent average atmospheric
current at the surface of the Earth in those latitudes was judged
to be from the south-west; or, rather, without particularizing one
exact point of the compass, and with allowance for great variations
in different localities, and at different times, we may better say
from south of west towards north of east.

To account for the component from the south in these
westerly winds of our middle latitudes, it came to be supposed,
for instance, by Leopold von Buch* prominently, as also by many
others, that the air departing for the northern hemisphere from
the top of the Equatorial Belt of buoyant air, while flowing

---

* Leopold von Buch, *Gesammelte Schriften*, Vol. III. Berlin, 1877, where there is
to be found his "Physikalische Beschreibung der Canarischen Inseln," Berlin, 1825,
chapter 2, "Bemerkung über das Klima der Canarischen Inseln," pp. 288, 289, and
290. A slightly abbreviated translation of the passage in question is given in
Dove's *Law of Storms*, Scott's translation, 1862, p. 39, in a chapter entitled "The
Upper Return Trade Wind."

northward still in the lofty regions of the atmosphere and over the trade-wind zone, soon becomes a current from the south-west, and continues after descending to the Earth's surface at the northern border of the trade-wind region still to move forward in continuation of its old course as a current from the south-west. But why in the lower regions a pole-ward motion should be maintained rather than a return flow towards the Equator, and how the return from higher to lower latitudes to compensate for this supposed pole-ward surface current should be accomplished, are questions which appear to have been scarcely mooted or to have been left enshrouded in vagueness.

Many examples might be cited indicating the wide currency which such conclusions attained to, but one or two may suffice. Thus, for instance, in Johnston's *National Atlas*, of date 1843, we have a map of the winds by Dr Heinrich Berghaus, of Berlin, on which the zone of south-westerly winds of middle latitudes is described in mysteriously poetic words more captivating to the imagination than satisfying to the reason, as *"Region of South-Westerly Currents of Air, or of the downward returning North-Eastern Trade Wind in Triumphal Conflict with the Northern Polar Currents."*

Herschel, in his *Astronomy**, of date 1850, gives an account for explanation of the south-west winds of middle latitudes substantially to the same effect as that of Leopold von Buch and Berghaus, and with like vagueness as to the return currents from higher towards lower latitudes.

But from the shelter of that prevalent vagueness, Maury, in 1855, stepped out and boldly offered a scheme of the general currents of atmospheric circulation which he supposed to prevail, in courses extending from Pole to Pole, and traversing in different ways the lower regions of the atmosphere next the surface of the Earth, and the upper regions which present themselves less directly to the observation of men. Fig. 2 is a copy of his diagram which, in conjunction with his printed explanations, sets forth his scheme of supposed circulation†.

That figure shows a hemisphere of the Earth's surface taken

---

* Third edition.

† Maury's *Physical Geography of the Sea*. The first and second editions appeared in 1855. The statements here made apply alike to his second and sixth editions, and presumably also to other editions.

from Pole to Pole. The continents and lands generally are not exhibited, and their disturbing effects on the atmospheric motions are left almost entirely out of consideration. The circulation imagined and described by Maury in connection with the diagram is meant to be a fair representation of what he would suppose likely to be realised in case of local and temporary disturbances and irregularities being only in a small degree effective. His

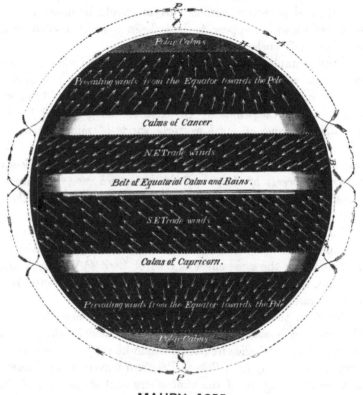

MAURY—1855.

Fig. 2.

supposed circulation may be thus described:—He supposes that the air entering the Belt of Equatorial Calms from the southern hemisphere rises there to the lofty regions of the atmosphere, and flows thence as an upper current to the Belt of the Calms of Cancer where it descends to the bottom, from whence it travels on as a south-west wind over the surface of the sea to the high

latitudes round the Pole; and that then ascending at and near
the Pole, it flows as an upper current out to the Calms of Cancer,
where it sinks again to the bottom of the atmosphere crossing
the current already mentioned as descending there, and then
passes along the surface of the sea as a bottom current forming
the north-east Trade Wind, and then enters the Belt of Equatorial
Calms, rises there, crossing the previously mentioned rising current
there, and thence departs as an upper current towards the Calms
of Capricorn, to go through a circulation in the Southern Hemisphere which is an exact counterpart of that already described
for the Northern Hemisphere. The supposed currents are further
indicated by arrows in the diagram, which, on inspection, may
easily be understood. It is to be understood that the diagram
shows a hemisphere of the surface of the Earth with the two
trade-wind zones exhibited one on each side of the Equator,
and separated by the Equatorial Belt of Calms and Rains, which
is often also called the Doldrum Belt. And that it also shows
the two Border Belts, or Calms of Cancer and Capricorn; and
also, in the Northern Hemisphere, the zonal region of wind
prevailing from south of west, and, in the Southern Hemisphere,
the corresponding zone of prevalent winds from north of west.
The arrows shown on the surface of the globe throughout these
various zones indicate the directions of motion of the bottom
currents of the atmosphere constituting the winds blowing on the
surface of the sea. Around the representation of the globe the
atmosphere is shown in section with arrows to indicate the north
and south, and up and down motions in the circulation, which has
just now been described in words.

In offering this scheme of atmospheric circulation, Maury
himself, in respect to the part of it which he propounds as
taking place in the regions between the trade-wind zones and
the Poles, confesses that it is "for some reason which does not
appear to have been very satisfactorily explained by philosophers"
that the currents he supposes do take place instead of their
contraries. In short, he admits that he does not think reason
has been found why in those regions the lower current should be
towards the Pole and the upper towards the Equator, instead of
what we might more obviously expect—namely, a flow towards
the Pole in the upper regions of the atmosphere, and a return
current towards the Equator in the lower regions close upon

the surface of the sea. He even describes the known prevalent motion of the bottom layers of the atmosphere towards the Pole in extra-tropical latitudes as being seemingly paradoxical as to its reason, and although he offers an argument for abatement of the paradox, that argument on the slightest consideration may readily be seen to be futile.

In 1856—the year following after the publication of Maury's scheme of circulation in his book entitled *The Physical Geography of the Sea*—quite a new theory was put forward by Ferrel in a paper on "The Winds and the Currents of the Ocean," published

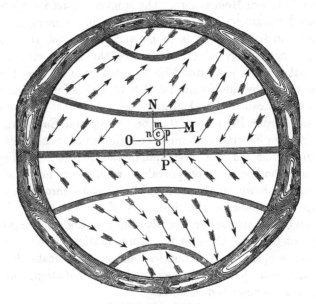

FERREL—1856.

Fig. 3.

in the *Nashville Journal of Medicine and Surgery**. The scheme of circulation which he then proposed and upheld by mathematical reasonings is illustrated in his paper by a diagram, from which fig. 3 here is taken as a copy. This scheme, as may be noticed by reference to the diagram, and as may be further ascertained by reference to the original paper, includes for each hemisphere

---

* October and November, 1856. This essay is to be found reprinted in *Professional Papers of the U.S.A. Signal Service*, No. 12, published by authority of the Secretary of War, Washington, Office of the Chief Signal Officer, 1882.

three zonal rings of atmosphere, making six in all, each having a separate circulation for itself, except that some small amount of commingling would necessarily take place at each narrow annular interface of meeting between two contiguous zonal rings. For either hemisphere one of these zonal rings of atmosphere covers the trade-wind region of that hemisphere, another covers the middle latitudes in which winds prevail from south of west in the northern hemisphere, and north of west in the southern, and the third covers the polar region.

Now, attention for simplicity being confined to the northern hemisphere, explanations of the scheme' may be continued as follows :—

In the trade-wind zonal ring the bottom current flows from the Calms of Cancer as the Trade Wind to the Equatorial Belt and rises there, and flows then in the upper regions of the atmosphere till it comes to a situation aloft nearly over the Calms of Cancer, and thence it descends obliquely to the Calms at bottom to flow again towards the Equator, and so to begin another circuit alike in character to the one now described. Next in the zonal ring of the middle latitudes, according to the scheme, the current of air taken as beginning at the Calms of Cancer advances in the lower regions over the surface of the sea as a wind from south of west till it comes to about the Arctic Circle where it ascends to the upper regions, to begin a return course proceeding southwards as an upper current till it comes to places aloft nearly over the Calms of Cancer, thence to descend to those Calms below, and so to complete its circulation from some part of that belt back again to the same belt. Next as to the supposed circulation in the zonal ring of the Arctic Regions, it may suffice to say briefly that the lower current is asserted to be from the Pole and the upper current towards the Pole, the ascent from the lower to the upper being at or near to the Arctic Circle, and the descent being in a region closely surrounding the Pole, all as may be seen by inspection of the diagram.

Ferrel, in setting forth in his paper his scheme of circulation and his theoretical reasonings on the subject, introduces as a fundamental principle in it the assertion that there must be a heaping up of the top layers of the atmosphere to a maximum height at about the parallel of 28° and a "depression" of them over the Equator, and also a "depression" of them at and around the Poles and in high latitudes generally, and his diagram is purposely drawn to represent these features.

What has now been said is enough to give a good general idea of Ferrel's scheme of Atmospheric Circulation of 1856. His assumptions, his reasonings, and his conclusions are, I may say with confidence, pervaded by impossibilities and incongruities. But notwithstanding this his paper is deserving of credit for the praiseworthy efforts it manifests towards a more complete consideration of important principles bearing on the subject, which had previously been unknown or neglected or imperfectly touched upon by others.

While I have told of this paper by Mr Ferrel at the present stage in order of dates, yet I deem it right to explain here that I had no knowledge of its existence, nor of any of its author's views, until some years after the publication of the new theory by myself, about which I have to tell forthwith in the present paper.

Through a paper* read before the Natural History and Philosophical Society of Belfast in 1856, by Mr Joseph John Murphy, of that town, interest was strongly aroused in my mind, in the question of what ought to be supposed to be the true state of the case as to the courses of atmospheric circulation in the zonal regions situated between the trade-wind zone and the Pole in each hemisphere. In that paper Mr Murphy brought under notice of the Society the scheme of currents of atmospheric circulation set forth by Maury, as the truth; and gave a theory or course of reasoning formed by himself, for explaining on dynamic principles how those supposed motions should be accounted for.

On the subject so presented for consideration, I had to judge that Mr Murphy's course of reasoning was not valid for sustaining Maury's theory of the atmospheric motions, and I had to judge moreover, that Maury's theory was itself, in so far as it dealt with the circulation outside of the trade-wind zones, entirely untenable and impossible.

Mr Murphy's course of reasoning, however, included within it one important element not limited in its scope to the application made of it in that particular course of reasoning. It was the supposition that the low barometric pressure of Polar regions

* On the *Circulation of the Atmosphere*, by Mr Joseph John Murphy, Belfast Natural History and Philosophical Society, 27th February, 1856.

and other high latitudes, already discovered as a fact, through observations of voyagers and others, was to be regarded as due to the centrifugal force of the air revolving from west to east throughout the great cap of atmosphere covering the middle and high latitudes.

Having rejected Maury's theory, and having got the benefit of the valuable suggestion just referred to in Mr Murphy's paper, I succeeded in framing a new theory for the circulation in the regions outside of the trade-wind zones. That new theory I put forward in a paper read by me at the meeting of the British Association, held at Dublin, in the following year, 1857; and a clear account of it is to be found in the Abstract of the paper published in the British Association volume for that year.

The verbal explanations given in the reading of that paper before the meeting were illustrated by a drawing showing the scheme of circulation described in the paper. Fig. 4, here given, is an accurate copy of that drawing, differing from it only in some unimportant matters, such as in the number of arrows shown, and in its being drawn with abatement of some exaggerations which were made in the original in order to render small features more readily visible at a distance in a large room. The full significance of the original in all respects is retained unchanged in the copy here.

In endeavouring to penetrate the mystery as to what the courses of circulation might be in the middle and higher latitudes, I was in preliminary ways fully satisfied that Hadley's theory* in its main features—those, in fact, which in the present paper I have already described with commendation—must be substantially true, and must form the basis of any tenable theory that could be devised.

Now, under Hadley's theory, when we come to consider what may be the courses of circulation that we should attribute to the

---

* Reference having been made in the text here to my paper read at the British Association Meeting for 1857, on the "Grand Currents of Atmospheric Circulation," and to the Abstract of it printed in the volume of the Association for that year, I have to mention as a correction that the theory here described and correctly designated as Hadley's Theory, was in the printed Abstract erroneously named as Halley's Theory. I was led into that mistake as to authorship of the commonly accepted explanation of the trade winds, through my finding it designated as "Halley's theory of the trade winds" by Maury, in his *Physical Geography of the Sea*, to whose newly proposed views in that book my attention was at the time specially applied.

atmosphere in the latitudes outside of the trade-wind zones, we
should naturally be led to expect (as I have pointed out in some
detail in an earlier part of the present paper in describing his
theory) that the great sheet of air floating out from the Equator
in the upper regions of the atmosphere towards either Pole, while
having a motion towards the east also, would gradually cool in

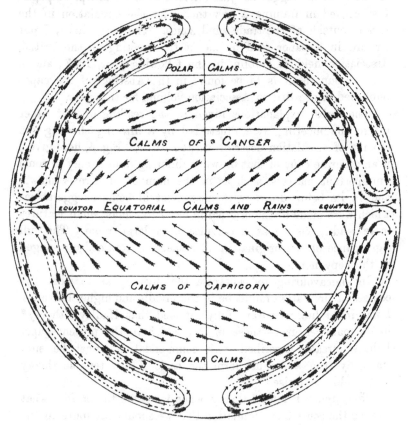

THOMSON—1857.

Fig. 4.

advancing to higher latitudes, and would therefore descend in
middle and high latitudes to the Earth's surface and would next,
as a bottom current, flow back towards the Equator while also
flowing eastward, and so would be a current towards the Equator,
not towards the Pole. But, on the other hand, it had been

brought out through accumulated observational results that the winds of middle latitudes while blowing towards the east, which so far is in agreement with Hadley's theory, do, in opposition to what would be expected under that theory, blow more towards the Pole than from the Pole. Thus the facts and theory seemed to be at variance. It then occurred to me that facts and theory could be reconciled by supposing that the great circulation brought into probability under Hadley's theory does actually occur, but occurs subject to this modification, that a thin stratum of air on the surface of the Earth in the latitudes higher than about 30°— a stratum in which the inhabitants of those latitudes have their existence, and of which the movements constitute the observed winds of those latitudes—being by friction and impulses on the surface of the Earth retarded with reference to the rapid whirl or vortex motion from west to east of the great mass of air above it, tends to flow towards the Pole, and actually does so flow under the indrawing influence of the partial void in the central parts of that vortex, due to the centrifugal force of its revolution. Thus it appeared to me that in temperate latitudes there are three currents at different heights:—That the uppermost moves towards the Pole and is part of a grand primary circulation between Equatorial and Polar Regions;—that the lowermost moves also towards the Pole, but is only a thin stratum forming part of a secondary circulation;—that the middle current moves from the Pole and constitutes the return current for both the preceding;— and that all these three currents have a prevailing motion from west to east in advance of the Earth. This was the substance of the new theory which I framed and which, in 1857, I submitted to the British Association at its Dublin meeting*. The atmospheric currents supposed under this theory are indicated by arrows in the diagram, fig. 4, and may be traced out readily on inspection. This drawing it is to be understood is not intended to offer any indications of supposed variations in height from bottom to top of the atmosphere in different latitudes.

I exhibited at the meeting, as an illustration, a simple experiment easily extemporizable on any occasion. It is mentioned in the printed abstract briefly in the following words:—"If a shallow circular vessel with flat bottom, be filled to a moderate depth with water, and if a few small objects, very little heavier than

---

* [*Supra*, p. 144.]

water, and suitable for indicating to the eye the motions of the water in the bottom, be put in, and if the water be set to revolve by being stirred round, then, on the process of stirring being terminated, and the water being left to itself, the small particles in the bottom will be seen to collect in the centre. They are evidently carried there by a current determined towards the centre along the bottom in consequence of the centrifugal force of the lowest stratum of the water being diminished in reference to the strata above, through a diminution of velocity of rotation in the lowest stratum by friction on the bottom. The particles being heavier than the water, must, in respect of their density, have more centrifugal force than the water immediately in contact with them; and must, therefore, in this respect have a tendency to fly outwards from the centre, but the flow of water towards the centre overcomes this tendency and carries them inwards; and thus is the flow of water towards the centre in the stratum in contact with the bottom palpably manifested."

The general hydraulic principle intended thus to be illustrated by the exhibition of an easily conducted simple case of it is, that if water were lying on a revolving flat-bottomed circular plate or tray, and were revolving at each part quicker than the tray immediately below that part, a flow would institute itself in the bottom layer towards the centre, and that this would occur alike for different speeds of revolution of the tray, and would still take place, likewise, in the case of the speed of revolution of the tray being abated to zero. The case of the non-rotative tray was taken for illustration of the more general proposition simply because of the facility which that particular case presents for being brought into visible manifestation, so as to form to an intelligent mind a help to the imagination in considering the action of the great cap of air lying on the middle and higher latitudes, and revolving prevalently at each part quicker than the Earth below that part does. I offer these explanatory remarks here because in a paper by Mr Ferrel, to be told of a little further on, my illustration by means of the non-revolving tray has been made a point of adverse criticism as to both the nature and the value of the theory I had offered.

Now, before passing quite away from the subject of the original framing of my own theory, I feel it right to make special reference to two considerations which were put forward by Mr Ferrel in his paper of October, 1856.

Firstly.—Ideas were put forward in that paper by Mr Ferrel
to the effect, that the low barometric pressure found observationally
to exist in polar regions and other high latitudes, is due to the
centrifugal force or tendency of the air of the surrounding middle
latitudes revolving from west to east quicker than does the earth
below: but his views on the matter being unknown to Mr Murphy
and to myself, did not happen to influence my considerations.

And *secondly*, Mr Ferrel in that paper adduced in connection
with other suppositions an idea which, taking it in a wider scope
than that in which he applied it, and with congruity in application
not pertaining to the case for which he adduced it, I may describe
as implying considerations to the effect that in an atmosphere
covering a zonal region such as that of the middle latitudes, and
having eastward motion relative to the Earth's surface or, what
is the same, having a speed of eastward revolution quicker than
that of the Earth below it, a layer at bottom retarded by friction
on the Earth's surface, and so having less centrifugal tendency than
has the quicker eastward-going air above will be caused to take,
along with its eastward motion, a motion also towards the Pole.

The principle is an important one in its applicability to
atmospheric circulation; but Mr Ferrel did not apply it to good
account.   He applied it only in reference to a system of motions
already assumed by him, but which in the actual atmosphere
are impossible as to causes for their origin and maintenance, and
are incongruous in their mutual relations.   His purpose in this
matter was to show reason for the bottom current flowing towards
the Pole while he had the upper current assumed as flowing
towards the Equator.   He assumed throughout the whole depth
from bottom to top in his zonal ring of the atmosphere a motion
eastward relative to the Earth, and thereby explained that the
frictionally resisted bottom part should flow towards the Pole.
But now we have to observe that the only reason why under his
theory he can be entitled to assume eastward motion in the lower
portion is because of that portion having been previously assumed
to flow towards the Pole; and as to the upper portion which he
assumes to flow from the Pole, that reason does not hold at all,
and the upper portion should rather be supposed, under his theory,
to flow westward than eastward.   Thus it comes out that he
explained the motion towards the Pole in the lower part of the
atmosphere by first assuming, for no valid reason, a motion towards

the Pole of that lower part. But now, for the primary assumption of that motion towards the Pole in the lower portion of the atmosphere, the reason which he assigned, and which I have just now treated as being not valid, was his supposed heaping up of the atmosphere at top, and consequent increased pressure at bottom at about the parallel of 28°; but, for the heaping up of the atmosphere there he needs in the upper region of the atmosphere over the middle latitudes a speed of revolutional motion greater than that of the Earth's surface immediately below, briefly a relative eastward motion, so that there may be the necessary centrifugal tendency for producing the heaping up, and that is incongruous with the flow in those upper regions taking place, as under his theory he made it do from higher to lower latitudes— from the Arctic Circle to about the parallel of 28°.

He has not thereby anticipated the new and, I think I may say, the true theory offered by me, in which the great body of the lower half of the atmosphere is already shown for good reason to have motion towards the Equator along with motion from west to east, but that a comparatively thin lamina at bottom of it, in virtue of frictional retardation of its eastward motion and consequent abatement of centrifugal tendency in it as compared to the air above, is caused to reverse what would otherwise be its motion towards the Equator, and to take its course towards the Pole instead.

The next publication to which I have to advert is a second paper by Mr Ferrel. It is entitled "The Motions of Fluids and Solids, relative to the Earth's Surface; comprising Applications to the Winds and the Currents of the Ocean*," and is dated at its close, "Cambridge, Mass., February, 1860," and is noted on its title-page as being "Taken from the First and Second Volumes of the *Mathematical Monthly*."

In that paper he offered a scheme of atmospheric circulation totally different from his previous one of 1856, and entailing a fundamentally altered theory. In fact, he there abandoned his arrangement of six zonal vortex rings of circulation, three for the northern hemisphere and three for the southern, and, instead, he adopted really the scheme that had been put forward by me

---

* New York: Ivison, Phinney, and Company. London: Trübner and Co., 1860. It appears that this paper was subsequently republished by the United States Signal Service in *Professional Papers*, No. VIII., with extensive notes giving the mathematical processes in detail by Professor Frank Waldo.

in 1857 with its two great currents of primary circulation, one flowing from equatorial to polar regions above, and the other flowing as a great return current from polar to equatorial regions below; together with the bottom subordinate current close on the surface of the Earth in middle latitudes or middle and higher latitudes, flowing pole-ward on account of the frictional retardation by the Earth's surface of its eastward relative motion and consequent diminution of centrifugal tendency.

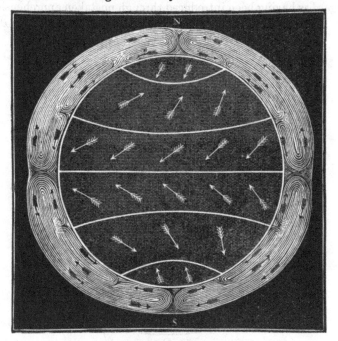

FERREL—1860.

Fig. 5.

These currents are shown distinctly by arrows in his diagram, notwithstanding some puzzling confusion introduced by lines which present the appearance of being meant to indicate average current lines, but which, in some parts, would suggest impossible courses, and which show signs of their having been put in without deliberate care. Fig. 5 here is a copy of that diagram*.

* The same diagram exhibiting his scheme of the winds is repeated in a subsequent paper by Ferrel of date 1861, which he offered as being more popular and less mathematical. It is to be found reprinted in *Professional Papers of the United States Signal Service*, No. XII.

The diagram retains two vestiges of his original scheme. Thus it exhibits distinctly a depression of the top of the atmosphere at the Equator making a place of minimum height for the atmosphere there; and it retains systems of arrows throughout the polar regions representing winds having, relatively to the Earth's surface, motion towards the west together with motion towards the Equator; and so in the polar region of the northern hemisphere representing north-east winds. Both of these features I regard as having been introduced through mistaken apprehension. In a later work, indeed, by Mr Ferrel, entitled, *A Popular Treatise*

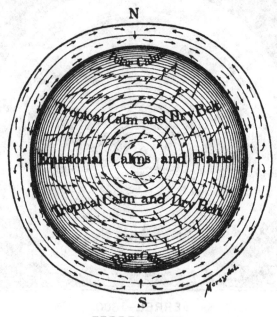

FERREL—1889.

Fig. 6.

on the Winds, 1889*, both these features of his former scheme of circulation are completely eliminated from his scheme and theory as there presented. This is shown by his diagram† taken in connection with the printed explanations by which it is accompanied. Fig. 6 is a copy of this diagram. The depression of the top of the atmosphere, or more strictly speaking, the depression

* London: Macmillan and Co.

† Ferrel's *Popular Treatise on the Winds*, 1889, § 105, p. 155.

of any isobaric interface in the very lofty regions to a minimum height over the Equator remains in the diagram, but it is expressly eliminated by words in the accompanying text. This diagram, when corrected according to Mr Ferrel's printed words is, as may readily be seen, essentially the same as my own.

In the closing passage of his second paper, 1860, Ferrel made mention of the theory given by me at the British Association meeting in Dublin, 1857, but he did this with erroneous representation of the theory, and with inadequate recognition of its importance and of the fundamental changes he had made from his own previous theory in adopting the main features of mine and incorporating them with some remnants of his own previous views or modes of consideration.

I proceed next to offer some considerations which, I think, may be of intrinsic interest in themselves, besides helping towards the development and elucidation of true theory in regard to atmospheric motions and other conditions.

I have to mention, at this stage, that it may sometimes be convenient, as an aid towards brevity and clearness in expression, to characterize air which has no eastward or westward motion relative to the Earth's surface as having *par*, or being *at par* of revolutional velocity and, likewise, to use the designation *over par* of revolutional velocity to signify eastward relative motion, and *under par* to signify westward relative motion.

(*a*) Recalling to notice the theory of Maury and the first theory of Ferrel given in his 1856 paper, and drawing attention to the confluence supposed, under both these theories, of two great upper currents of the atmosphere meeting aloft over the belt called the Calms of Cancer in the Northern Hemisphere, and of other two currents likewise meeting over the belt called the Calms of Capricorn in the Southern, I think it is well to remark that, if such a confluence were to take place of two currents, one coming from higher latitudes and the other from lower to a zonal belt of meeting, the current from the higher latitudes would have a rapid westward motion relatively to the Earth below, that is, a revolutional velocity greatly under par, and the current from lower latitudes would have a rapid relative motion eastward, or, in other words, a revolutional velocity greatly over par. They would meet one another obliquely with a velocity of each relative to the other very great because of its having had no frictional mitigating resistance such as the Earth's surface would afford to currents meeting in like manner at bottom of the atmosphere. Thus the belt of meeting aloft would be a place of extraordinary commotion, and this commotion would be propagated with the two descending currents down to the surface of the Earth below; and thus, instead of the Calms of Cancer or Capricorn we ought to expect to find there a belt of wild and varying storms. This very simple, and, I think, very obvious principle, is one of the numerous objections which might singly or conjointly have checked both Maury and Ferrel in the early inception of their theories, and might reasonably have prevented them from propagating views so fallacious.

(b) Next we may raise questions, and proceed to solve them more or less completely, as to what must be the general character of the motions of the air at various places in the trade-wind zone, both in the lower great current approaching to the Equatorial or Doldrum Belt*, and in the upper great current departing from that Equatorial Belt and flowing aloft over the trade-wind zone to pass over the Border Belt and thence into what, for want of a better name, we may for the present call the middle latitudes. In doing this, we shall have to consider and bring to light some features of the motions of the atmosphere in the middle latitudes more fully in detail than hitherto in the present paper.

Let us accompany in thought the progress of a point advancing with the current along an average stream line, or rather an average current course in the great under-current from Polar to Equatorial Regions. For simplicity, let us confine attention to the Northern Hemisphere. Let us begin the course somewhere within the middle latitudes. To help imagination we may fix on a point of commencement situated vertically over New York. The moving point may, if we please, be idealized as being a small balloon

* This equatorial belt of rising air may also well be called the Medial Belt, while the Calms of Cancer and Capricorn may be referred to as the Border Belts, this last especially when it is wished to speak of either of these two indifferently, without distinction as to whether it be in the northern or southern hemisphere. Also, either of these border belts may very well be described, or may be named when desirable, as the Belt of Offturn Parting or briefly as the Offturn Parting. The reason for this name will be seen readily by inspection of the diagram, where the great return current towards the Equator is parted into two currents, one going on southwards, and the other turning off towards the north.

constrained by frictionless guidance to keep in an average current course while being propelled along that average course by the more or less varying motions of the surrounding air. Now, during the progress of the travelling point in its course making way both eastward and southward, so long as the bottom lamina close to the surface of the Earth directly beneath the travelling point is blowing eastward with over-par revolutional velocity, the air above the bottom lamina there must be going forward with still greater over-par velocity. The reason for this statement is, that the only cause for maintenance of eastward relative motion in the frictionally restrained bottom lamina is, that the air above in virtue of revolutional momentum brought from equatorial regions, and not yet exhausted, is blowing with over-par revolutional velocity, and driving forward the resisted lamina below. Also, as long as the over-par velocity is existing in the frictionally resisted bottom lamina under our travelling point, a flow pole-ward also must exist in that bottom lamina at the place, for the time being, directly below the travelling point. This is for reasons fully explained in the account already given of my own theory. When further, the travelling point, in making its way southward, arrives at a stage where that eastward bottom over-par motion no longer exists directly below, and the travelling point then goes on making progress further south, it comes to places where the bottom lamina at the place then below it is moving equatorward because of indraught thither, and because all reason for that lamina's going northward has ceased. Thereafter in the trade-wind region now entered upon the surface of the Earth is dragging the bottom air forward revolutionally, and so is helping it briskly towards the Equator through increasing its centrifugal tendency.

Then we have to notice that the air, during its course equatorwards and back again through the trade-wind zone, receives forward revolutional momentum through the frictional forward drag applied to it by the Earth's surface, and it loses no revolutional momentum, as the vacuum above the atmosphere can take none from it. So in departing northwards, as the grand upper current, it must carry with it far more revolutional momentum than it had in entering, as the great under current from the north across the Border Belt; but that great under current in entering was either at par, or partly at par, and partly at over-par, of revolutional velocity; consequently the grand upper current must depart across the Border Belt with great over-par of revolutional velocity.

It follows from this, as a corollary, that the top of the atmosphere, or any isobaric interface near the top, must have a declivity in approaching the Border Belt from the top of the Equatorial Belt; and the Border Belt must not have a maximum height with declivity thence to a minimum at the Equator.

The foregoing demonstration seems also likely to give help towards the proper interpretation to be put on observational results recorded by the Krakatoa Committee, and to render highly improbable any suggestions such as seem to be conveyed in some parts of the report, to the effect that in the very lofty regions of the atmosphere—at such elevations as 13 miles above the sea level—a velocity such as 70 miles per hour from east to west has been indicated in the atmosphere, through the phenomena manifested after the great Krakatoa eruption.

(c) In connection with the reasoning or demonstration I have just given, there is another element which I regard as forming part of the whole truth, and which must, I think, form an important element towards the development of the theory more completely. I have already indicated in the demonstration just now offered, that the bottom lamina of the atmosphere in the trade-wind region is especially helped to advance towards the Equator by the increased absolute centrifugal tendency superimposed on it by the forward revolutional drag it there receives from the Earth's surface; and which communicates to it, throughout its course towards the Equator, new accessions of revolutional momentum, and prevents it from getting into under-par of revolutional velocity, so much as does the air above it in the great under current towards the Equator. For ready apprehension of this, it is well to notice that the under-par and increased under-par of revolutional velocity imply westward relative motion in the bottom lamina, and quicker westward relative motion in the air next above within the great under current towards the Equator.

This greater abatement of absolute revolutional velocity below par, or increase of relative velocity westward, constitutes a condition opposing flow towards the Equator in the main body of the great under current, and we may reasonably suppose that the principal flow towards the Equator takes place in the bottom lamina—the lamina whose motions constitute the winds noticeable by action on the sails of ships. So we may suppose that the main body of that great under current blows nearly due westward with only a small component of motion equator-ward. I could not venture, through theoretical considerations alone, to form an opinion as to the velocity of westward relative motion which might thus be attained to in the main body of the great under current, or the velocity of westward relative motion which might remain in some parts of the upper current proceeding from the Equator before it has made much advance in latitude to places importantly nearer the Earth's axis. The complications involved in the frictional conditions attendant on the flow of sheets of air with others below and with others above going at very different velocities render the question practically unsolvable by theory alone. But I have to point out emphatically that the Doldrum air, deadened as it is to the condition commonly spoken of as equatorial calm, is very approximately at par of revolutional velocity, and when it rises to the top, or to the very high regions, of the atmosphere, it will have scarcely any westward relative motion, and therefore will not be able to make its way thence as an upper current pole-ward except by flowing as we may say down hill, or as we may better say, among isobaric interfaces down-sloping forward. The lower part of this sheet of deadened air departing aloft pole-ward, and which lower part is much below the top of the atmosphere, and is in close contiguity with the current of westward relatively moving air (already just now mentioned) commencing to move pole-ward without ever having attained to par of revolutional velocity, will we may suppose, by buffeting and commingling between it and that westward relatively moving air, be dragged forward from the Equator, even among up-sloping isobaric interfaces, in a manner that may be likened to being dragged up hill.

I might at present extend the explanations and reasonings on this matter somewhat further, but I abstain from doing so in order not to prolong unduly the present paper. I prefer to leave the subject over for further consideration and exposition by myself, perhaps, and probably by others.

(d) I propose next to offer some considerations in respect to the atmosphere of the polar regions. For simplicity of expression

I shall speak, in particular, of the polar regions of the Northern Hemisphere; and I intend that in this, as indeed throughout nearly all I have said in the present paper, the complications introduced into atmospheric motions by local distinctions of the Earth's surface into land and sea are to be, primarily at least, disregarded.

I consider that we should take as one element of our theories the principle that we have to suppose a stagnation of impounded air around the Pole over a great extent of the Polar Regions, this impounded air being maintained by the influx along the surface of the Earth of air frictionally deprived of the over-par of revolutional velocity which is possessed by the great cap of air higher up above the surface of the Earth. This impounded air lags, I affirm, in the Polar Regions, being unable, for want of revolutional momentum, and accompanying want of centrifugal inertial tendency, to take part readily in the great circulation between polar and equatorial regions. In fact, it cannot get out from its imprisonment there except by being dragged away through gradual entanglement with the comparatively rapidly revolving air arriving by the great upper current from regions having more rapid revolutional motion and passing away in the great middle return current towards the Equator.

(e) Further, I may now offer some considerations as to whether, according to theory, we should expect very clear skies to prevail in the Polar Region of impounded deadened air. I think we must suppose the great upper atmospheric current converging towards the Pole and having over-par of revolutional velocity must be already very dry, owing to its greatly reduced pressure and cold temperature. So, when its air descends in level to return towards the Equator, that air must, I think, be greatly under its saturation point with water-substance; or, in other words, must be far from ready to form clouds, or to precipitate rain or snow. We have to recollect that descending air is generally very rainless.

On the other hand the bottom flow along the surface of the land and sea converging towards the Pole I affirm to be moist. It will be from lower latitudes and generally warmer climates, and will carry moisture with it from sea and land. This bottom current will supply water-substance for cloud and snow in the impounded deadened polar air. The cold of radiation out to interstellar space, coupled with expansion in ascending before it can join the great middle current of return towards the Equator, will cause clouds and snow.

I will now conclude this paper by offering a sketch of a contemplated experimental apparatus for affording practical illustration of the theory of Atmospheric Circulation which I have propounded.

The apparatus would consist mainly of a horizontal circular tray kept revolving round a vertical axis through its centre. The tray would be filled to some suitable depth with water. Heat would be applied round its circumference at bottom, and cold would be applied or cooling would be allowed to proceed in and around the central part at or near the surface. Under these circumstances I would expect that motions would institute themselves, which would be closely allied to those of the great general currents supposed under the theory to exist in either hemisphere of the Earth's atmosphere. The motions of the water, I would propose, should be rendered perceptible to the eye by dropping in small particles of aniline dye, and perhaps by other contrivances. Great variations would be available in respect to the velocity of rotation given to the tray, and in respect to the depth of water used, and the intensity of the heating and cooling influences applied. By various trials with variations in these respects I think it likely that the phenomena expected could be made manifest.

CONGELATION AND LIQUEFACTION

29. THEORETICAL CONSIDERATIONS ON THE EFFECT OF PRESSURE IN LOWERING THE FREEZING POINT OF WATER.

[From *Cambridge and Dublin Mathematical Journal*, November 1850; taken, with some slight alterations made by the author, from the *Transactions of the Royal Society of Edinburgh* (Jan. 2, 1849), Vol. XVI. Part 5, 1849, p. 575.

This fundamental paper and the one next succeeding were reprinted in Sir William Thomson's (Lord Kelvin's) *Mathematical and Physical Papers*, Vol. I. pp. 156–164 and 165–169.]

SOME time ago my brother, Professor William Thomson, pointed out to me a curious conclusion to which he had been led, by reasoning on principles similar to those developed by Carnot, with reference to the motive power of heat. It was, that *water at the freezing point may be converted into ice by a process solely mechanical, and yet without the final expenditure of any mechanical work.* This at first appeared to me to involve an impossibility, because water expands while freezing; and therefore it seemed to follow, that if a quantity of it were merely enclosed in a vessel with a moveable piston and frozen, the motion of the piston, consequent on the expansion, being resisted by pressure, mechanical work would be given out without any corresponding expenditure; or, in other words, a perpetual source of mechanical work, commonly called a perpetual motion, would be possible. After farther consideration, however, the former conclusion appeared to be incontrovertible; but then, to avoid the absurdity of supposing that mechanical work could be got out of nothing, it occurred to me that it is necessary farther to conclude, that *the freezing point becomes lower as the pressure to which the water is subjected is increased.*

The following is the reasoning by which these conclusions are proved.

First, to prove that water at the freezing point may be converted into ice by a process solely mechanical, and yet without the final expenditure of any mechanical work:—Let there be supposed to be a cylinder, and a piston fitting water-tight to it, and capable of moving without friction. Let these be supposed to be formed of a substance which is a perfect non-conductor of heat; also, let the bottom of the cylinder be closed by a plate, supposed to be a perfect conductor, and to possess no capacity for heat. Now, to convert a given mass of water into ice without the expenditure of mechanical work, let this imaginary vessel be partly filled with air at 0° C.*, and let the bottom of it be placed in contact with an indefinite mass of water, a lake for instance, at the same temperature. Now, let the piston be pushed towards the bottom of the cylinder by pressure from some external reservoir of mechanical work, which, for the sake of fixing our ideas, we may suppose to be the hand of an operator. During this process the air in the cylinder would tend to become heated on account of the compression, but it is constrained to remain at 0° by being in communication with the lake at that temperature. The change, then, which takes place is, that a certain amount of work is given from the hand to the air, and a certain amount of heat is given from the air to the water of the lake. In the next place, let the bottom of the cylinder be placed in contact with the mass of water at 0°, which is proposed to be converted into ice, and let the piston be allowed to move back to the position it had at the commencement of the first process. During this second process, the temperature of the air would tend to sink on account of the expansion, but it is constrained to remain constant at 0° by the air being in communication with the freezing water, which cannot change its temperature so long as any of it remains unfrozen. Hence, so far as the air and the hand are concerned, this process has been exactly the converse of the former one. Thus the air has expanded through the same distance through which it was formerly compressed; and since it has been constantly at the same temperature during both processes, the law of the variation of its pressure with its volume must have been the same in both. From this it follows, that the hand has received back exactly the same amount of mechanical work in the second process as it gave out in the first. By an analogous reason it is easily shown that the air also has received again exactly the same amount of heat as it gave out during its compression; and, hence, it is now left in a condition the same as that in which it was at

* The centigrade thermometric scale is adopted throughout this paper.

the commencement of the first process. *The only change which has been produced then, is that a certain quantity of heat has been abstracted from a small mass of water at 0°, and dispersed through an indefinite mass at the same temperature, the small mass having thus been converted into ice.* This conclusion, it may be remarked, might be deduced at once by the application, to the freezing of water, of the general principle developed by Carnot, that no work is given out when heat passes from one body to another without a fall of temperature; or rather by the application of the converse of this, which of course equally holds good, namely that no work requires to be expended to make heat pass from one body to another at the same temperature.

Next, to prove that the freezing point of water is lowered by an increase of the pressure to which the water is subjected:—Let the imaginary cylinder and piston employed in the foregoing demonstration, be again supposed to contain some air at 0°. Let the bottom of the cylinder be placed in contact with the water of an indefinitely large lake at 0°; and let the air be subjected to compression by pressure applied by the hand to the piston. A certain amount of work is thus given from the hand to the air, and a certain amount of heat is given out from the air to the lake. Next, let the bottom of the cylinder be placed in communication with a small quantity of water at 0°, enclosed in a second imaginary cylinder similar in character to the first, and which we may call the water cylinder, the first being called the air cylinder; and let this water be, at the commencement, subject merely to the atmospheric pressure. Let, however, resistance be offered by the hand to any motion of the piston of the water cylinder which may take place. Things being in this state, let the piston of the air cylinder move back to its original position. During this process, heat becomes latent in the air on account of the increase of volume, and therefore the air abstracts heat from the water, because the air and water, being in communication with one another, must remain each at the same temperature as the other, whether that temperature changes or not. The first effect of the abstraction of heat from the water must be the conversion of a part of the water into ice, an effect which must be accompanied with an increase of volume of the mass enclosed in the water cylinder. Hence, on account of the resistance offered by the hand to the motion of the piston of this cylinder, the internal pressure is increased, and work is received by the hand from the piston. Towards the end of this process, let the resistance offered by the hand gradually decrease, till, just at the end it becomes nothing, and the pressure within the water cylinder thus becomes again equal to that of the atmosphere. The temperature of the mass of partly frozen water must now be 0°, and the air in the other cylinder, being in communication with this, must have the same temperature. The air is therefore at its original temperature, and it has its original volume, or, in other words, it is in its original state. Farther, let the ice be converted, under atmospheric pressure, into water; the requisite heat being transferred to it from the lake by the mechanical process already pointed out, which involves no loss of mechanical work. Thus, now at the conclusion of the operation, the whole mass of water is left in its original state; and likewise, as has already been shown, the air is left in its original state. Hence no work can have been developed by any change on the air and water, which have been used. But work has been given out by the piston of the water cylinder to the hand; and therefore an equal quantity* of work must have been given from the hand to the air piston, as there is no other way in which the work developed could have been introduced into the apparatus. Now, the only way in which this can have taken place is by the air having been colder, while it was expanding in the second process, than it was while it was undergoing compression during the first. Hence it was colder than 0° during the course of the second process; or, in other words, *while the water was freezing, under a pressure greater than that of the atmosphere, its temperature was lower than* 0°.

The fact of the lowering of the freezing point being thus demonstrated, it becomes desirable, in the next place, to find what is the freezing point of water for any given pressure. The most obvious way to determine this would be by direct experiment with freezing water. I have not, however, made any attempt to do so in this way. The variation to be appreciated is extremely small, so small in fact as to afford sufficient reason for its existence never having been observed by any experimenter. Even to detect its existence, much more to arrive at its exact amount by direct experiment, would require very delicate apparatus which

---

* In saying " an equal quantity" I, of course, neglect infinitely small quantities in comparison to quantities not infinitely small.

would not be easily planned out or procured. Another, and a better mode of proceeding has, however, occurred to me : and by it we can deduce, from the known expansion of water in freezing, and the known quantity of heat which becomes latent in the melting of ice, together with data founded on the experiments of Regnault on steam at the freezing point, a formula which gives the freezing point in terms of the pressure ; and which may be applied for any pressure, from nothing up to many atmospheres. The following is the investigation of this formula.

Let us suppose that we have a cylinder of the imaginary construction described at the commencement of this paper ; and let us use it as an ice-engine analogous to the imaginary steam-engine conceived by Carnot, and employed in his investigations. For this purpose, let the entire space enclosed within the cylinder by the piston be filled at first with as much ice at 0° as would, if melted, form rather more than a cubic foot of water, and let the ice be subject merely to one atmosphere of pressure, no force being applied to the piston. Now, let the following four processes, forming one complete stroke of the ice-engine, be performed.

PROCESS 1. Place the bottom of the cylinder in contact with an indefinite lake of water at 0°, and push down the piston. The effect of the motion of the piston is to convert ice at 0° into water at 0°, and to abstract from the lake at 0° the heat which becomes latent during this change. Continue the compression till one cubic foot of water is melted from ice.

PROCESS 2. Remove the cylinder from the lake, and place it with its bottom on a stand which is a perfect non-conductor of heat. Push the piston a very little farther down, till the pressure inside is increased by any desired quantity which may be denoted, in pounds on the square foot, by p. During this motion of the piston, since the cylinder contains ice and water, the temperature of the mixture must vary with the pressure, being at any instant the freezing point which corresponds to the pressure at that instant. Let the temperature at the end of this process be denoted by $-t°$ C.

PROCESS 3. Place the bottom of the cylinder in contact with a second indefinitely large lake at $-t°$, and move the piston upwards. During this motion the pressure must remain constant at p above that of the atmosphere, the water in the cylinder increasing its volume by freezing, since if it did not freeze, its pressure would diminish, and therefore its temperature would increase, which is impossible, since the whole mass of water and ice is constrained by the lake to remain at $-t°$. Continue the motion till so much heat has been given out to the second lake at $-t°$, as that if the whole mass contained in the cylinder were allowed to return to its original volume without any introduction or abstraction of heat, it would assume its original temperature and pressure. This, if Carnot's principles be admitted, as they are supposed to be throughout the present investigation, is the same as to say,—Continue the motion till all the heat has been given out to the second lake at $-t°$, which was taken in during Process 2, from the first lake at $0°$ *.

PROCESS 4. Remove the cylinder from the lake at $-t°$, and place its bottom again on the non-conducting stand. Move the piston back to the position it occupied at the commencement of Process 1. At the end of this fourth process the mass contained in the cylinder must, according to the condition by which the termination of Process 3 was fixed, have its original temperature and pressure, and therefore it must be in every respect in its original physical state.

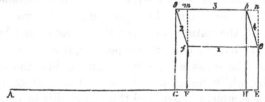

By representing graphically in a diagram the various volumes and corresponding pressures, at all the stages of the four processes which have just been described, we shall arrive, in a simple and easy manner, at the quantity of work which is developed in one complete stroke by the heat which is transferred during that stroke from the lake at $0°$ to the lake at $-t°$. For this purpose,

* [This is reprinted from *Camb. and Dub. Math. Journal*, Nov. 1850. In the original paper, *Trans. R.S. Edin.*, Jan. 1849, the last two sentences did not appear except as regards the last three lines beginning with "continue"; and the following footnote appeared.

"This step, as well as the corresponding one in Carnot's investigation, it must be observed, involves difficult questions, which cannot as yet be satisfactorily answered, regarding the possibility of the absolute formation or destruction of heat as an equivalent for the destruction or formation of other agencies, such as mechanical work; but, in taking it, I go on the almost universally adopted supposition of the perfect conservation of heat."]

let E be the position of the piston at the beginning of Process 1 ; and let some distance, such as EG, represent its stroke in feet, its area being made a square foot, so that the numbers expressing, in feet, distances along EG may also express, in cubic feet, the changes in the contents of the cylinder produced by the motion of the piston.   Now, when 1·087 cubic feet of ice are melted, one cubic foot of water is formed.   Hence, if EF be taken equal to ·087 feet, F will be the position of the piston when one cubic foot of water has been melted from ice, that is, the position at the end of Process 1, the bottom of the cylinder being at a point A distant from F by rather more than a foot.   Let FG be the compression during Process 2, and HE the expansion during Process 4.   Let ef be parallel to EF, and let Ee represent one atmosphere of pressure ; that is, let the units of length for the vertical ordinates be taken such that the number of them in Ee may be equal to the number which expresses an atmosphere of pressure.   Also let gh be parallel to EF, and let fm represent the increase of pressure produced during Process 2.

Then the straight lines ef and gh will be the lines of pressure for Processes 1 and 3 ; and for the other two processes, the lines of pressure will be some curves which would extremely nearly coincide with the straight lines fg and he.   For want of experimental data, the natures of these two curves cannot be precisely determined ; but, for our present purpose, it is not necessary that they should be so, as we merely require to find the area of the figure $efgh$, which represents the work developed by the engine during one complete stroke, and this can readily be obtained with sufficient accuracy.   For, even though we should adopt a very large value for fm, the change of pressure during Process 2, still the changes of volume gm and hn in Processes 2 and 4 would be extremely small compared to the expansion during the freezing of the water ; and from this it follows evidently that the area of the figure $efgh$ is extremely nearly equal to that of the rectangle $efmn$, but fe is equal to FE which is ·087 feet.   Hence the work developed during an entire stroke is ·087 $\times p$ foot-pounds.   Now this is developed by the descent from 0° to $-t°$ of the quantity of heat necessary to melt a cubic foot of ice ; that is, by 4925 thermic units, the unit being the quantity of heat required to raise a pound of water from 0° to 1° centigrade.   Next we can obtain another expression for the same quantity of work ; for, by the tables deduced in the preceding paper* from the experiments of Regnault, we find that the quantity of work developed by one of the same thermic units descending through one degree about the freezing point, is 4·97 foot-pounds. Hence, the work due to 4925 thermic units descending from 0° to $-t°$ is $4925 \times 4·97 \times t$ foot-pounds. Putting this equal to the expression which was formerly obtained for the work due to the same quantity of heat falling through the same number of degrees, we obtain

$$4925 \times 4 \cdot 97 \times t = ·087 \times p.$$

Hence $\qquad t = ·00000355\, p$ ....(1).

This, then, is the desired formula for giving the freezing point $-t°$ centigrade, which corresponds to a pressure exceeding that of the atmosphere by a quantity p, estimated in pounds on a square foot.

To put this result in another form, let us suppose water to be subjected to one additional atmosphere, and let it be required to find the freezing point. Here $p =$ one atmosphere $= 2120$ pounds on a square foot; and therefore, by (1),

$$t = ·00000355 \times 2120, \text{ or } t = ·0075.$$

That is, the freezing point of water, under the pressure of one additional atmosphere, is $-·0075°$ centigrade; and hence, if the pressure above one atmosphere be now denoted in atmospheres†, as units, by n, we obtain t, the lowering of the freezing point in degrees centigrade, by the following formula,

$$t = ·0075\, n \text{ ....(2).}$$

(The phenomena predicted by the author, in anticipation of any direct observations on the freezing point of water, have been fully confirmed by experiment. See a short paper published in the *Proceedings of the Royal Society of Edinburgh* (Jan. 1850), and republished in the *Philosophical Magazine* for August 1850, under the title "The Effect of Pressure in Lowering the Freezing Point of Water experimentally demonstrated. By Prof. William Thomson." [Reprinted *infra*, No. 30.])

* [An Account of Carnot's *Theory of the Motive Power of Heat*; with Numerical Results deduced from Regnault's *Experiments on Steam*. By William Thomson. From the *Transactions of the Royal Society of Edinburgh*, XVI. 1849, p. 541. Read Jan. 2, 1849. Reprinted in Sir William Thomson's *Mathematical and Physical Papers*, Vol. I. pp. 113—155, Art. XLI.]

† The atmosphere is here taken as being the pressure of a column of mercury of 760 millimetres; that is 29·92, or very nearly 30 English inches.

30. THE EFFECT OF PRESSURE IN LOWERING THE FREEZING
POINT OF WATER EXPERIMENTALLY DEMONSTRATED. By
Professor WILLIAM THOMSON.

[From the *Proc. R. S. E.* Jan. 1850; *Phil. Mag.* XXXVII. 1850; *Annal. de
Chimie,* XXXV. 1852; *Journ. de Pharm.* XVIII. 1850; *Poggend. Annal.*
LXXXI. 1850. Reprinted here from Sir William Thomson's (Lord
Kelvin's) *Mathematical and Physical Papers,* Vol. I. p. 165, Art. XLV.]

ON the 2nd of January 1849, a communication entitled
"Theoretical Considerations on the Effect of Pressure in Lowering
the Freezing Point of Water, by James Thomson, Esq., of
Glasgow," was laid before the Royal Society, and it has since been
published in the *Transactions,* Vol. XVI. part 5*.   In that paper
it was demonstrated that, if the fundamental axiom of Carnot's
Theory of the Motive Power of Heat be admitted, it follows, as a
rigorous consequence, that the temperature at which ice melts
will be lowered by the application of pressure; and the extent of
this effect due to a given amount of pressure was deduced by a
reasoning analogous to that of Carnot from Regnault's experimental determination of the latent heat, and the pressure of
saturated aqueous vapour at various temperatures differing very
little from the ordinary freezing point of water.   Reducing to
Fahrenheit's scale the final result of the paper, we find

$$t = n \times 0.0135;$$

where "t" denotes the depression in the temperature of melting
ice produced by the addition of n "atmospheres" (or n times
the pressure due to 29·922 inches of mercury) to the ordinary
pressure experienced from the atmosphere.

In this very remarkable speculation, an entirely novel physical
phaenomenon was *predicted* in anticipation of any direct experiments on the subject; and the actual observation of the phaenomenon was pointed out as a highly interesting object for
experimental research.

To test the phaenomenon by experiment without applying
excessively great pressure, a very sensitive thermometer would be
required, since for ten atmospheres the effect expected is little

---

* It will appear also, with some slight alterations made by the author, in the
Cambridge and Dublin Mathematical Journal, November 1850. [See *supra,* p. 196.]

more than the tenth part of a Fahrenheit degree; and the thermometer employed, if founded on the expansion of a liquid in a glass bulb and tube, must be protected from the pressure of the liquid, which, if acting on it, would produce a deformation, or at least a compression of the glass that would materially affect the indications. For a thermometer of extreme sensibility, mercury does not appear to be a convenient liquid; since, if a very fine tube be employed, there is some uncertainty in the indications on account of the irregularity of capillary action, due probably to superficial impurities, and observable even when the best mercury that can be prepared is made use of; and again, if a very large bulb be employed, the weight of the mercury causes a deformation which will produce a very marked difference in the position of the head of the column in the tube according to the manner in which the glass is supported, and may therefore affect with uncertainty the indications of the instrument. The former objection does not apply to the use of any fluid which perfectly wets the glass; and the last-mentioned source of uncertainty will be much less for any lighter liquid than mercury, of equal or greater expansibility by heat. Now the coefficient of expansion of sulphuric aether at $0°$ C. being, according to Mr I. Pierre[*], $\cdot00151$, is eight or nine times that of mercury (which is $\cdot000179$, according to Regnault), and its density is about the twentieth part of the density of mercury. Hence a thermometer of much higher sensibility may be constructed with aether than with mercury, without experiencing inconvenience from the circumstances which have been alluded to. An aether thermometer was accordingly constructed by Mr Robert Mansell of Glasgow, for the experiment which I proposed to make. The bulb of this instrument is nearly cylindrical, and is about $3\frac{1}{2}$ inches long and 3/8ths of an inch in diameter. The tube has a cylindrical bore about $6\frac{1}{2}$ inches long: about $5\frac{1}{2}$ inches of the tube are divided into 220 equal parts. The thermometer is entirely inclosed, and hermetically sealed in a glass tube, which is just large enough to admit it freely[†]. On

---

[*] See Dixon, *On Heat*, p. 72.

[†] Following a suggestion made to me by Professor Forbes of Edinburgh, I have in subsequent experiments with this thermometer used it with enough of mercury introduced into the tube in which it is hermetically sealed to entirely cover its bulb; as I found that, without this, if the experiment was conducted in a warm room, the indications of the thermometer were frequently deranged by the portion of the water which was left free from ice becoming slightly elevated in temperature.

comparing the indications of this instrument with those of a thermometer of Crichton's with an ivory scale, which has divisions corresponding to degrees Fahrenheit of about 1/25th of an inch each, I found that the range of the aether thermometer is about 3° Fahrenheit; and that there are about 212 divisions on the tube corresponding to the interval of temperature from 31° to 34°, as nearly as I could discover from such an unsatisfactory standard of reference. This gives 1/71 of a degree for the mean value of a division. From a rough calibration of the tube which was made, I am convinced that the values of the divisions at no part of the tube differ by more than 1/30th of this amount from the true mean value; and, taking into account all the sources of uncertainty, I think it probable that each of the divisions on the tube of the aether thermometer corresponds to something between 1/68 and 1/75 of a degree Fahrenheit.

With this thermometer in its glass envelope, and with a strong glass cylinder (Œrsted's apparatus for the compression of water), an experiment was made in the following manner:

The compression vessel was partly filled with pieces of clean ice and water: a glass tube about a foot long and 1/10th of an inch internal diameter, closed at one end, was inserted with its open end downwards, to indicate the fluid pressure by the compression of the air which it contained; and the aether thermometer was let down and allowed to rest with the lower end of its glass envelope pressing on the bottom of the vessel. A lead ring was let down so as to keep free from ice the water in the compression cylinder round that part of the thermometer tube where readings were expected. More ice was added above; so that both above and below the clear space, which was only about two inches deep, the compression cylinder was full of pieces of ice. Water was then poured in by a tube with a stopcock fitted in the neck of the vessel, till the vessel was full up to the piston, after which the stopcock was shut.

After it was observed that the column of aether in the thermometer stood at about 67°, with reference to the divisions on the tube, a pressure of from 12 to 15 atmospheres was applied, by forcing the piston down with the screw. Immediately the column of aether descended very rapidly, and in a very few minutes it was below 61°. The pressure was then suddenly removed, and immediately the column in the thermometer began to rise rapidly. Several times pressure was again suddenly applied, and again suddenly removed, and the effects upon the thermometer were most marked.

The fact that the freezing point of water is sensibly lowered by a few atmospheres of pressure was thus established beyond all doubt. After that I attempted, in a more deliberate experiment, to determine as accurately as my means of observation allowed me to do, the actual extent to which the temperature of freezing is affected by determinate applications of pressure.

In the present communication I shall merely mention the results obtained, without entering at all upon the details of the experiment.

I found that a pressure of, as nearly as I have been able to estimate it, 8·1 atmospheres produced a depression measured by 7½ divisions of the tube on the column of aether in the thermometer; and again, a pressure of 16·8 atmospheres produced a thermometric depression of 16½ divisions. Hence the observed lowering of temperature was 7½/71, or ·106° F. in the former case, and 16½/71, or ·232° F. in the latter.

Let us compare these results with theory. According to the conclusions arrived at by my brother in the paper referred to above, the lowering of the freezing point of water by 8·1 atmospheres of pressure would be 8·1 × ·0135, or ·109° F., and the lowering of the freezing point by 16·8 atmospheres would be 16·8 × ·0135, or ·227° F. Hence we have the following highly satisfactory comparison, for the two cases, between the experiment and theory:

Observed pressures	Observed depressions of temperatures	Depressions according to theory, on the hypothesis that the pressures were truly observed	Differences
8·1 atmospheres	·106° F.	·109° F.	− ·003° F.
16·8 atmospheres	·232° F.	·227° F.	+ ·005° F.

It was, I confess, with some surprise, that, after having completed the observations under an impression that they presented great discrepancies from the theoretical expectations, I found the numbers I had noted down indicated in reality an agreement so remarkably close, that I could not but attribute it in some degree to chance, when I reflected on the very rude manner in which the quantitative parts of the experiment (especially the measurement of the pressure, and the evaluation of the division of the aether thermometer) had been conducted.

I hope before long to have a thermometer constructed, which shall be at least three times as sensitive as the aether thermometer I have used hitherto; and I expect with it to be able to perceive the effect of increasing or diminishing the pressure by less than an atmosphere, in lowering or elevating the freezing point of water.

If a convenient *minimum* thermometer could be constructed, the effects of very great pressures might easily be tested by hermetically sealing the thermometer in a strong glass, or in a metal tube, and putting it into a mixture of ice and water, in a strong metal vessel, in which an enormous pressure might be produced by the forcing-pump of a Bramah's press.

In conclusion, it may be remarked, that the same theory which pointed out the remarkable effect of pressure on the freezing point of water, now established by experiment, indicates that a corresponding effect may be expected for all liquids which expand in freezing; that a reverse effect, or an elevation of the freezing point by an increase of pressure, may be expected for all liquids which contract in freezing; and that the extent of the effect to be expected may in every case be deduced from Regnault's observations on vapour (provided that the freezing point is within the temperature-limits of his observations), if the latent heat of a cubic foot of the liquid, and the alteration of its volume in freezing be known.

31. ON THE PLASTICITY OF ICE, AS MANIFESTED IN GLACIERS.

[From the *Proceedings of the Royal Society*, Vol. VIII. 1856—7, p. 455.

Received April 1st, 1857, read May 7th, 1857.]

THE object of this communication is to lay before the Royal Society a theory which I have to propose for explaining the plasticity of ice at the freezing point, which is shown by observations by Professor James Forbes, and which is the principle of his Theory of Glaciers.

This speculation occurred to me mainly in or about the year 1848. I was led to it from a previous theoretical deduction at which I had arrived, namely, that the freezing point of water, or the melting point of ice, must vary with the pressure to which the water or the ice is subjected, the temperature of freezing or melting being lowered as the pressure is increased. My theory on that subject is to be found in a paper by me, entitled "Theoretical Considerations on the Effect of Pressure in Lowering the Freezing Point of Water," published in the *Transactions of the Royal Society of Edinburgh*, Vol. XVI. part 5, 1849 *. It is there inferred that the lowering of the freezing point, for one additional atmosphere of pressure, must be ·0075° centigrade; and that if the pressure above one atmosphere be denoted in atmospheres as units by n, the lowering of the freezing point, denoted in degrees centigrade by t, will be expressed by the formula $t = ·0075\,n$.

The phenomena which I there predicted, in anticipation of direct observations, were afterwards fully established by experiments made by my brother, Professor William Thomson, and described in a paper by him, published in the *Proceedings of the Royal Society of Edinburgh* (Jan. 1850) under the title "The Effect of Pressure in Lowering the Freezing Point of Water Experimentally Demonstrated†."

The principle of the lowering of the freezing point by pressure being laid down as a basis, I now proceed to offer my explanation, derived from it, of the plasticity of ice at the freezing point as follows :—

If to a mass of ice at 0° centigrade, which may be supposed at the outset to be slightly porous, and to contain small quantities of liquid water diffused through its substance, forces tending to change its form be applied, whatever portions of it may thereby be subjected to compression will instantly have their melting point lowered so as to be below their existing temperature of 0° cent. Melting of those portions will therefore set in throughout their substance, and this will be accompanied by a fall of temperature in them on account of the cold evolved in the liquefaction. The liquefied portions being subjected to squeezing of the compressed mass in which they originate, will spread themselves out through the pores of the general mass, by dispersion from the regions of greatest to those of least fluid pressure. Thus the fluid pressure

---

* [*Supra*, p. 196.]   † [*Supra*, p. 204.]

is relieved in those portions in which the compression and liquefaction of the ice had set in, accompanied by the lowering of temperature. On the removal of this cause of liquidity—the fluid pressure, namely—the cold which had been evolved in the compressed parts of the ice and water, freezes the water again in new positions, and thus a change of form, or plastic yielding of the mass of ice to the applied pressures, has occurred. The newly formed ice is at first free from the stress of the applied forces, but the yielding of one part always leaves some other part exposed to the pressure, and that, in its turn, acts in like manner; and on the whole, a continual succession goes on of pressures being applied to particular parts—liquefaction in those parts—dispersion of the water so produced in such directions as will relieve its pressure,—and recongelation, by the cold previously evolved, of the water on its being relieved from this pressure. Thus the parts recongealed after having been melted must, in their turn, through the yielding of other parts, receive pressures from the applied forces, thereby to be again liquefied, and to enter again on a similar cycle of operations. The succession of these processes must continue as long as the external forces tending to change of form remain applied to the mass of porous ice permeated by minute quantities of water.

Postscript received 22nd April, 1857.

It will be observed that in the course of the foregoing communication, I have supposed the ice under consideration to be porous, and to contain small quantities of liquid water diffused through its substance. Porosity and permeation by liquid water are generally understood, from the results of observations, and from numerous other reasons, to be normal conditions of glacier ice. It is not, however, necessary for the purposes of my explanation of the plasticity of ice at the freezing point, that the ice should be at the outset in this condition; for, even if we commence with the consideration of a mass of ice perfectly free from porosity, and free from particles of liquid water diffused through its substance, and if we suppose it to be kept in an atmosphere at or above 0° centigrade, then, as soon as pressure is applied to it, pores occupied by liquid water must instantly be formed in the compressed parts in accordance with the fundamental principle of the explanation which I have propounded—the lowering, namely, of the freezing or melting point by pressure, and the fact that ice cannot exist at 0° cent. under a pressure exceeding that of the atmosphere. I would also wish to make it distinctly understood that no part of the ice, even if supposed at the outset to be solid or free from porosity, can resist being permeated by the water squeezed against it from such parts as may be directly subjected to the pressure, because the very fact of that water being forced against any portions of the ice supposed to be solid will instantly subject them to pressure, and so will cause melting to set in throughout their substance, thereby reducing them immediately to the porous condition.

Thus it is a matter of indifference as to whether we commence with the supposition of a mass of porous or of solid ice.

[The following paragraph was added to the paper as read in August 1857 at the Dublin meeting of the British Association. See *Report*, Part II. p. 40.]

Mr Thomson then referred to an experiment made by Prof. Christie, late Secretary to the Royal Society, showing the plasticity of ice in small hand specimens, and also to more recent experiments by Prof. Tyndall to the same effect, and very interesting on account of the striking way in which they exhibit the phenomena. He also stated that another very important quality of ice was brought forward by Faraday in 1850 (see *Athenaeum*, No. 1181). It was that two pieces of moist ice will consolidate into one on being laid in contact with one another, even in hot weather. The theory he had just propounded, he said, afforded a clear explanation of this fact as follows:—The two pieces of ice, on being pressed together at their point of contact, will, at that place, in virtue of the pressure, be in part liquefied and reduced in temperature, and the cold evolved in their liquefaction will cause some of the liquid film intervening between the two masses to freeze. It is thus evident, he added, that by continued pressure fragmentary masses of ice may be moulded into a continuous mass ; and a sufficient reason is afforded for the reunion, found to occur in glaciers, of the fragments resulting from an ice cascade, and for the mending of the crevasses or deep fissures which result occasionally from their motion along their uneven beds.

32. CORRESPONDENCE WITH PROF. FARADAY ON REGELATION*.
(*Hitherto unpublished.*)

To Professor Faraday.

> 6, FRANKLIN PLACE,
> BELFAST, 30*th July* 1858.

SIR,

At the request of my brother Professor William Thomson of Glasgow College, I take the liberty of sending to you the enclosed abstract of a Paper "On the Effect of Pressure in Lowering the Freezing Point of Water and on the Plasticity of Ice†."

The theory explained in this paper, as you will observe by a clause near the conclusion of the abstract‡, I consider affords a satisfactory explanation of the interesting property of ice to which you directed attention, that separate masses of ice laid in contact with one another will, even in hot weather, unite or freeze firmly together.

> I am Sir Your obedient servant
> JAMES THOMSON.

To Professor James Thomson.

> ROYAL INSTITUTION,
> 4 *August* 1858.

MY DEAR SIR,

I receive it as very kind of you that you should send me the Belfast report of your paper on Ice etc. I knew of, and have been very much interested with, your views respecting ice, and the effect generally of pressure on the fusing point. Prof. W. Thomson told me of them long ago.

All the reasoning in the report I accept as truth; though I may hesitate in supposing that it contains the *whole* of what concerns the assumption of the solid or the liquid state by a particle of water under given circumstances.

---

* [As regards this and some of the following sections reference should be made to Faraday's *Experimental Researches in Chemistry and Physics*, 1859, as follows:—

"On Certain Conditions of Freezing Water," a Royal Institution discourse of 1850, pp. 372—4; "On Ice of Irregular Fusibility," a letter to Professor Tyndall, Dec. 9, 1857, pp. 374—7; and a note "On Regelation," pp. 377—82 written for the reprint in that volume. All are referred to in this correspondence.]

† [Read before the Belfast Natural History and Philosophical Society, 2nd Dec., 1857. It comprised the substance of papers Nos. 29, 31, *supra*, pp. 196 and 208.]

‡ [*Supra*, p. 211.]

It is curious and interesting to observe how much the general question has drawn attention. Forbes is thinking about it—so is Tyndall, and others also in Paris. The peculiar supposition of a *stickiness* of the ice, at the freezing or solidifying point, is interesting; but one wants some better proof than, or additional proof besides, the fact of regelation; I have not worked at the subject of late; but I could not make ice stick to gold or metal at 32°, and I do not think that Forbes' shilling is any proof of it.

Being sure that your principle is correct, which requires pressure and its variations;—and admitting stickiness, simply because I am not prepared to deny it;—I am at present strongly of opinion that there is another efficient cause of regelation to which I think I have referred in the old *Athenaeum* report:—but I have not that here, and therefore cannot clearly say. It seems to me that a particle of water, touching ice on one side and water at the same temperature on the other, is not so apt to change its state for that of ice as another touching ice on both sides;—and this, not as the consequence of any very limited peculiarity in water and ice, but in subordination to a far more general law, if I may so call it, that in bodies of the same kind the particles tend to retain the state of those which surround them. Thus water may be cooled many degrees below 32° F. but a particle of the water in the midst of the cooled mass remains *as water* though colder than ice:—and yet a warmer body than itself, as for instance a spicule of ice, touching it on one side and so breaking up the continuous liquid contact around it, instantly makes it solidify.

In the same manner I can conceive that a particle of ice in the middle of ice may be raised with the mass to a *higher* temperature, before it become water, than a particle of the same ice at the surface can,—this change from ice to water and water to ice being independent of any effect of *pressure*.

We have an illustration of this effect in water also, when it changes from the liquid to the gaseous state, instead of from the liquid to the solid state. It is given us in Donny's beautiful results; where he shows that water, freed from air, may be heated to I think 300° F. under the pressure of one atmosphere only, and yet not boil or be converted into vapour within, though the introduction of the minutest bubble of air will cause it to explode:— a given liquid particle in the mass not being able at this high temperature to change its state into that of vapour whilst in contact on every side with liquid particles like itself, though if once the contact of these particles be broken in any point, they will burst with violence into the new state.

Great numbers of other cases of this kind may be found amongst bodies able to change their state; but they will readily occur to you. I am about to reprint my report from the *Athenaeum* in a volume of Experimental Researches on Physics; I think I shall arrange these views into some kind of form so as to give them a place beside your views and those of Forbes and others. You have them, uncorrected by any experiments except those of former time, for I have made none lately though I have seen Tyndall's.

> Believe me to be
> My dear Sir
> Most truly yours
> M. FARADAY.

To Professor Faraday.

> 2 DONEGALL SQUARE WEST, BELFAST,
> *9th Nov.* 1858.

MY DEAR SIR,

I thank you very much for having so kindly written to me in August last on the subject of the freezing and melting point of water and ice.

I intended to write to you at the time, thanking you for your letter; but, through illness in my family, and repeated absences from home on business, I have hitherto found myself detained from reverting to the subject.

I recollect that in the old *Athenaeum* report of your Paper—to which however I have not had access of late—mention is made of your supposition that there is a tendency for a film of water between two surfaces of ice to become ice, other particles touching ice only on one side becoming water at the same time, and supplying the cold necessary to freeze the water between the two surfaces of ice. Having been long aware of that supposition, and now having farther had your letter of Aug. 4 to consider, I still incline to think that the freezing of the film between the two masses of ice is due simply to the melting by *pressure* of portions of the ice pressed against one another at places of contact.

Soon after the time when I received your letter, I met with a paper by Prof. Forbes of Edinburgh (from the *Proceedings of the Royal Society of Edinburgh* dated the 19th of April 1858) " On Properties of Ice near its Melting Point." This I presume to be the paper in which his shilling experiment is mentioned to which you alluded in your letter to me. In that paper Forbes states an experiment as proving that two masses of ice placed extremely close together, and having a film of water intervening, but free from pressure against one another, will unite even in a moderately warm atmosphere. I am not, however, prepared to admit the validity of his proof in this matter. He *had* slight springs pressing the masses of ice together: but beyond this I conceive that the capillary attraction of the intervening water would draw the masses against one another with a force quite notable;—and also I conceive that the film of water between the two masses of ice would be sustained almost entirely by capillary attraction at its upper surface, and would therefore exist (like mercury in a barometer) under a pressure *less* than that of the atmosphere. This diminished fluid pressure would *raise* the freezing point of the film of water, and would produce a tendency to its freezing, even by contact with such parts of the adjacent ice as exist under atmospheric pressure,—and much more by contact with the parts of the two masses of ice in contact with one another and pressing against one another by the forces of the spring used to maintain contact and of the capillary attraction of the fluid pulling the masses against one another. Thus I still incline to think that mere proximity is not enough to cause the two pieces of ice to unite.

In your letter to me you make reference, as bearing on this subject, to what appears to be a very general law, namely that, in bodies of the same kind, the particles tend to retain the state of those which surround them:—as, for instance, that water may be cooled much below 32° F. without freezing, but that a spicule of ice touching it will instantly make it solidify: and that water may be raised under atmospheric pressure to a temperature far above 212° F. without boiling, but that a bubble of air introduced will cause it to explode. It seems to me, however, that these phenomena are essentially distinct from what can occur with a film of water touching ice on one side, or on two opposite sides, because, in the phenomena you adduce, the tendency is for a particle to *retain* a state in which it already exists, that state being the same as the state of the surrounding ones: but I do not think they show a tendency for a particle differing in its state from the surrounding particles to *assume* their state. It is certain that, in numerous cases, every particle shows a great resistance to change from a state in which it and all adjacent particles exist. In the case of a film of water between two masses of ice, however, the difficulty of *making a beginning* either of melting or freezing does not exist, as both water and ice are present together.

I do not see in the principle I have proposed any insufficiency to explain the regelation of fractured ice, and the plasticity of ice; and, according to the views I have just now submitted, I do not see that we ought to suppose the occurrence of another efficient cause of regelation, such as you have suggested.

With great deference I beg to offer the above remarks to you.

I am, with thanks, yours most truly,

James Thomson.

P.S.   It may be well to mention that I conceive the powerful tendency to adhesion manifested between flannel and ice in a warm atmosphere, is to be attributed partly to the fibres of the flannel being drawn against the ice by capillary attraction of the liquid films, and thus being made to apply a pressure to the ice which must slightly lower its melting point; and again partly to the *diminished fluid pressure* of the liquid films produced by capillary attraction, which must *raise* the freezing point of those films.   J. T.

To Professor James Thomson.

Albemarle Street, W.
15 *Nov.* 1858.

My dear Sir,

I am much obliged by your note.   I happen to be occupied in collecting my papers (not electrical) into a volume, and on reprinting the notice on ice from the *Athenaeum* have added a further development of my views:—it will come out in the course of the winter I dare say, but that is as the printer likes.

You represent that my view gives no account of the beginning of the regelation.   That is not so great a difficulty to my mind as to yours; because, since the year 1833, if not before that time, I have been obliged to admit that particles cannot be so exclusively engaged, *even by a combining chemical action,* as to be indifferent to, or without relation to, those alongside of them.

I will mention a difficulty as regards this *beginning* of regelation which occurs to me under your view, *if the particles be supposed to be without this external relation.* You admit with me that bodies tend to retain that state which they for the time possess; and that against a change of temperature of *many degrees of heat;*—then how can the small change of temperature not amounting to the $\frac{1}{100}$th of a degree, due to difference of pressure in many of the regelation experiments, cause that change from solid to fluid, or fluid to solid, which many degrees of temperature change, applied in the common way, will not effect?

It seems that in ice the melting temperature is irregular, i.e. that certain portions of the ice tend to melt before other portions. I have given my view of this irregularity in a note, which Tyndall has added to his last paper in the *Phil. Trans.* :—

If ice of *equal purity* should, either from crystalline arrangement or some other cause, prove to be a mixture of particles having differences in their fusibility, then it would be very easy to build up a fourth theory of regelation. But time, reconsideration, new thoughts and new experiments, will no doubt clear up all these matters.

Ever my dear Sir,

Very truly yours,

M. FARADAY.

To Professor Faraday.

2 DONEGALL SQUARE WEST, BELFAST,
24th Nov. 1858.

MY DEAR SIR,

I am much obliged for your letter of the 15th inst.; and I presume I have to thank you and Prof. Tyndall jointly for a copy of his paper containing the note from you appended, which he has kindly sent to me and of which the contents interest me much.

There was hard frost here last night: and having happened to meet with slabs of ice about $1\frac{1}{2}$ inch thick, I have repeated

Forbes' experiment of suspending two slabs of ice on a horizontal glass rod thus :—

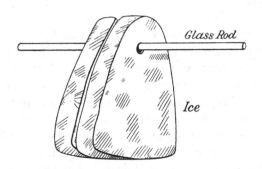

Glass Rod

Ice

I found that in a few hours (during which water was constantly dropping from them) the two slabs were stuck fast together—and I did not find it necessary to apply springs to produce any pressure at all. I merely pressed them gently with my hands for a few seconds at first, to make them keep close together by an incipient cohesion or a cohesion of some minute spots. The cohesion then went on increasing till in a few hours it became very strong. Although there is no spring *necessary*, yet the capillary attraction due to the films of water between the two plates, alluded to in my last letter to you, must keep up a steady force, drawing the plates together and causing pressure at some parts. I could see many square inches of films of water between the two plates or slabs of ice.

This water, being situated as at *a, a, a, a, a, a*, in the figure, must exist, by virtue of capillary attraction, under less than atmospheric pressure *at all parts of itself.* This diminished pressure must tend to cause the water to freeze even at a temperature slightly above the ordinary freezing point. The diminished pressure in the liquid film, extending over several square inches, will cause the external atmospheric pressure to force the two slabs against one another with quite a notable force.

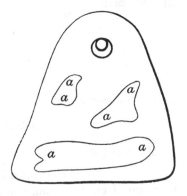

It seems to me that you have somewhat misconceived my meaning in respect to a "difficulty of making a beginning" as you say in your last letter:—

"I will mention a difficulty as regards this *beginning* of regelation which occurs to me under your view, *if the particles be supposed to be without this external relation.* You admit with me that bodies tend to retain that state which they for the time possess; and that against a change of temperature of *many degrees of heat*;—then how can the small change of temperature not amounting to the $\frac{1}{100}$th of a degree, due to the difference of pressure in many of the regelation experiments, cause that change from solid to fluid, or fluid to solid, which many degrees of temperature change, applied in the common way, will not effect?"

With reference to this I would say, that I think the resistance to change of state from liquid to solid, or solid to liquid, or liquid to gaseous, or gaseous to liquid, under consideration, occurs only when the substance is not present in the two states already (as for instance when water is cooled below 32° F., without freezing, there is no ice present) but when both water and ice are in contact, as is the case in the regelation experiments, it is a part of my theory to suppose that there is no tendency for water to remain water, or ice to remain ice, against any change of temperature however slight tending to change the state of the water or the ice.

I shall look with much interest to the ideas given in your note annexed to Prof. Tyndall's paper, and to the suggestion you give in your last letter to me of a "fourth theory of regelation" as being perhaps possible to be built up.

I am, My dear Sir,

Very truly yours,

JAMES THOMSON.

33. ON TUBULAR PORES IN ICE FROZEN ON STILL WATER.

[A paper on this subject was read before the Belfast Natural History and
Philosophical Society on the 20th April 1864, of which no report has
been preserved. The following letter to Professor William Thomson,
written on the 19th Feb. 1862, gives the substance of his explanation,
with possible application to similar structures in rock.]

2 DONEGALL SQUARE WEST,
BELFAST, 19th *Feb.* 1862.

MY DEAR WILLIAM,

Perhaps you may have noticed in slabs of ice frozen on
the surface of a pond or of any vessel of water, great numbers of
small vertical pipes or tubes usually smaller than a pin or a needle.
In fig. 1 I have represented these pipes.

Surface of the Ice

Fig. 1.

They seem to me to begin from a very small size at top and
usually to increase somewhat in diameter downwards.

The question occurred to me how these pipes or tubes originated. They seemed to contain air; and
I judged that they are produced by
bubbles of air expelled from the water
during its congelation:—that each pipe
is produced by one bubble which is
small at first, and which floats up against
the under surface of the ice:—that the
ice is prevented from growing at the
spot where the bubble touches it and so
excludes the water from touching it:
or that in fact the water cannot freeze to the ice at a spot where

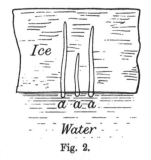

Ice

a a a

Water

Fig. 2.

no water is in contact with it, but where the air of the bubble is alone in contact with it:—that thus as the ice grows a tubular opening is left behind the advancing round front of the bubble as shown in fig. 2 or better in fig. 3 on a larger scale:—and that as the ice goes on growing and the bubble in the water leaves a tail of itself behind in the ice, new supplies of air are continually expelled from the freezing water and added to the various bubbles; so keeping up the supply of air to them during their growth which keeps pace with the thickening ice.

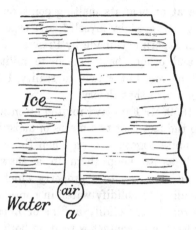

Fig. 3.

Dr Wyville Thomson happened to point out to me yesterday some elongated vesicles in amygdaloidal trap rock. He remarked on them as being of great interest, and supposed them to be formed by streams of ascending bubbles. I think it very possible that they may be explained on the same principle as the tubes in the ice. What do you think of all this? Have you ever seen the tubes in the ice described or explained?

Your affectionate brother,

JAMES THOMSON.

34. ON RECENT THEORIES AND EXPERIMENTS REGARDING ICE AT OR NEAR ITS MELTING POINT.

[From the *Proceedings of the Royal Society*, Vol. x. 1859, pp. 152—160. Read November 24, 1859.]

MY object in the following paper is to discuss briefly the bearings of some of the leading theories of the plasticity and other properties of ice at or near its melting point, on speculations on the same subject advanced by myself*, and especially, to offer an explanation of an experiment made by Professor James D. Forbes, which to him and others has seemed to militate against the theory proposed by me, but which, in reality, I believe to be in perfect accordance with that theory.

In the year 1850, Mr Faraday† invited attention, in a scientific point of view, to the fact that two pieces of moist ice, when placed in contact, will unite together, even when the surrounding temperature is such as to keep them in a thawing state. He attributed this phenomenon to a property which he supposed ice to possess, of tending to solidify water in contact with it, and of tending more strongly to solidify a film or a particle of water when the water has ice in contact with it on both sides than when it has ice on only one side.

In January 1857, Dr Tyndall, in a paper (by himself and Mr Huxley) read before the Royal Society and in a lecture delivered at the Royal Institution, adopted this fact as the basis of a theory by which he proposed to explain the viscosity or plasticity of ice, or its capability of undergoing change of form, which was previously known to be the quality in glaciers in virtue of which their motion down their valleys is produced by gravitation. Designating Mr Faraday's fact under the term "regelation," Dr Tyndall described the capability of glacier ice to undergo changes of form, as being not true viscosity, but as being the result of vast numbers of successively occurring minute fractures, changes of

---

* *Proceedings of Royal Society*, May 1857. Also *British Association Proceedings*, Dublin Meeting, 1857. Also *Proceedings of Belfast Literary and Philosophical Society*, Dec. 2, 1857. [See *supra*, No. 31, p. 208.]

† Lecture by Mr Faraday at the Royal Institution, June 7, 1850; and Report of that Lecture, *Athenaeum*, 1850, p. 640. [Footnote, *supra*, p. 212.]

position of the fractured parts, and regelations of those parts in their new positions. The terms *fracture* and *regelation* then came to be the brief expression of his idea of the plasticity of ice. He appears to have been led to deny the applicability of the term viscosity through the idea that the motion occurs by starts, due to the sudden fractures *of parts in themselves not viscous or plastic.* The crackling, he pointed out might, according to circumstances, be made up of separate starts distinctly sensible to the ear and to the touch, or might be so slight and so rapidly repeated as to melt almost into a musical tone. He referred to slight irregular variations in the bending motion of the line marked by a row of pins on a glacier, by Prof. Forbes, as being an indication of the absence of any quality that could properly be called viscosity, and of the occurrence of successive fractures and sudden motions in a material not truly viscous or plastic. I can only understand his statements on this subject by supposing that he conceived the material between the cracks to be rigid, or permanent in form, when existing under strains within the limit of its strength, or when strained less than to the point of fracture.

This theory appeared to me to be wrong*; and I then published,

---

* While the offering of my own theory as a substitute for Professor Tyndall's views seems the best argument I can adduce against them, still I would point to one special objection to his theory. No matter how fragile, and no matter how much fractured a material may be, yet if its separate fractured parts be not possessed of some property of internal mobility, I cannot see how a succession of fractures is to be perpetuated. A heap of sand or broken glass will either continue standing, or will go down with sudden falls or slips, after which a position of repose will be attained; and I cannot see how the addition of a principle of reunion could tend to reiterate the fractures after such position of repose has been attained. When these ideas are considered in connection with the fact that while ice is capable of standing, without immediate fall, as the side of a precipitous crevasse, or of lying without instantaneous slipping on a steeply sloping part of a valley, it can also glide along, with its surface nearly level, or very slightly inclined, I think the improbability of the motion arising from a succession of fractures of a substance having its separate parts devoid of internal mobility will become very apparent. If, on the other hand, any quality of internal mobility be allowed in the fragments between the cracks, a certain degree at least of plasticity or viscosity is assumed in order to explain the observed plasticity or viscosity. That fractures— both large and exceedingly small—both large at rare intervals, and small, momentarily repeated—do, under various circumstances, arise in the plastic yielding of masses of ice, is, of course, an undoubted fact: but it is one which I regard not as the cause, but as a consequence, of the plastic yielding of the mass in the manner supposed in my own theory. It yields by its plasticity in some parts, until other parts are overstrained and snap asunder, or perhaps also sometimes slide suddenly past one another.

in a paper communicated to the Royal Society, a theory which had occurred to me mainly in or about the year 1848, or perhaps 1850; but which, up till the date of the paper referred to, had only been described to a few friends verbally. That theory of mine may be sketched in outline as follows :—If to a mass of ice at its melting point, pressures tending to change its form be applied, there will be a continual succession of pressures applied to particular parts—liquefaction occurring in those parts through the lowering of the melting point by pressure—evolution of the cold by which the so melted portions had been held in the frozen state—dispersion of the water so produced in such directions as will afford relief to the pressure—and recongelation, by the cold previously evolved, of the water on its being relieved from this pressure : and the cycle of operations will then begin again; for the parts recongealed after having been melted, must in their turn, through the yielding of other parts, receive pressures from the applied forces, thereby to be again liquefied and to proceed through successive operations as before.

Professor Tyndall, in papers and lectures subsequent to the publication of this theory, appears to adopt it to some extent, and to endeavour to make its principles cooperate with the views he had previously founded on Mr Faraday's fact of so-called "regelation *."

Professor James D. Forbes adopts Person's view, that the dissolution of ice is a *gradual*, not a *sudden* process, and so far resembles the tardy liquefaction of fatty bodies or of the metals, which in melting pass through intermediate stages of softness or viscosity. He thinks that ice must essentially be colder than water in contact with it; that between the ice and the water there is a film varying in local temperature from side to side, which may be called plastic ice, or viscid water; and that through this film heat must be constantly passing from the water to the ice, and the ice must be wasting away, though the water be what is called *ice-cold*.

* I suppose the term regelation has been given by Professor Tyndall as denoting the second, or mending stage in his theory of "*Fracture* and *regelation*." Congelation would seem to me the more proper word to use after fracture, as *regelation* implies previous melting. If my theory of *melting by pressure and freezing again on relief of pressure* be admitted, then the term regelation will come to be quite suitable for a part of the process of the union of the two pieces of ice, though not for the whole, which then ought to be designated as the process of *melting* and *regelation*.

There is a manifest difficulty in conceiving the possibility of the state of things here described: and I cannot help thinking that Professor Forbes has been himself in some degree sensible of the difficulty; for in a note of later date by a few months than the paper itself, he amends the expression of his idea by a statement to the effect that if a small quantity of water be enclosed in a cavity in ice, it will undergo a gradual *"regelation"*; that is, that the ice will in this case be gradually increased instead of wasted. In reference to the first case, I would ask,—What becomes of the cold of the ice, supposing there to be no communication with external objects by which heat might be added to or taken from the water and ice jointly considered? Does it go into the water and produce viscidity beyond the limit of the assumed thin film of viscid water at the surface of the ice? Precisely a corresponding question may be put relatively to the second case— that of the large quantity of ice enclosing a small quantity of water in which the reverse process is assumed to occur. Next, let an intermediate case be considered—that of a medium quantity of water in contact with a medium quantity of ice, and in which no heat, nor cold, practically speaking, is communicated to the water or the ice from surrounding objects. This, it is to be observed, is no mere theoretical case, but a perfectly feasible one. The result, evidently, if the previously described theories be correct, ought to be that the mixture of ice and water ought to pass into the state of uniform viscidity. Prof. Forbes's own words distinctly deny the permanence of the water and ice in contact in their two separate states, for he says, "bodies of different temperatures cannot continue so without interaction. The water *must* give off heat to the ice, but it spends it in an insignificant thaw at the surface, *which therefore wastes even though the water be what is called ice-cold.*" Now the conclusion arrived at, namely, that a quantity of viscid water could be produced in the manner described, is, I am satisfied, quite contrary to all experience. No person has ever, by any peculiar application of heat to, or withdrawal of heat from, a quantity of water, rendered it visibly and tangibly viscid. We even know that water may be cooled much below the ordinary freezing point and yet remain fluid.

Professor Forbes regards Mr Faraday's fact of regelation as being one which receives its proper explanation through his theory described above; and, in confirmation of the supposition

T   15 that ice has a tendency to solidify a film of water in contact with it, and in opposition to the theory given by me, that the regelation is a consequence of the lowering of the melting point in parts pressed together, he adduces an experiment made by himself, which I admit presents a strong appearance of proving the influence of the ice in solidifying the water, to be not essentially dependent on pressure. This experiment, however, I propose to discuss and explain in the concluding part of the present paper.

Professor Forbes accepts my theory of the plasticity of ice as being so far correct that it points to *some* of the causes which may reasonably be considered, under peculiar circumstances, to impart to a glacier a portion of its plasticity. In the rapid alternations of pressure which take place in the moulding of ice under the Bramah's press, it cannot, he thinks, be doubted that the opinions of myself and my brother Prof. William Thomson are verified*.

Mr Faraday, in his recently published *Researches in Chemistry and Physics*, still adheres to his original mode of accounting for the phenomenon he had observed, and for which he now adopts the name "regelation"; or, at least, while alluding to the views of Prof. Forbes as possibly being admissible as correct, and to the explanation offered by myself as being probably true in principle, and possibly having a correct bearing on the phenomena of regelation, he considers that the principle originally assumed by himself may after all be the sole cause of the effect. The principle he has in view, he then states as being, when more distinctly expressed, the following:—"In all uniform bodies possessing cohesion, *i.e.* being either in the liquid or the solid state, particles which are surrounded by other particles having the like state with themselves tend to preserve that state, even though subject to variations of temperature, either of elevation or depression, which, if the particles were not so surrounded, would cause them instantly to change their condition." Referring to water in illustration, he says that it may be cooled many degrees below 32° Fahr., and still retain its liquid state; yet that if a piece of the same chemical substance—ice—at a higher temperature be introduced, the cold water freezes and becomes warm. He points out that it is

* Forbes, "On the Recent Progress and Present Aspect of the Theory of Glaciers," p. 12 (being introduction to a volume of Occasional Papers on the Theory of Glaciers), February 1859.

certainly not the change of temperature which causes the freezing; for the ice introduced is warmer than the water; and he says he assumes that it is the difference in the condition of cohesion existing on the different sides of the changing particles which sets them free and causes the change. Exemplifying, in another direction, the principle he is propounding, he refers to the fact that water may be exalted to the temperature of 270° Fahr., at the ordinary pressure of the atmosphere, and yet remain water; but that the introduction of the smallest particle of air or steam will cause it to explode, and at the same time to fall in temperature. He further alludes to numerous other substances—such as acetic acid, sulphur, phosphorus, alcohol, sulphuric acid, ether, and camphine—which manifest like phenomena at their freezing or boiling points, to those referred to as occurring with the substance of water, ice, and steam; and he adverts to the observed fact that the contact of extraneous substances with the particles of a fluid usually sets these particles free to change their state, in consequence, he says, of the cohesion between them and the fluid being imperfect; and he instances that glass will permit water to boil in contact with it at 212° Fahr., or by preparation can be made so that water will remain in contact with it at 270° Fahr. without going off into steam; also that glass can be prepared so that water will remain in contact with it at 22° Fahr. without solidification, but that an ordinary piece of glass will set the water off at once to freeze.

He afterwards comes to a point in his reasoning which he admits may be considered as an assumption. It is "that many particles in a given state exert a greater sum of their peculiar cohesive force upon a given particle of the like substance in another state than few can do; and that as a consequence a water particle with ice on one side, and water on the other, is not so apt to become solid as with ice on both sides; also that a particle of ice at the surface of a mass (of ice) in water is not so apt to remain ice as when, being within the mass there is ice on all sides, temperature remaining the same." This supposition evidently contains two very distinct hypotheses. The former, which has to do with ice and water present together, I certainly do regard as an assumption, unsupported by any of the phenomena which Mr Faraday has adduced. The other, which has to do with a particle of ice in the middle of continuous ice, and which assumes that it will not so readily change to water, as another particle of ice in contact with water, I think is to be accepted as probably true. I think the general bearing of all the phenomena he has adduced is to show that the particles of a substance when existing all in one state only, and in continuous contact with one another, or in contact only under special circumstances with other substances, experience *a difficulty of making a beginning of their change of state*, whether from liquid to solid, or from liquid to gaseous, or probably also from solid to liquid : but I do not think anything has been adduced showing a like difficulty as to their undergoing a change of state, when the substance is present in the two states already, or when a beginning of the change has already been made. I think that when water and ice are present together, their freedom to change their state on the slightest addition or abstraction of heat, or the slightest change of pressure, is perfect. I therefore cannot admit the validity of Mr Faraday's mode of accounting for the phenomena of regelation.

Thus the fact of regelation which Prof. Tyndall has taken as the basis of his theory for explaining the plasticity of ice, does in my opinion as much require explanation as does the plasticity of ice which it is applied to explain. The two observed phenomena, namely the tendency of the separate pieces of ice to unite when in contact, and the plasticity of ice, are indeed, as I believe, cognate results of a common cause. They do not explain one another. They both require explanation ; and that explanation, I consider, is the same for both, and is given by the theory I have myself offered.

I now proceed to discuss the experiment by Prof. Forbes, already referred to as having been adduced in opposition to my theory. He states that mere *contact* without pressure is sufficient to produce the union of two pieces of moist ice* ; and then states, as follows, his experiment by which he supposes that this is proved :—"Two slabs of ice, having their corresponding surfaces ground tolerably flat, were suspended in an inhabited room upon a horizontal glass rod passing through two holes in the plates of ice, so that the plane of the plates was vertical. Contact of the even surfaces was obtained by means of two very weak pieces of watch-spring. In an hour and a half the cohesion was so complete,

* "On some Properties of Ice near its Melting-Point," by Prof. Forbes, *Proceedings of the Royal Society of Edinburgh*, April 1858.

that, when violently broken in pieces, many portions of the plates (which had each a surface of twenty or more square inches) continued united. In fact it appeared as complete as in another experiment where similar surfaces were pressed together by weights." He concludes that the effect of pressure in assisting "regelation" is principally or solely due to the larger surfaces of contact obtained by the moulding of the surfaces to one another.

I have myself repeated this experiment, and have found the results just described to be fully verified. It was not even necessary to apply the weak pieces of watch-spring, as I found that the pieces of ice, on being merely suspended on the glass rod in contact, would unite themselves strongly in a few hours. Now this fact I explain by the capillary forces of the film of interposed water as follows:—Firstly, the film of water between the two slabs —being held up against gravity by the capillary tension, or contractile force, of its free upper surface, and being distended besides, against the atmospheric pressure, by the same contractile force of its free surface round its whole perimeter, except for a very small space at bottom, from which water trickles away, or is on the point of trickling away—exists under a pressure which, though increasing from above downwards, is everywhere, except at that little space at bottom, less than the atmosphere pressure. Hence the two slabs are urged towards one another by the excess of the external atmospheric pressure above the internal water pressure, and are thus pressed against one another at their places of contact by a force quite notable in its amount. If, for instance, between the two slabs there be a film of water of such size and form as might be represented by a film one inch square, with its upper and lower edges horizontal, and with water trickling from its lower edge, it is easy to show that the slabs will be pressed together by a force equal to the weight of half a cubic inch of water. But so small a film as this would form itself even if the two surfaces of the ice were only very imperfectly fitted to one another. If, again, by better fitting, a film be produced of such size and form as may be represented by a square film with its sides four inches each, the slabs will be urged together by a force equal to the weight of half a cube of water of which the side is four inches; that is, the weight of 32 cubic inches of water or 1·15 pound, which is a very considerable force. Secondly, the film of water existing, as it does, under less than atmospheric pressure, has its freezing point raised in virtue of the reduced pressure; and it would therefore freeze even at the temperature of the surrounding ice, namely the freezing point for atmospheric pressure. Much more will it freeze in virtue of the cold given out in the melting by pressure of the ice at the points of contact, where, from the first two causes named above, the two slabs are urged against one another.

The freezing of ice to flannel or to a worsted glove on a warm hand is, I consider, to be attributed partly to capillary attraction acting in similar ways to those just described; but in many of the observed cases of this phenomenon there will also be direct pressures from the hand, or from the weight of the ice, or from other like causes, which will increase the rapidity of the moulding of the ice to the fibres of the wool.

[This paper was also read at the British Association meeting at Aberdeen, September 1859.]

35. NOTE ON PROFESSOR FARADAY'S RECENT EXPERIMENTS ON "REGELATION."

[From the *Proceedings of the Royal Society*, April 25, 1861, Vol. XI. p. 198. Also *Phil. Mag.* Fourth Series, Vol. XXIII. (1862), pp. 407, 411.]

SOME time ago*, Principal James D. Forbes showed that two slabs of ice, having each a face ground tolerably flat, and being both suspended in an atmosphere a little above the freezing point upon a horizontal rod of glass passed through two holes in the plates of ice, so that the plates might hang vertically and in contact with one another, would unite gradually so as to adhere strongly together. This interesting experiment Principal Forbes adduced as being in opposition to the theory offered by me† of the plasticity of ice, and of the tendency of pieces of thawing ice to unite when placed in contact.

He thought it showed that pressure was not essential to the union of the two pieces of ice. I pointed out, in reply‡, that the film of water between the two slabs, being held up against gravity

* "On some Properties of Ice near its Melting-Point," by Prof. Forbes, *Proceedings of the Royal Society of Edinburgh*, April 1858.
† *Proceedings of Royal Society*, May 1857, and *British Association Reports*, 1857. [No. 31, *supra*, p. 208.]
‡ *Proceedings of Royal Society*, Nov. 24, 1859, Vol. x. p. 159. [No. 34, *supra*, p. 229.]

by the capillary tension or contractile force of its free upper surface, and being distended besides against the atmospheric pressure, by the contractile force of its free surface round its whole perimeter —except for a very small space at bottom, from which water trickles away, or is on the point of trickling away,—exists under a pressure which, though increasing from above downwards, is everywhere, except at that little space near the bottom, less than atmospheric pressure:—that hence the two slabs are urged against one another by the excess of the external atmospheric pressure above the internal water pressure, and are thus pressed against one another by a force quite notable in amount;—that, further, the film of water existing as it does, under less than atmospheric pressure, has its freezing point raised in virtue of the reduced pressure; and would therefore freeze even at the temperature of the surrounding ice, which I took to be the freezing point for atmospheric pressure; and would still more strongly be impelled to freeze by the joint action of this condition with the cold given out in the melting by pressure of the ice at the points of contact where the two slabs are urged against one another.

To this explanation of Principal Forbes's experiment I still adhere as mainly correct, though admitting of some further development and slight modification in reference to a point to which I shall have to make further allusion in what follows, and which seems to me to be as yet rather obscure:—the influence, namely, of the tension in the ice due to its own weight, which makes it *not* be subject internally to simple atmospheric pressure:— and though I shall also, in what follows, point out some additional conditions, almost necessarily present in the experiment, which, under my general view of the plasticity of ice, would act in conjunction with those already adduced, and would increase the rapidity of the union.

Professor Faraday, holding it in view to remove the ground on which my explanation of Principal Forbes's experiment was founded, has contrived and carried out a set of new and very beautiful experiments from which the capillary action referred to has been completely eliminated, and he has still found the union of the ice to occur, and to increase with time, and has met with a curious additional phenomenon of "flexible adhesion*."

---

* *Proceedings of Royal Society*, April 26, 1860, Vol. x. p. 440. [Faraday's *Researches in Chemistry and Physics*, pp. 373, 378.]

In these experiments, when two pieces of ice, rounded so as to be convex at their points where mutual contact is to be allowed, arc placed in water, and are either anchored so as to be wholly under water, or are placed floating when so formed that they can touch one another only under water, and that, at the water surface, there shall be a wide space between them so that there shall be no capillary action drawing them together, he showed that the pieces of ice, in either of these cases, if brought gently into contact, will adhere together; unless indeed the movement bringing them into contact be so directed as to introduce forces capable of tearing them apart again by obliquity of action, by agitation of the water, or by other disturbances. He showed also that, if when the two pieces of ice have become attached at their point of contact, a slight force, such as may be given by one or two feathers, be applied, tending to separate them, at one side of their point of contact, they will roll round one another with a seemingly flexible adhesion; or that, if the point of a floating wedge-shaped piece of ice is brought under water against the side of another floating piece, it will stick to that piece like a leech.

He showed that if the pieces be allowed to remain for a few moments in contact, their adhesion will become rigid, so that on a force being applied sufficient to break through the joint, the rupture will occur with a crackling noise, though the pieces may still continue to hold together, rolling on one another with the flexible adhesion. He made some other experiments nearly the same as these, but in which he showed the flexible and rigid adhesion to occur while there is constantly a decided tensile force applied externally tending to pull the pieces asunder, instead of any external force tending to press them together. He thinks that the phenomena of flexible and rigid adhesion "under tension" go towards showing that pressure is not necessary to "regelation." He then gives his own idea of the flexible and rigid adhesion in the following words :—"Two convex surfaces of ice come together; the particles of water nearest to the place of contact, and therefore within the efficient sphere of action of those particles of ice which are on both sides of them, solidify; if the condition of things be left for a moment, that the heat evolved by the solidification may be conducted away and dispersed, more particles will solidify, and ultimately enough to form a fixed and rigid junction, which will remain until a force sufficiently great to break through it is applied. But if the direction of the force resorted to can be relieved by any hinge-like motion at the point of contact, then I think that the union is broken up amongst the particles on the opening side of the angle, whilst the particles on the closing side come within the effectual regelation distance; regelation ensues there and the adhesion is maintained, though in an apparently flexible state. The flexibility appears to me to be due to a series of ruptures on one side of the centre of contact, and of adhesion on the other,—the regelation, which is dependent on the vicinity of the ice surfaces, being transferred as the place of efficient vicinity is changed. That the substance we are considering is as brittle as ice, does not make any difficulty to me in respect of the flexible adhesion; for if we suppose that the point of contact exists only at one particle, still the angular motion at that point must bring a second particle into contact (to suffer regelation) before separation could occur at the first; or if, as seems proved by the supervention of the rigid adhesion upon the flexible state, many particles are concerned at once, it is not possible that all these should be broken through by a force applied on one side of the place of adhesion, before particles on the opposite side should have the opportunity of regelation, and so of continuing the adhesion."

The interpretation thus put by Prof. Faraday on his experiments is not convincing to me; but, on the contrary, I think the experiments are in perfect accordance with my own theory, and tend to its confirmation. My view of the phenomena of these experiments is as follows:—

The first contact of the two pieces of ice cannot occur without impact and consequent pressure; and, small as the total force may be, its intensity must be great, as the surface of contact must be little more than a geometrical point. This pressure produces union by the process of melting and regelation described by me in previous papers. On the application of the forces from the two feathers, at one side of the point of contact, tending to cause separation, the isthmus of ice formed by the union of the two pieces comes to act as a tie or fulcrum subject to tensile force; and consequently a corresponding pressure will occur at the side of the isthmus, far from the feathers; and that pressure will effect the union of the ice at the side where it occurs.

The tensile force, it may readily be supposed, tends to preserve the isthmus, internally at least, in the state of ice, whatever may be its influence on the external molecules of the isthmus, and to solidify such water as, having occupied pores in the interior during previous compression, may now, by the linear tension or pull, be reduced in cubical pressure or hydrostatic pressure, because the melting point of wet ice is raised by diminution of pressure of the water in contact with it*.   The pull applied to the isthmus

* How the *surface* of a bar of ice immersed in ice-cold water, as distinguished from the *interior* of the bar, may, in respect to tendency either to melt away, or to solidify to itself additional ice from the water, be influenced by the application of linear tension to the bar, I am not quite prepared to say positively. The application of tension, whether linear, superficial, or cubical (that is, whether simply in one direction, or in two directions crossing one another, or in three directions crossing one another), to a piece of ice immersed in water at any given pressure, atmospheric for instance, is very distinct from the application of what might be called cubical tension, that is, diminution of hydrostatic pressure, to the surrounding water. In the former case the pressure of the water at the external surface of the ice will not be reduced by the application of the tension to the ice; though that of the water in the internal pores may, or probably in many of them must, be so; but in the second case, the diminution of cubical pressure in the external water effects the same diminution of pressure in the ice, and also in the water occupying pores in the ice.   The theory and quantitative calculation which I originally gave (*Trans. Roy. Soc. Edin.* Vol. XVI. Part 5, 1849, and *Cambridge and Dublin Math. Journ.* Nov. 1850) [No. 29, *supra*, p. 196] of the effect of increase of pressure in lowering the freezing point of water, and of course also of the effect of diminution of pressure in raising it, applied solely to effects of pressure communicated to the ice *through the water*, and therefore equal in all directions, and equally occurring in the ice and the water; but when changes of pressure in one or more directions are applied to the ice as distinguished from the water, the theory does not apply in any precise way to determine the conditions of the melting of the ice, or of its growth by the freezing of the adjacent water to its surface.   There seems to me to be yet a field open for much additional theoretical and experimental investigation in this respect†; but so far as I have applied the principle of the lowering of the freezing point of water by pressure in developing or sketching out a theory of the plasticity of ice, I think I have done so correctly.   I perceived that the application of pressures tending to change the form of the ice must necessarily produce volume-compression in some parts of the mass, accompanied by the occurrence of increased fluid pressure in the pores which might already exist in those parts, or which would arise in them as a consequence of the pressure; and this I thought was a sufficient basis on which to rest the theory, even without precise knowledge of all the varying influences on the melting or freezing of the ice or water, of all the possible varieties of pressures or stresses that could be applied to the ice, and of fluid pressure that could occur in the water contemporaneously with those stresses in the ice.   Some additional developments of this part of the subject, which have occurred to me, may, I hope, form the subject of a future paper. [See No. 36, *infra*, p. 236.]

† [For the analytical solution of this problem, see Willard Gibbs' classical memoir (1876) 'On the Equilibrium of Heterogeneous Substances,' reprinted in his *Collected Papers*, Vol. I. pp. 185—218, especially the footnote, p. 197, giving references to Prof. Thomson's work.]

thus appears to put it out of the condition in which my theory
has clearly indicated a cause of plasticity, and I presume makes it
cease, or almost entirely cease, to be plastic.

I believe no plastic yielding of ice to tension has been discovered by observation in any case, and I think there are
theoretical reasons why ice should be expected to be very brittle
in respect to tensile forces. The isthmus then being supposed
devoid of plasticity at its extended side, ultimately breaks at that
side, when the opening motion caused by the feathers has arrived
at a sufficient amount to cause fracture, and the ice newly formed
on the compressed side comes now to act as a tie instead of the
part which has undergone disruption, and holds together the two
pieces of ice, or serves as a fulcrum under tension to communicate
a compressive force to the points of the two pieces of ice immediately beyond it; and so the rolling action with a constant
union at the point of contact goes forward.

It is to be observed that the leverage of the forces applied
by the feathers is so great, compared with the distance from the
fulcrum or tensile part of the isthmus, to the compressed part in
process of formation at the other side, as that the compression
may usually be considered almost equal to the tension: and the
tension in the extended part cannot be of small intensity, being
sufficient to break that side of the isthmus. In the experiments
which gave flexible adhesion *seemingly* under tension, it is not to
be admitted that tension was really the condition under which
the ice existed at the places where the union was occurring.

To apply a simple disruptive force to the whole isthmus of
ice, it would be necessary to take very special precautions in
order to arrange that the line of application of the disruptive
forces should pass through the point of contact of the two pieces.
If that were done, and the forces were gradually increased till the
cohesive strength of the isthmus were overcome, it is clear that
the two pieces of ice would separate altogether, and there would
be no flexible adhesion; but the flexible adhesion, when it occurs,
is essentially dependent on the existence of an intense pressure
at the side of the isthmus remote from the line of the externally
applied disruptive forces, or of the single force applied in some
of the experiments to one only of the pieces, and resisted by the
inertia of the other.

It is further to be observed that tremors and slight agitations to which the two pieces of ice, united at their point of contact, may be subject, arising from undulations imparted to the water in which the ice is immersed, by manipulation of the experimenter, —from the tread of people on adjacent floors,—from the passage of vehicles on neighbouring streets,—from convective movements of the water,—and from other causes,—will be sources of power or energy operative in bringing about an increase of adhesion with time; that is to say, in changing gradually the flexible into the rigid adhesion.

It will now of course be obvious that the conditions involved in the explanation just offered of Prof. Faraday's experiments must also usually be present in the experiment of Principal Forbes. Their incidental occurrence, however, as additional causes increasing the rapidity of the union of the two slabs of ice, does not overthrow the particular explanation of Principal Forbes's experiment which I had offered as a perfect answer to the objection raised by that experiment against my general view of the plasticity of ice; and as indicating clearly and certainly the occurrence of all the conditions required for the union of the two pieces of ice under my theory. The contingent occurrence of the additional conditions now specially brought forward, was indeed from the first somewhat familiar to my mind, but was left out of the explanation as being unessential and not perhaps quite so clearly apparent. Their occurrence has, however, now become essential to the explanation of Prof. Faraday's new experiments:— and by it I consider these are shown not to militate against my general theory of the plasticity of ice, but to corroborate it strongly, and to confirm its application to the various observed cases of the union of two pieces of moist ice when placed in contact.

36. ON CRYSTALLIZATION AND LIQUEFACTION, AS INFLUENCED BY STRESSES TENDING TO CHANGE OF FORM IN THE CRYSTALS.

[From the *Proceedings of the Royal Society*, December 5, 1861, Vol. xi. p. 473. Also *Phil. Mag.* Fourth Series, Vol. xxiv. (1862), pp. 395—401.]

IN a paper submitted to the Royal Society, and printed in the *Proceedings* for April 25th, 1861, I directed attention in a note [page 234] to the question of how the *surface* of a bar of ice in ice-cold water, as distinguished from the interior of the bar, may, by the application of tension to the bar, be influenced in respect to tendency either to melt away, or to solidify to itself additional ice from the water; but did not then venture to offer a positive answer. I pointed out as a matter deserving of special attention, and as affording scope for much additional theoretical and experimental investigation, the distinction between the application to ice in ice-cold water, of stresses tending to change its form, the stresses not being participated in by the water; and the application directly to the water, and through that to the ice, of cubical or hydrostatic pressures or tensions, these being participated in by the water and the ice alike; and I pointed out that the theory and quantitative calculation which I had originally given* of the effect of pressure in lowering the freezing point of water, or of diminution of pressure in raising it, applied solely to effects of pressure communicated to the ice *through the water*, and therefore equal in all directions, and equally occurring in the ice and the water; but that when changes of pressure in one or more directions are applied to the ice as distinguished from the water, the theory does not apply in any precise way to determine the conditions of the melting of the ice, or of its growth by the freezing of the adjacent water to its surface; and I expressed the hope that I might subsequently communicate to the Society some further developments of the subject.

On following up various considerations which had then occurred to me, I soon formed positively the opinion that *any stresses whatever, tending to change the form of a piece of ice in ice-cold water* (whether these stresses be of the nature of pressures or tensions, that is pushes or pulls, and whether they be in one direction alone, or in more directions than one), *must impart to the ice a tendency to melt away, and to give out its cold, which will tend to generate, from the surrounding water, an equivalent†quantity of ice free from the applied stresses.* I came also to the more general inference that *stresses tending to change the form of any crystals in the saturated solutions from which they have been crystallized must give them a tendency to dissolve away, and to generate, in substitution for themselves, other crystals free*

---

* *Transactions Roy. Soc. Edin.* Vol. xvi. Part 5, 1849; and *Cambridge and Dublin Math. Journ.* Nov. 1850. [No. 29, *supra*, p. 196.]

† [The following note in Prof. Thomson's hand is found written in a copy of this paper:—P.S. The quantity here not explicitly enough referred to as an equivalent quantity is more accurately the same quantity minus that which would be melted by the energy of the straining which disappears in the melting.]

from the applied stresses or any equivalent stresses. In the month of May last, I tested this inference by applying stresses to crystals of common salt in water saturated with salt dissolved from the crystals themselves; and found the crystals to give way gradually, with a plastic yielding, like the yielding of wet snow, but very much slower. The crystals, with the brine in which they were immersed, were, in the first set of experiments, placed in a glass tube, like a test-tube, and a glass piston, or rammer, fitting the tube loosely, so as not to be water-tight, was placed on the top of the salt which lay like fine sand in the bottom, and the piston was loaded with weights. The piston went on descending from day to day through spaces, which, though small, and though diminishing as the crystals became more compacted against one another, were still distinctly visible. When the rate of descent became very slow, I added more weights, and found that the rate of descent increased, as was to be expected. I afterwards procured a strong brass cylinder with a loosely fitted, not water-tight piston, or rammer, and in this I subjected crystals of common salt in their saturated brine to very heavy stresses, and thus compressed them rapidly and easily into a hard mass like rock-salt. The top surface presented a perfect impression of the tool marks on the bottom of the piston, such as might have been made in wax. The expulsion of the minute quantities of brine remaining in pores in the salt when it has become very closely compacted, appears to be a slow and difficult process; as, after the pressure had been continued for about a fortnight, I still found a slight oozing of brine from a pore which happened to exist in the side of the cylinder.

Experiments by the application of tensile stresses, or of any other stresses than those mixed and chiefly compressive ones which arise when the crystals are pressed in a close vessel by a rammer, would probably not be very easily carried out; and I have not as yet tried any except those by pressure. I feel quite convinced, however, that melting, or dissolving, must result from all kinds of stresses tending to change of form. I think the following statement may be assumed as a general physico-mechanical principle or axiom, and I think it involves the truth of the opinion just expressed :—

If any substance, or any system of substances, be in a condition in which it is free to change its state (whether of molecular arrangement, or of mechanical relative position and connexion of its parts, or of rest or motion), and if mechanical work be applied to it (or put into it) as potential energy, in such a way as that the occurrence of the change of state will make it lose (or enable it to lose) (or be accompanied by its losing) that mechanical work from the condition of potential energy, without receiving other potential energy as an equivalent; *then the substance or system will pass into the changed state.* The consideration of a few cases, in some of which there is not freedom for the substance or system to change its state, and in others of which there is freedom, will render the meaning of this more clear.

Gunpowder may be cited as an example of a substance in a condition *not free to change its state,* although when it is made to explode by a spark, it passes to an altered condition, and, in doing so, even gives out a great amount of mechanical work. That is to say, that *on the whole* it is more than free to change to the exploded state, or it tends so to change, but there is some kind of obstacle at ordinary temperatures, to the change, which either vanishes at a high temperature, or requires the application of mechanical work to begin the overcoming of it. When the change is once begun, the requisite help is given to the succeeding parts by those which have gone off first.

Again, water confined in a high reservoir is not free to go to a lower one; although a siphon, primarily filled with water, may help the parts successively over the obstacle by lending to each the requisite mechanical work in advance, which it afterwards pays to the parts which are to follow, besides that it gives out in its fall a great additional amount of power or energy applicable otherwise. Two reservoirs of water, on the same level, and having an opening between them under the water surface, would represent the case of *perfect freedom for change of state*; and two on a level with one another, but separated by a partition, would represent the case in which no mechanical work would finally be either given out or absorbed by the change, but in which there is not perfect freedom to change, until a siphon or other means of help is applied.

A bell hung from an axle and then turned up, and left resting against a stop a little beyond its position of unstable equilibrium, is not free to go down, but a slight pull will bring it over this position and make it free to swing, which the work stored as potential energy in the raising of it from its low or hanging position, will cause it to do; its fall till it comes to the bottom being essentially accompanied by the loss of that potential energy, as such, though not as actual energy, out of the system of which it and the earth are the two parts, and in which change of their distance asunder constitutes change of their potential energy.

If in an atmosphere of steam resting on water at its boiling temperature for the pressure of the steam; as, for instance, in the inside of a boiler partly filled with water, and partly with steam, an inverted cup, or bell-shaped vessel, be suspended, and if it then, being full of steam, be forced down under the water, mechanical work will be imparted as potential energy to the system of which the steam and water in the boiler form one part, and the earth is the other part; though, for brevity of expression, the work may be spoken of as applied to the steam and water. In this case there is perfect freedom for the steam forced under the water to condense and cause by communication of its latent heat the generation of an equal quantity of steam at the surface of the water under which the bell was sunk. The occurrence of this change of state will enable the system to lose the potential energy which had been imparted to it by the submersion of the steam, or will release that energy which had been stored, and the system *will pass into the changed state*; that is to say, a certain part of the steam will change to water, and, instead of that, a different part of the water will be changed to steam; and this change will be accompanied by a transmission of heat from the part condensing to the part evaporating. This is all in accordance with the axiom; and we know otherwise that it must take place, as the steam being pressed when submerged must condense and give its latent heat to the water, and that heat must generate an equal quantity of steam at the surface of the water, where the pressure is less. Thus the truth of the axiom is confirmed.

If a quantity of ice and water be enclosed in a cylinder with a water-tight piston, and if this be put into a completely closed vessel filled with other ice and water, and if the piston then be pulled with any given force and fixed in its new position (which might be done in many ways, as for instance, by the use of an axle passing air-tight through the side of the outer vessel), mechanical work will be introduced as potential energy into the system consisting of all the things enclosed in the outer vessel. But there is perfect freedom for the water enclosed in the cylinder to proceed to freeze, obtaining the requisite cold from the ice in the water confined around the cylinder and within the outer vessel. The occurrence of this change would be accompanied by the system's losing or giving up the potential energy which had been stored in it. According to the axiom, then, the change ought to occur. But we know otherwise that it must occur; because the diminution of hydrostatic pressure in the cylinder raises the freezing point of the enclosed water, and makes it freeze by the cold of the surrounding mixture of ice and water, which, besides, by being itself subjected to increased pressure, tends to give out cold by the lowering of its freezing point. Thus the truth of the axiom is again confirmed.

Lastly, if a bar of ice in ice-cold water be subjected to any stress (a pull for instance) tending to change its form, it will receive mechanical work from the force, or forces, applied, and that work will be stored as potential energy in the elasticity of the ice. Now, if there be another piece of ice in juxtaposition with this piece, seeing that, at the beginning, both these pieces were free from externally applied forces, and were both in the state in which either was perfectly free to melt and cause an equal quantity of water to freeze to the other*, it will follow according to the axiom, now supposed to be established, that the application of the stress *will cause this action to occur*.

The case of crystals in their solutions might be stated almost in the same words as the case of ice in ice-cold water; but it is to

* The supposition here assumed, however, of there being perfect freedom for either of two pieces of ice, which are immersed in the same water, and are alike free from stresses, to melt, and, by giving out its cold, to cause an equal quantity of water to freeze to the other, will probably not meet with assent at present from all, as it appears to be a prevailing opinion that water and ice in contact are *not* in a state of perfect indifference as to retaining or interchanging their conditions. It is supposed that ice has a property of tending to solidify water in contact with it, and the more so if there be ice on both sides of the water than if on only one side. Again, it is supposed that ice is essentially colder than water in contact with it, and that the water must continually be giving off heat to the ice. Both these opinions are inconsistent with the supposition here assumed. I conceive, however, that that supposition is amply confirmed by the fact that it was involved essentially throughout the reasoning, by which I was led to conclude that the freezing point is lowered by increase of pressure, and to calculate the amount of the lowering. That reasoning led to true results and I believe it could not have done so unless the supposition were true, that when water and ice are present together their freedom to change their state on the slightest addition or abstraction of heat, or on the slightest application of mechanical work tending to the change, is perfect.

T.   16

be observed that, in their case, the necessity for the translation of
one chemical substance through another (the salt through the
dissolving liquid), and not of heat or cold alone, causes a great
slowness of the process, as compared with that of the yielding of
the ice, in ice-cold water, to applied stresses.

At an early stage of the considerations which led to the
opinions on the influence of stresses on crystallization and liquefaction described in the present paper, the question arose to me:—
Is a spiculum or single crystal of ice, which has solidified itself in
the interior of water, and is therefore not colder than the water,
plastic ? Or would it, when in the water, and attached by one
end, as for instance to a crust of ice lining the containing vessel,
gradually bend upwards by its own buoyancy in the heavier water?
My idea is that it is not plastic. I cannot conceive of the growth
of a crystal proceeding with one continuous or uninterrupted
structural arrangement, if during its growth the part already
formed undergoes permanent change of form, such as would be
due to any plastic or ductile yielding. I think we must suppose
the molecules in the interior of one crystal to be so locked into
one another, by the forces of crystalline cohesion, that any one of
them, or set of them, would experience a difficulty in making a
beginning of the change of state from solid to liquid. I have not
succeeded even in forming any clear conception of continuous
crystalline structure admitting of what may be called ductile or
malleable bending (that is, bending beyond limits of elasticity
such as occurs in lead, copper, tin, and many other metals), and
still remaining of the nature of one continuous crystal. What in
soft or malleable crystals of copper or other metals, deposited in
the electrotype process, may be the nature of the change of
molecular arrangement induced by bending them, I cannot say ;
but I suppose that, in their yielding, their crystalline structure is
materially altered, and rendered discontinuous where, before, it
was continuous.

In a mass of plastic ice, I incline to think that the internal
melting, to which I attribute the plasticity, must occur at the
surfaces of junction of separate crystals or fragments of crystals ;
though probably pores, formed through melting by pressures or
stresses, may penetrate crystals by entering them from their
moistened surfaces or their junctions with other crystals. It now
becomes clear, I think, that the influence of stresses affecting the

ice, and tending to make it melt without there being necessarily any consequent pressure applied to the water in contact with the ice, must come to be taken into account in any theory of the plasticity of ice approaching to completeness.

This view does not, however, I think, supersede the theory of the plasticity of ice sketched out by myself in former papers, but rather constitutes an amendment, and further development of it. Any complete theory of the plasticity of ice, and of the nature of glacier motion, must comprise the conditions as to fluid pressure and structural arrangement of the water and air included in the ice, and must so explain the lamination of the glacier, seen as blue and white veins. My brother, Professor William Thomson, in papers in the *Proceedings of the Royal Society* for February [January] 25 and April 22, 1858*, endeavoured to follow up my previously published views on the plasticity of ice with an explanation of the laminated structure, based on the same principles. The explanation he then offered, I think, cannot fail to assist in suggesting the direction in which the true solution is ultimately to be sought for; yet I feel confident that no full and true solution has as yet been found†.

In the foregoing part of the present paper, I have shown reason why stresses applied to crystals when in contact with the liquid from which they have been produced, should be expected to cause them to melt or dissolve away. The following line of reasoning to show that stresses applied to a crystal will cause a resistance to the deposition of additions to it from the liquid, or, in other words, a resistance to its growth, will, I think, prove to be correct. When a crystal grows, the additions, it seems to me, must lay themselves down in a state of molecular fitting, or regular interlocking with the parts on which they apply themselves; or, in other words, they must lay themselves down so as to form one continuous crystalline structure with the parts already crystallized. It thus seems to me that, if a crystal grows when

* [See Sir William Thomson (Lord Kelvin), *Math. and Phys. Papers*, Vol. v. pp. 47 and 50.]

† I have my brother's authority for stating that, although he believes the physical principles suggested in his papers here referred to, to be capable of being developed into a true explanation of the phenomena, yet he considers further investigation necessary, and does not feel confident as to the correctness of that part of the explanation he offered, in which the mutual action of two vesicles in a line oblique to that of maximum pressure is considered.

under a stress, the new crystalline matter must deposit itself in the same state of stress as the part is in on which it lays itself. If, then, we consider a spiculum of ice growing in water, and if we apply any stress, a pull for instance, to it while it is thin, and then fix it in its distended state, and if then by the transference* to the water beside it of cold taken from any other ice at the freezing point we cause it to grow, which it may do if there be no other crystal of ice beside it more free than it to receive accessions, then the additional matter will, I think, lay itself down in the same state of tensile stress as the original spiculum was put into by the applied pull. The contractile force of the crystal will thus be increased in proportion to the increase of its cross-sectional area. If it now be allowed to contract and relax itself, it will give out, in doing so, more mechanical work than was applied to the ·original spiculum during distention. Hence there would be a gain of mechanical work without any corresponding expenditure; or we could theoretically have a means of perpetually obtaining mechanical work out of nothing, unless it were the case that greater cold is required to freeze water into ice on the stressed crystal than on a crystal free from stress. Hence we must suppose that a greater degree of cold will be required to cause the stressed crystal to grow. The reasoning just given has been for brevity stated somewhat in outline; but I trust the full meaning can readily be made out, and that what has been said may suffice.

I wish now to suggest as an important subject for investigation, The Effect of Change of Pressure (hydraulic pressure) in changing the Crystallizing Temperatures of Saline or other Solutions of given Strengths,—as I feel sure that such effect must exist, but am not aware that it has been hitherto discussed or experimented on, and as it is intimately connected with the matters under consideration in the present paper and with subjects discussed in previous papers, which I have submitted to the Royal Society, on Ice†.

* A theoretic air-engine for making such transferences of heat or cold was used in the reasoning by which I determined theoretically the lowering of the freezing point by pressure, and the same is admissible here.

† [For further developments see Willard Gibbs, *loc. cit.*, *supra*, p. 234.]

37.   CORRESPONDENCE WITH W. THOMSON ON THE INFLUENCE OF STRESS ON CRYSTALLIZATION.

KILMICHAEL, BRODICK,
BY ARDROSSAN.
October 18, 1861.

MY DEAR JAMES,

I cannot quite make up my mind as to the stress, or non stress, of new crystalline matter deposited on a crystal under stress. I rather think there will not be stress, but a crystalline discontinuity of some kind. Still perhaps your paper may stand as it is. You have time enough to think of it however and it is easy to send to Stokes either a qualifying sentence, which will make there be nothing to recant if it should turn out that there is not stress, or a request to cancel that part of the paper. As far as I can see it is at least a legitimate hypothesis that there is stress; but I think it is also a legitimate hypothesis that there is not stress which you reject perhaps too decidedly. I have been thinking on writing a mathematical investigation determining the amount of lowering of the freezing point, on the supposition that your hypothesis holds. Your hypothesis is simply that the addition to or subtraction from a crystal, kept elongated, is a reversible operation. There is also a mathematical course open, to investigate the thermal result of the work done by first elongating a crystal and then letting it melt. I think I see how to do this perfectly without either of the above hypotheses, but my hands are too full just now to think of writing it out just at present.

*   *   *   *   *   *   *   *   *   *

W. THOMSON.

2 DONEGALL SQUARE WEST,
BELFAST.
October 19, 1861.

MY DEAR WILLIAM,

* * * You will see on looking over my R. S. paper that I do state the idea rather as a probable hypothesis than as a confidently offered theory. You will see I say that the reasoning will "*I think prove to be correct*," and twice again I use the words "*it seems to me.*" If you still think I ought to state the matter with more doubt perhaps I may do so on hearing from you again. Still I have a pretty strong opinion that the hypothesis or theory offered will turn out to be correct: and I think the notes which I enclose to you support it pretty strongly.

I shall be very glad if when your hands are less full you can make out the mathematical investigations as to melting and liquefaction which you propose. I would think there would be some data deficient such as the modulus of elasticity or the coefficient of rigidity of the crystal of ice, salt, or other substance. * * *

JAMES THOMSON.

October 19, 1861.

NOTES ON CRYSTALLIZATION AND LIQUEFACTION.

If a crystal be subjected to stress the molecules at its surface tend to dissolve away and the stored potential energy, making its escape in conjunction with their dissolving, affords a power to drive them into the fluid state even against a resistance. We may take, as the resistance opposed, a slight reduction of temperature in the case of a spiculum of ice in water, or of any crystal in its saturated solution, when the solubility increases with the temperature. Now as there is energy driving the molecules at the surface into the fluid state even against a resistance, let the diminution of temperature constituting the resistance be applied perfectly gradually. It must at first, when infinitely small, and afterwards while extremely small, be unable to hold in check the force or energy tending to produce melting or dissolving of the crystalline molecules at the surface. But that tendency must be overcome before the process of dissolving can be reversed, and converted into a process of deposition of new crystalline matter from the solution, and so the lowering of temperature must be increased to some definite amount before the reversal can take place. The only plausible objection that I can see to this reasoning is that perhaps it might be urged that, although to commence the reversal of the process a real lowering of temperature (not infinitely small) would be required; yet that the first film of new crystalline matter deposited might be supposed to lay itself down free from stress and that then growth might subsequently proceed by farther deposition on this unstressed surface. The objection does not however seem to me to be valid.

In answer to the suggested objection I would say that if its first step held good it would merely amount to this, that a new crop of crystals would grow over the surface of the first one from great numbers of points where deposition might first begin, while the spaces on the surface of the original crystal between those points would proceed to dissolve away, and the new crop of crystals would be undermined or cut off at their roots by this dissolving process;—and so no real covering of unstressed crystalline matter would be thrown over the stressed surface of the original crystal. That I am justified in saying that between the points where deposition of unstressed matter would first begin (on the supposition that it can begin at all) the spaces on the surface of the original crystal would proceed to dissolve away, is evident from this;—that on the commencement of deposition of unstressed matter in proximity to a stressed part of the original crystal, there would instantly be a rise of temperature to the ordinary melting or dissolving temperature; and the cold which previously checked the dissolving of parts of the original crystal exposed to the fluid and under stress would vanish by spending itself in producing the growth of unstressed new crystals. But, we find it to be the fact that a crystal *can* grow although to some extent stressed; for instance stressed by its own buoyancy or by its own heaviness relatively to the surrounding fluid;—for if its continuous growth were interrupted by the stress, we could have no crystals at all but only perhaps some sort of powdery or muddy depositions of incipient crystals.

Hence on the whole I think the view I have put forward is confirmed and is very likely to be correct; namely that additional depositions on a stressed crystal will be likewise stressed; and that the stress in the original crystal will give a resistance to the deposition of more crystalline matter in continuous structural molecular arrangement with itself.

*　*　*　*　*　*　*　*　*　*

JAMES THOMSON.

KILMICHAEL.
October 31, 1861.

MY DEAR JAMES,

* * * I think I see some more light about deposition on stressed surfaces but have not time to write to-night except to say that, in supposing a crystal pulled artificially you do not take into account the molecular condition of the fixtures or appliances at the ends. It will depend on how these are dealt with during the deposition whether the added matter is under stress or not. But with a spiculum of ice growing vertically upwards, no doubt the whole deposited matter will be under uniform stress.

*   *   *   *   *   *   *   *   *   *

W. THOMSON.

31 WELLINGTON ROAD,
DUBLIN.
November 13, 1861.

MY DEAR WILLIAM,

* * * Now I think the action at the middle of a crystal may be considered properly by itself, as merely dependent on the stress in it and the temperature of the adjacent fluid; or its temperature and strength of solution if the crystal be not of the same chemical composition as the fluid. Because we may suppose such degree of cold or of strength of solution to be applied at the ends or points of fixture, as will prevent any interfering action from being propagated from the ends to the part under consideration at the middle, so that the molecules of fluid and crystal at the middle may be left free to act for themselves in the deposition of the additional matter.

One view which may illustrate my idea would be to consider the deposition at the middle of an infinitely long crystal so that no effect could be propagated from the ends to the part under consideration. No doubt there will be much remaining to be examined into both theoretically and by observation and experiment. If one part in the length of a long crystal be subjected to melting while another part is subjected to the freezing or crystallising of additional matter to itself and if the whole be subject to stress; there may be much to be found out as to what will go on at the boundary between the part melting and the part receiving accessions. I am quite at a loss to understand such things as the transverse striae on rock crystal, and the minute facets of geometrical forms such as triangles often visible on the nearly plane pyramidal faces of the termination of the rock crystal.

Your affectionate brother
JAMES THOMSON.

P.S. I may add that if a crystal, say of ice, were pulled longitudinally, and if the fixtures at the ends were such as to expose the ends to more intense stress than the middle, and if then the cold were made such as to cause some deposition of additional ice to occur; I would suppose that the deposition would occur at the middle; although the intensely stressed parts at the ends might be, at the same time, melting away.

GLASGOW,
November 20, 1861.

MY DEAR JAMES,

I have not had time to think with any effect about the ice problem all this time, and therefore rather than wait longer to reply to your last, I shall write one case for objection, or question, with reference to your views. Suppose *CD, EF* to

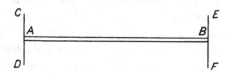

be fixed ends to a vessel of water. (Imagine them perfectly rigid, and kept at a fixed distance from one another.) Now if the spiculum or bar of ice *AB*, fixed to them at its ends, be in a state of stress, such as would be produced by initially pulling the two ends asunder, and leaving them at an invariable distance, do you imagine that ice deposited in the way of augmentation on *AB* will all lie down in equal stress, and will therefore cause an increasing force pulling the ends together? I can imagine that there would be some increase on the force pulling the ends together, produced by the augmentation of the ice, but not that it would go on indefinitely. I think it would be necessary to go on applying artificially means of stretching (by clasps successively applied for instance) to the newly deposited ice, to make it have the same stress as the original spiculum. It seems to me most probable that without such artificial management, the newly deposited ice would gradually have less stress, and in fact be gradually heterogeneous to some extent in its crystalline arrangement, although not necessarily discontinuous like a distinct crystalline deposit. If A is a clamped end of a spiculum initially pulled out longitudinally, I think it probable that augmentations might deposit on it somewhat according to the dotted lines, and that these deposits would take off something from the stress of the portions GH of the bar.

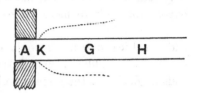

This would bring the end section K to bear rather more than its original stress, because the GH portion would contract a little. The end section K would therefore probably melt away, and ultimately break off and relieve the whole bar from tension. The inner and outer portions in the bar that remains would be in states of relative constraint then of a somewhat complicated character.

I am by no means confident that all the above is correct, but I suggest it for your consideration. Still I *cannot* see that an initial stress of two or three grains weight for instance on a fine spiculum, could originate a powerful action tending to draw in the sides of the vessel, and (provided these do not yield at all) increasing without limit as the crystal grows, which I think according to your supposition, as first stated at least, would be the case. We were glad to have so much better accounts from you last time. I hope the next will be still better, and will soon come. I am getting on fairly with my work, and my leg I think is not suffering.

Your affectionate brother,
WILLIAM THOMSON.

FROM A LETTER TO WILLIAM THOMSON OF 29 NOV. 1861
WRITTEN BY MRS JAMES THOMSON.

James thinks when he next writes to you that he will be able to answer satisfactorily the remarks you make (in letter of Nov. 20) as to the growth of a crystal under stress. In the meantime he wishes me to tell you that it is not necessary for his case to maintain the *indefinite* growth in thickness of a long spiculum when under stress. You point out that there might be a tendency to melt away at the extremities or points of attachment of the crystal when pulled longitudinally until at last breakage might occur at one extremity, but James says it is enough for his case to consider what goes on at the middle of the crystal where it is growing thicker during the time previous to the supposed breakage and if it grows when under stress he thinks each thin lamina of the fresh ice will form itself subject to like stress with that of the lamina on which it lays itself down. He also thinks that he is entitled theoretically to suppose that the cold producing the additional freezing may be designedly so distributed as to prevent any one part of the spiculum, whether at the end or at the middle, from falling behind the rest in growth. That is to say, a little extra cold applied at the parts where you suppose the crystal to thin away would prevent it from thinning and would make it grow equally with the rest or even more so if sufficient cold were applied. James wishes you to observe that spicula of ice growing as they actually do in the middle of a mass of water frozen externally as a crust, are subject to cross strains in consequence of their own tendency to flotation. These strains are no doubt exceedingly small, but he thinks if a force were applied upwards at the middle of one of them so as to bend it decidedly with an elastic bending, the crystal might still continue to grow if sufficient cold were supplied to the adjacent water to produce the solidification of new ice under like stress with that of the part of the crystal on which the new ice would grow. James would not be surprised if it should turn out that stresses in a straight spiculum might have something to do with the origination of branch crystals growing off from it; the branches being at the commencement produced at points where accidentally the crystals might grow a little quicker than at intervening points and so being subject to less strains than those intervening points of the original spiculum. However he does not wish to be positive on this part of the matter.

P.S.

November 30, 1861.

My dear William,

I did not manage to get the foregoing letter ready in time to send off to you last night. On reading it over to-day along with your letter I think it forms my full answer to your letter. You will see that it is of no consequence for my case to maintain the *indefinite* growth of the crystal by uniform laminar additions. I require only the admission that at any moment an infinitely thin lamina added to the previous surface will have the same stress as that previous surface has, and that consequently to cause a crystal of ice to grow when under stress a greater cold will be requisite than to cause one to grow when free from stress. It will then follow as suggested in the foregoing letter that at the most stressed points the growth will be resisted but will be left more free at any points of the crystals of ice which may be free from stresses. This I think may tend to cause original straight spicula to grow out into branches thus:—though I have not sufficiently observed the actual growth of crystals to venture to say much positively on this point. I am feeling better to-day and was teaching at the College for three hours, and out walking afterwards.

Your affectionate brother,
JAMES THOMSON.

W. Thomson replies to this in a letter of Dec. 10th, 1861, as follows :—

"I think your last on the growth of crystals is satisfactory, but it is a very difficult subject, and I must think more about it before I write again on it."

38. CORRESPONDENCE WITH H. C. SORBY, F.R.S., ON GEOLOGICAL EFFECTS OF PRESSURE.

BROOMFIELD,
SHEFFIELD.
February 5, 1862.

MY DEAR SIR,

I make no apology in doing what I should be glad for anyone to do to me, and therefore write to you, although not known to you personally.

I have just received the *Proceedings of the Royal Society,* and read your paper on crystallization etc. p. 473* with the greatest pleasure; since it has been published most opportunely, and serves most admirably to clear up a curious geological difficulty. The fact is, I had come to very nearly the same conclusions as you

* [No. 36, supra, p. 236.]

have, from an entirely different course of reasoning. I had made out certain facts, and, at a meeting of our Chemical Society, had argued that, if pressure would cause a crystal in a saturated solution to be dissolved and crystallize again in some other part where there was not the pressure, the whole of the difficulties would be removed. I brought forward the subject in order to learn if anyone had made any experiments on the subject, and based some of my arguments on the case of ice, on which you have written, and in relation to which you have made such interesting discoveries. I had planned to myself the trying of just such experiments as you suggest, but other things have prevented my doing so.

It will perhaps be well to give you a short sketch of what I have done, which has led me to form my conclusions; and I should esteem it a great favour if you will give me your opinion as to the accuracy of my explanations.

In the Nagelflue in Switzerland are certain conglomerates of limestone pebbles; and, when these pebbles are in contact and have apparently been forcibly pressed against each other, one has often made its way into the other; thus in Section, *A*. I have prepared various thin sections of these pebbles to examine with the microscope, according to the plan described in my various published papers, and I have thus succeeded in proving most completely that the impressed pebble did not yield as a *plastic* or as a *rigid* body to the impressing pebble, but the depression has been due to the *actual removal* of the limestone, and that not *mechanically* but *chemically*. Crystalline carbonate of lime is found on the exterior of the pebble, and in cracks where there would be no pressure; and, if it be correct to form the conclusion you have printed in italics about the middle of p. 474* of your paper, i.e. if the pressure would cause the carbonate of lime to be dissolved from the part where one pebble pressed against the other, and to crystallize out in another part, where free from pressure, the whole, or nearly the whole, of the difficulties would be removed. I may say that this question has been much studied in Germany and France, and the best geologists there have failed to give a satisfactory explanation of these curious impressed pebbles; but my methods

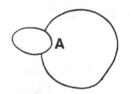

* [See foot of p. 237.]

of microscopical investigation appear likely to remove all the doubts. The only thing not entirely cleared up, if the principles you advocate be correct, is that very often the pebble which makes the impression has not itself been acted on; but perhaps we may suppose that its nature was such that its exterior resisted the pressure in such a manner that the calcareous part was not subjected to the same amount of pressure as in the case of the other, for both are impure Limestone, and sometimes they contain a large proportion of sand etc. It is curious that your paper should have appeared so opportunely, for I was going to commence writing a paper on the subject, to be published in Germany, as soon as I had finished a few things I now have in hand; and I am glad I had not written it sooner.

Trusting that the great interest, which, as you see, I take in the subject of your paper, will be a sufficient excuse for my thus writing to you, and only wishing I could show you my objects and talk the matter over with you here,

I remain, Yours very truly,

H. C. SORBY.

To H. C. SORBY, ESQ., F.R.S.

2 DONEGALL SQUARE WEST,
BELFAST.
March 14, 1862.

MY DEAR SIR,

I hope you will not attribute to any want of interest in the subject of your letter, as to limestone pebbles having been penetrated into by others, my having allowed so long time to elapse with little more than a brief acknowledgment of receipt of your letter. The phenomena you describe appear to me to be of very great interest; and I think it most likely that you are right in suggesting that the theoretical views in my paper to the R. S. on Crystallization and Liquefaction as influenced by Stresses in the Crystals, may probably be coupled with your observed phenomena as affording the theory of their explanation and as confirming the supposition you had been finding yourself led to, from quite different grounds, to the effect that pressure might probably cause a crystal in a saturated solution to be dissolved and to crystallize again in some other part where there was not the pressure, since, as you argue, if this supposition were correct the whole of the difficulties as to your observed phenomena would be removed.

With reference to one point which you mention, the fact namely that "very often the pebble which makes the impression has not itself been acted on." I think we are quite entitled to account for this by supposing nearly as you suggest, that the two may not be of exactly equal solubility or that the one may, from a slight difference in chemical composition, or from some other slight difference, be not quite so easily soluble as the other.

I shall look forward with great interest to the farther development of the line of observation and reasoning you have entered on in respect to this matter, and it would give me great pleasure if opportunity should permit of my seeing your objects and conversing with you on the subject. I think all you have described in your letter tallies perfectly well with my views.

Believe me,

Yours very truly,

JAMES THOMSON.

BROOMFIELD,
SHEFFIELD.
March 22, 1862.

MY DEAR SIR,

You will doubtless be pleased to learn that, in the same manner as your prediction about the thawing of ice was proved by experiment, so also now does there appear every probability of your prediction about the solution of salts under pressure being established. I have been making some experiments, and though they are not anything like so complete as I intend to make them, and do not establish the fact beyond all doubt, yet they so far clearly prove that it is uncommonly well worth while to carry them out more thoroughly. I have experimented with a solution of common salt and portions of rock salt, since it is soluble to the same extent, or at all events sufficiently nearly so, as not to be influenced by any small change in temperature. I have proved that the effects of the heat of the hand, that might influence the result, in examination and preparing for the experiment, are not appreciable; nor was there any solution when the crystal was exposed in the tube used for the experiment for 24 hours. However, when in the same tube and in the same concentrated solution exposed for 15 minutes to a pressure of 100 lbs. to the square inch, then for 10 minutes to 150, and at last for 5 minutes to 200 lbs. per square inch, though the amount dissolved was only small, there was no kind of doubt whatever about it, as observed by myself and two friends who are used to experiments.

I am going to try a much longer action, so as to be more certain, but in the meantime thought you might like to know that so far the result is so satisfactory.

Yours very truly,

H. C. SORBY.

To Mr Sorby's letter of 22nd March, 1862, I reply on the 24th thanking him for writing: and I add:—

"As I know you have done much in investigating slaty cleavage in rocks, I wish to suggest that it seems to me that perhaps some interesting results in the formation of slaty cleavage might be arrived at by pressing between two flat plates (such as those of a screw letter copying press) Roman or Portland Cement or some other quick or slow setting hydraulic limes when worked with water to the pasty or clayey consistency. The screw of the press I think should be turned at regular short intervals during the solidification (crystallizing?) of the stony substance of the cement. I offer this only as a mere casual suggestion. You may be better able to judge, than I, of the prospect of any interesting results being obtainable in that way."

In a P.S. to same letter I add some recommendations that the cement should be kept under water and I give this figure [omitted].

BROOMFIELD,
SHEFFIELD.
August 20, 1862.

MY DEAR SIR,

Before leaving here for the Highlands I thought I would write a few lines to report the progress I have made in my researches. The subject has grown famously. I have proved beyond all doubt that when a salt occupies *less* space when it dissolves it is *more* soluble, and when *more* space *less* soluble, under pressure. I have devised a plan of keeping up a pressure

of from 100 to 200 atmospheres for weeks and months together
without any trouble. So far I have not made out the law of the
increased or decreased solubility, but it varies *directly* as the
pressure for chloride of sodium, and I think the facts show that
it does not vary merely as the decreased space when various salts
are compared together. This is however not all, for I find that
mechanical pressure modifies chemical decomposition, and that
certain changes go on half as fast again under a pressure of
100 atmospheres. Much remains to be done, but facts seen in
studying the microscopical structure of rocks clearly show that
mechanical pressure has in some cases been the governing
principle in chemical changes, and this, combined with the
experiments hitherto made, lead me to conclude provisionally
that there is a correlation between chemical action and mechanical
forces, and that in the same way that we can get mechanical force
from chemical action, so we may also get chemical action out of
mechanical force. Will not this be a good step in advance?
I of course look upon it with the eye of a geologist, and can
clearly perceive that it will enable us to explain in the most
simple manner a whole lot of facts which have hitherto been
nearly incomprehensible.

I hope you will be at Cambridge in October, for I shall take
a lot of my specimens with me, but shall not read a paper on the
subject.

Yours very truly,

H. C. SORBY.

[Cf. H. C. Sorby, "On the Direct Correlation of Mechanical and Chemical
Forces," Bakerian Lecture, 30 April 1863, *Proc. Roy. Soc.* Vol. XII. p. 538.
Also, "On Impressed Limestone Pebbles as Illustrating a New Principle in
Chemical Geology," *Proc. Geological and Polytechnic Soc. of the West Riding
of Yorkshire*, 22 Nov. 1865, Vol. IV. p. 458.]

39. ON THE DISINTEGRATION OF STONES EXPOSED IN BUILDINGS
AND OTHERWISE TO ATMOSPHERIC INFLUENCE.

[From the *Report of the British Association*, Section A: p. 35
Cambridge, 1862.]

THE author having first guarded against being understood as
meaning to assign any one single cause for the disintegration of
stones in general, gave reasons to show:

1st. That there may frequently be observed cases of disintegration which are not referable to a softening or weakening of the stone by the dissolving away or the chemical alteration of portions of itself, but in which the crumbling is to be attributed to a disruptive force possessed by crystalline matter in solidifying itself in pores or cavities from liquid permeating the stone.

2nd. That in the cases in question the crumbling away of the stones, when not such as is caused by the freezing of water in pores, usually occurs in the greatest degree at places to which, by the joint agency of moisture and evaporation, saline substances existing in the stones are brought and left to crystallize.

3rd. That the solidification of crystalline matter in porous stones, whether that be ice formed by freezing from water, or crystals of salts formed from their solutions, usually produces disintegration—not, as is implied in the views commonly accepted on this subject, by expansion of the total volume of the liquid and crystals jointly, producing a fluid pressure in the pores—but, on the contrary, by a tendency of crystals to increase in size when in contact with a liquid tending to deposit the same crystalline substance in the solid state, even where, to do so, they must push out of their way the porous walls of the cavities in which they are contained, and even though it be from liquid permeating these walls that they receive the materials for their increase.

40. On Ground or Anchor Ice and its Effects in the St Lawrence.

[Printed from MS. of paper read at Natural History and Philosophical Society, Belfast, 7th May, 1862.]

[An abstract by myself was printed in the *Whig* and reissued in the Proceedings. J. T. It was reprinted in the *Philosophical Magazine*, Vol. XXIV. 1862, pp. 241—4.]

IF we watch the progress of the freezing and subsequent thawing, of a pond or lake of still water, we may find no indications of powerful action in the ice. Everything seems dead and passive. The stiff covering of ice forms itself and disappears without perhaps breaking a bulrush or a blade of grass which it may have held in its firm grasp, or disturbing perceptibly a pebble which it may have enclosed ; and, when a few days or weeks have elapsed, and the water has resumed its mobility, no apparent trace may remain to indicate the previous occurrence of so different a. condition of the lake surface.

Often, however, in other circumstances ice manifests itself as an agent of almost irresistible power, effecting important geological changes, and taxing severely, and often baffling the constructive skill of men. We see ice gathered in mountain valleys as masses of length and breadth which may be measured in miles, and depth in hundreds of feet; and we find these masses grinding and excavating the hard rocky beds over which they pass, and protruding down among cultivated fields, and overwhelming bridges, villages, or any other structures which may, too incautiously, have been built within the limits of their possible extension. We see icebergs swimming freely at sea, ready to crush the strongest ships; and we find them conveying huge masses of rock, and depositing them in the ocean at hundreds of miles distance from their previous place.

The ice of great rivers, in climates where the winter-cold is severe, produces striking geological effects; and it is by far the most formidable difficulty to be encountered in the construction of bridges, and other engineering works, in the river channels, or adjacent to their banks.

The ice of rivers forms itself in two very distinct modes, one as *surface ice* like that formed on the surface of still water, and the other as ice of a spongy or soft kind adhering to the bottom. The ice formed at the bottom is designated by the various names *ground ice, anchor ice* and *frasil ice.*

When a lake or pond of still water at any temperature above 40° F. is exposed to cold at its surface, the portions cooled are diminished in bulk; and, on account of the increase of their specific gravity, they sink to the bottom, and so a circulation is maintained until the temperature of the whole has arrived at 40° F. the point at which water attains its maximum density. Any farther cooling beyond that point expands the water to which it is applied, and causes it to float. Thus the previous circulation is brought to an end; and the coldest stratum of water, lying at the top, soon begins to freeze. The crust of ice when formed, and especially when it has become thick, opposes great resistance to the downward passage of the cold; and the cold which, in consequence of this resistance, arrives slowly at the under side of the crust spends itself immediately in freezing the first water it meets with. In this way, the deep and still water is protected from the cold, and retains approximately the temperature of its maximum density 40° F.; and never freezes, nor even approaches near to the freezing point.

Such being the conditions of the congelation of lake water, it has long been regarded by geologists and others as a singular circumstance that ice should ever be found growing at the bottom of a river, and many speculations have been advanced to account for the phenomenon. As yet, however, it would appear, none has been put forward which has met with general acceptance as at all satisfactory. Thus Sir Charles Lyell in a recent edition of his *Principles of Geology*, after referring to the discussions on this subject and treating of its geological bearings, leaves the origin of the ground ice as being still an unsettled point.

From among the suggestions which at various times have been offered, we may now set out by accepting as quite correct the view that the essential difference between the circumstances of the freezing of lake and river water, is that in the former case the water is left undisturbed to the action of the cold, and is allowed to adjust itself in strata in which the coldest parts, being also the lightest, float to the top,—while in the rivers the whole water is, by mixing due to its rapid flow, brought to a uniform temperature at the freezing point from top to bottom, and is thus brought into a condition in which it is ready to freeze at any part where additional cold may be applied.

It remains however still to be asked why is it that, in many cases, the freezing occurs first and chiefly at the bottom, or that the ice is found first to collect in masses there?

To account for this it was suggested more than 30 years ago that radiation of heat from the bottom, through the water, up to the cold atmosphere or sky, might keep the bottom cooled so as to effect the freezing of ice there from the water already ice cold. Arago rejected this supposition entirely and assigned two other reasons for the growth of the ice at the bottom. Firstly he supposed that there would be a peculiar aptitude to the formation of crystals of ice on the stones and asperities at the bottom (the water being throughout at the freezing point), like as there is found to be a special readiness to the formation of crystals on rough bodies in saline solutions:—and secondly, he supposed that the existence of less motion of the water at the bottom would favour the growth of the crystals there. Neither of these supposed reasons appears to me to be tenable now, when the principles of crystallization are better understood than they were at the time when this distinguished French naturalist discussed the subject:— for, first, the water of a rapid river when freezing has abundance of small spicula or fragments of ice floating and diffused through it, every one of which offers at least as free a point for the reception of new ice as can be presented by asperities on the bottom:— and secondly, the slower motion at the bottom would not favour the occurrence of freezing of new ice there, rather than at the top. Indeed, on the contrary, the greater fluid friction at the bottom, and the heavier pressure there, are causes slightly tending to oppose the freezing of new ice at the bottom.

Shortly after Arago's writings on this subject, which were published in France in 1833, the same subject was elaborately discussed in the Royal Society of London in a paper by the Rev. James Farquharson*. He maintained the radiation theory, and supposed that the ground ice grows under water much in the same way as hoar frost does on land. His paper is certainly very valuable as affording information from his own careful observations on the growth and appearances of the ground ice, in rivers where he had often seen it in England.

In his description of the growth, the appearance, and the qualities of this ice he states that when it first begins to form on the bottoms of the streams, it presents a rudely symmetrical appearance, which may be compared to little hearts of cauliflowers, fixed on the bottom, having a circular outline and a protuberance in the centre, with coral-like projections, and having a shining silvery aspect; and that these increase in numbers and in size until the whole bottom is covered with them. He mentions also that this ground ice is a cavernous mass, of variously sized pieces of ice, all small, adhering in an apparently irregular manner by their sides, angles, or points promiscuously, and having their interstices occupied with water.

These characters of the ground ice agree, I think, perfectly with the view which I have to propose for its origin, and tend to confirm that view, rather than to establish the supposition that the ice grows like hoar frost by crystallization at the bottom.

* *Phil. Trans.* 1835.

My own view, which I may at this stage submit, is that the crystals of ice are frozen from the water at any part of the depth of its stream;—whether the top, the middle, or the bottom; where cold may be introduced, either by contact or radiation; and that they may also be supplied in part by snow or otherwise; and that they are whirled about in currents and eddies until they come in contact with some fixed objects to which they can adhere, and which may perhaps be rocks or stones, or may be pieces of ice accidentally jammed in crevices of the rocks or stones; or may be ground ice already grown from such a beginning.

That pieces of ice under water have the property of adhering to one another with a continually increasing firmness, and this even when the surrounding water is above the freezing temperature, has been shown in a set of very interesting experiments by Professor Faraday, of which I had the pleasure of submitting my view of the reasons to this Society in the course of last Session*. I think too that the ready adhesion to the bottom, or to ice already anchored there, may possibly be increased by the effects of radiation, but I am confident that the anchor ice is not formed by crystallization at the place where it is found adhering.

That loose crystals of ice in a river might by their union form ground ice, is not at present quite a new suggestion; for I find in a Report by Mr Keefer, one of the Engineers who were engaged in the surveys for the project of a railway bridge over the St Lawrence at Montreal, the idea stated that snow, falling into the water, may be converted into frasil or anchor ice. This is, I believe, correct so far as it goes, but it is far from a full explanation, more especially when unconnected with any experiments to show a capability in crystals of ice to unite themselves when in contact under water; and it does not appear to have been generally accepted as satisfactory by people on the spot at Montreal, where the phenomena of the ground ice are met with on a grand scale.

The view of Mr Hodges, the Managing Engineer of the Contractors for the great bridge since built there across the St Lawrence, is quite different. His view is given incidentally in a graphic description of the ice of the St Lawrence, in his large and valuable work on the *Construction of the Victoria Bridge* published since the completion of that great undertaking. From the opportunities and necessities he met with, for observing

* [See *supra*, p. 231.]

and studying the ice phenomena of the St Lawrence, during the six years or more in which he was engaged in the building of the bridge, and in which the ice difficulties were the greatest with which he had to contend, his description of the phenomena may readily be accepted as valuable, and as worthy of much confidence. It relates, not merely to the ground ice, but to the ice phenomena in general of the St Lawrence. It describes the surface ice, and the formidable 'shovings' which occur, when the ice is beginning to fix itself across the surface of the river in winter, and afterwards when it is breaking up in spring; and it is so interesting that I trust I may be permitted to quote it with but little abbreviation.

In submitting his description, I shall, however, have to dissent entirely from his theoretical view of the formation of the ground ice.

"The ice," he says, "begins to form in the St Lawrence about the beginning of December. Then along the shores, and in the shallow quiet places where the current is least strong, a thin ice begins to make its appearance, gradually showing signs of increasing strength and thickness. Soon after, pieces of ice begin to come down from the lakes above; and then, as winter advances, anchor or ground ice comes down in vast quantities, thickening the otherwise clear water of the river. A word as to the ' Anchor Ice.' It appears to grow in rapid currents, and attaches itself to the rocks forming the bed of the river, in the shape of a spongy substance not unlike the spawn of frogs. Immense quantities form in an inconceivably short space of time, accumulating until the mass is several feet in depth. A very slight thaw, even that produced by a bright sunshine at noon, disengages it, when, rising to the surface, it passes down the river with the current. This description of ice appears to grow only in the vicinity of rapids, or where the water has become aerated by the rapidity of the current. It may be that the particles or globules of cold air are whirled by eddies until they come into contact with the rocky bed of the river, to which they attach themselves, and being of a temperature sufficient to produce ice, become surrounded with the semi-fluid substance of which anchor ice is formed. ' Anchor Ice ' sometimes accumulates at the foot of rapids in such quantities as to form a bar across the lake (similar to bars of sand at mouths of rivers) of some miles in extent, lifting the water in its locality several feet above its ordinary level. This frequently happens at the foot of

the Cedar Rapids at the head of Lake St Louis, where a branch of
the Ottawa empties itself into the St Lawrence. Upon such occasions the water at this point is dammed up to such a height as to
change its course, and run into the Ottawa, at the rate of some
four or five miles per hour. From thence it finds its way back
into the St Lawrence by the Rapids of St Anne, after performing
a circuit of some ten or twelve miles. The accumulation of ice
continues, probably for several weeks, till the river is quite full, and
so thickened as to make the current sluggish and cause a general
swelling of the waters. The pieces, too, become frozen together,
and form large masses, which by grounding, and diminishing the
sectional area of the river, cause the waters to rise still more (there
being always the same quantity of water coming over the rapids).
Then the large masses float and move further down the river,
where, uniting with accumulations previously grounded, they offer
such an obstruction to the semi-fluid waters that the channels
become quite choked, and what is called a ' jamb' takes place.

"The surface ice, arrested in its progress, packs into all sorts
of imaginable shapes; and, if the cold is very intense, a crust is
soon formed, and the river becomes frozen over till many square
miles extent of surface packed ice is formed. As the water rises,
the jamb against which this field rests, if not of sufficient strength
to hold it in place, gives way; when the whole river, after it is
thus frozen into one immense sheet, moves *en masse* down stream,
causing the 'shovings' so much dreaded by the people of Montreal.
The edges of the huge field moving irresistibly onwards, plough
into the banks of the river, in some instances to a depth of several
feet, carrying away everything within reach. In places the ice
packs to a height of 20 or 30 feet, and goes grinding and
crushing onwards till another jamb takes place, which, aided by
the grounded masses of packed ice upon the shoals and shores,
offers sufficient resistance to arrest in its progress the partially
broken up field.

"As the winter advances and the cold increases, the field of
packed ice becomes stronger, and as the lakes above become frozen
over, the ice from thence, which had hitherto tended so much to
choke the channel, ceases to come down, and the water in the river
gradually subsides, till it assumes its ordinary winter level, some
12 feet above its height in summer. The ' Ice Bridge,' i.e. the
complete and solid condition of the ice in the river, now be-

comes permanently formed for the winter, and this generally takes place about the first or second week in January. The thickest virgin ice seldom exceeds three feet. Upon the clear blue waters of the St Lawrence it is perfectly transparent.

"By the middle of March, the sun becomes very powerful at midday, which with the warm heavy rains, so affects the ice as to make it rotten, or, as it is usually called 'honeycombed'; and when it is in this state, a smart blow from any sharp-pointed instrument will cause a block, even though three feet thick, to fall into thousands of pieces, as if it was composed of millions of crystallised reeds placed vertically.

"The ice when it becomes thus weakened, is easily broken up by the winds, particularly in places where from the great depth of water in the lakes, they do not entirely freeze over. This ice, coming over the rapids, thickens the water, and causes a rise in the river as in early winter. The weakened fields of ice then begin to break up, and in a few days the river becomes free, excepting upon the wharves and some particular parts of the shore, where shovings may have taken place. In these places ice may be seen for many weeks. When the lake ice comes down before that in the river and its lower basins become rotten, great 'shovings' take place resulting in jambs, and the consequent rise of the water level."

The anchor ice then, you will observe, from Mr Hodges's description, appears to grow in rapid currents and attaches itself to the rocks forming the bed of the river in the shape of a spongy substance not unlike the spawn of frogs. Immense quantities form in an inconceivably short space of time, accumulating till the mass is several feet in depth. It sometimes accumulates at the foot of rapids in such quantities as to form a bar across the river, and if the river at the foot of the rapids enters a lake, as the St Lawrence does a little way above Montreal, this bar comes to resemble the bars of sand formed at the mouths of rivers; only that the ice bar collects and adheres together to the extent of damming up the river for several feet above its ordinary level which the bars of sand from want of tenacity will scarcely do.

Mr Hodges thinks that small bubbles of cold air are whirled by the eddies till they come in contact with the rocky bed of the river, to which they attach themselves, and then communicating their cold to the water at the bottom which is already at the freezing point, convert it into the anchor ice. With reference to this speculation it may suffice to point out that the cold which could be conveyed down into the water by small bubbles would be quite inadequate to produce the results in question. The bubbles would indeed be raised to the temperature of the surrounding water long before reaching the bottom, and so if they were capable of freezing any appreciable quantity of water the freezing would occur at the top and middle rather than at the bottom.

The whole of the facts as described by various observers I think bear out the new view which I have already offered, to the effect that the growth of the ice at the bottom of a river is due to the adhesion of spicula or other small pieces of ice formed elsewhere by cold; the adhesion being due to the property of ice at the freezing point now usually designated as 'regelation,' and of which in former papers I have stated my views at length.

P.S. I intended to say (but forgot in writing the paper) that the rapid growth of ground ice as a bar at the foot of rapids is much more like the accumulation there of ice formed elsewhere, than the crystallising of new ice there.

[The following note on this paper has been kindly supplied by Prof. H. T. Barnes, F.R.S., of Montreal, who has made extensive practical investigations on the mode of occurrence of, and the problems presented by, Ground Ice and Anchor Ice. An account of earlier discussions, including papers by Arago, Elsdale, and Farquharson, has been published by Prof. Barnes in *Trans. R. S. Canada*, 1906.

IT has been shown within recent years, by accurate measurements of water temperatures, that a river is subject to variations of temperature during winter months, depending on the meteorological conditions. With the advance of cold weather, and the air temperature falling below the freezing point, the water soon produces a quantity of ice. In the case of a lake, or river flowing very gently—one mile per hour or less—surface ice quickly forms, and increases in thickness as the winter progresses.

In an open river flowing too swiftly for surface ice to form, the effect of the cold air is to produce minute crystals throughout the mass of the river, chiefly by surface contact. These minute crystals are carried down below the surface by the eddies in the swiftly moving water.   In the absence of sun, and with the action of a cold wind or of excessive radiation at night, the temperature of the river falls a few thousandths of a degree below the freezing point, and during this time, the water and bodies immersed in it become super-cooled.   The flowing ice crystals under such circumstances readily adhere to the bottom or to objects immersed in the water.   The effect of the sun in offsetting the super-cooling of the water is very marked; and during the time of sunny weather no ice is formed below the surface, no matter how cold the air may be, and such ice as has been formed during the previous condition of super-cooling is very rapidly melted away and disintegrated.   The action of the sun is to cause a rise in the water temperature of the order of one-hundredth of a degree,—or more, if there are many days of sunny weather and a long stretch of open water.   It is evident that Prof. Thomson regarded the sticking of the ice below the surface of the water as being due to regelation.   That, however, can play no part in the phenomena; but the whole matter becomes simple and easy of explanation in view of the condition of super-cooling, which has been observed time and again in a river like the St Lawrence.

Prof. Thomson remarks that he is confident that anchor ice is not formed by crystallization at the place where it is found adhering.   While there is a great quantity of anchor ice produced from the sticking of the surface-formed crystals, yet there are undoubtedly times when the nocturnal radiation is so strong during a clear cold night that great quantities of anchor ice are produced *in situ*.   Both methods of growth are found in operation, but undoubtedly the condition of super-cooling greatly accelerates the formation.   In Canada, the surface-formed crystals go by the name of 'frazil,' and it is quite an easy matter to distinguish true frazil from the snow which has fallen into the water during a storm.]

41. On Conditions affording Freedom for Solidification to Liquids which tend to Solidification, but experience a Difficulty of making a Beginning of their Change of State.

[Printed from "memoranda of Communication which I made to the Natural History and Philosophical Society, Belfast, yesterday evening the 20th April, 1864. I was led to make this communication at that time, yesterday evening, though it had not been previously announced, through the subject being touched on in my paper for the evening 'On the Action of Frost in Disintegrating Earth and Stones*,' and because there was time available for the Society to receive extra communications. I had been, independently of this immediate inducement, in the intention of giving sometime a paper on the subject. J. T. 21st April, 1864."]

I REFERRED to my taking a different view from Faraday as to such phenomena as that water may be cooled below the freezing point without freezing, but that the introduction of a spiculum of ice will instantly cause solidification to set in. Faraday thought that this is an indication that ice in contact with water has a tendency to induce or constrain the water to assume the same physical condition as the ice itself. I mentioned my having come to a different or contrary opinion, that I think that when water and ice are in contact together the freedom of each to change to the other's state on the slightest possible cause being applied tending to such change, is perfect†.

The new matters on this subject, which I communicated, were in reference to the well known experiment with a strong solution of sulphate of soda warmed in a bottle, corked, and allowed to cool, and its solidifying then, often by the simple withdrawal of the cork,—or if not by that, its solidification on being touched with an iron nail or perhaps any of numerous other substances. My idea is that some quality of the nail affords the requisite disturbance for allowing the particles to go into the state to which they tend, but for the commencement of the change to which a difficulty exists which must be got over. I found that withdrawal of the cork could not of itself be the cause; also that the introduction of atmospheric pressure could not be the cause, for often it did not

* [See No. 42 next page.]
† [For development of this fundamental idea see J. Willard Gibbs' great memoir on the "Equilibrium of Heterogeneous Substances," 1876–8, reprinted in his *Scientific Papers*, Vol. I. p. 320.]

go off; shaking the bottle would not do it; pouring to another vessel would not do it; but I suppose there often is a partial vacuum and that particles of dust or little crystals of sulphate of soda are blown in. I found that I could drop in pieces, or crumbs, of effloresced sulphate of soda from the cork without setting it off. I therefore suppose the sulphate of soda must be in the crystalline state to produce the effect.

I found that numerous objects such as iron nails, common pins (brass tinned?) etc. if wet in the warmed solution and left in it to cool *did not cause* solidification, but that a dry nail afterwards inserted did.

I found that solidification *would* pass through a bladder. So the pores of a bladder do not suffice to prevent solidification in them; and this I mentioned as being an exception to the phenomenon which I had been pointing to as very usual, and exemplified in the case of ice freezing from some kinds of moist earth in my other paper. [See next paper.]

42. DISINTEGRATION OF EARTH BY COLUMNAR GROWTH OF ICE.

MS. Notes. 7th March 1864. After meeting of Literary Soc.

[Foundation of Paper "On the Action of Frost in Disintegrating Earth and Stones," *Belfast Nat. Hist. and Phil. Soc.* 20 April 1864.]

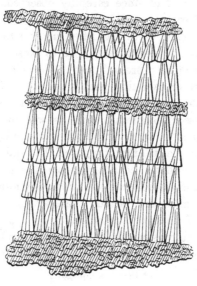

Rather than freeze within the pores of the moist earthy substance the water prefers to rise out of the moist earth, particle by particle, and to push before it upwards the load of columnar ice, gravel, etc. above it, and to draw up after it the water held hanging below it through capillary tension or pull—or, in other words, drawn up subject to less than atmospheric pressure by negative head of from zero up to about 4 ins. or more.

Thus the remarkable phenomenon showed itself of water passing from a region of less than atmospheric pressure into a place and condition, in the base of the columns of ice, where it was subject to more than atmospheric pressure, being loaded with the ice, etc., above it:—and this action went on rather than that the water would freeze within the pores of the moist earthy bottom.

Observations on *Water in Frost Rising against Gravity rather than Freezing in the Pores of Moist Earth.*

[Partly from MS. notes for paper read at Brit. Assoc. Edinburgh, 1871; and partly from abstract printed in *Brit. Assoc. Volume*, Part II, p. 34.]

In this paper Professor Thomson, in continuation of a subject which he had brought before the British Association at the Cambridge Meeting in 1862*, on the Disintegration of stones exposed to Atmospheric Influences, adduced some remarkable instances which he had since carefully observed. One of these was the case of earthy mud between paving stones in a paved drain at Belfast. The earth was lifted up and to a great extent detached from the paving stones, so as to present an appearance as if the stones had been pressed down by some extraordinary load rolled over them, or ramming with a pounder. On picking out some pieces of the raised frozen mud, and cutting them in various directions, Professor Thomson found, as was to be anticipated, that they contained ice frozen in such a way as might be supposed to result if the ice in forming itself had taken or received additional particles from the adjacent porous mud, while these in interposing themselves between the previously formed ice and the adjacent porous mud had pushed that porous substance out of their way, and if finally the mud on being deprived of a great portion of the water by which it had previously been made very wet had hardened by the freezing of the remaining moisture in its

* [See *supra*, p. 257.]

pores. The frozen mud showed the appearance of great numbers of stratified laminae of deposition from water. These laminae were, in many places, separated by interposed laminae of transparent ice, which no doubt must have intruded itself molecularly and forced open the laminar space for itself.

In another instance, observed by him in February 1864, he showed that water from a pond in a garden had in time of frost raised itself to heights of from four to six inches above the water surface level of the pond by permeating the earth bank, formed of decomposed granite, which it kept thoroughly wet, and out of the upper surface of which it was made to ascend by the frost, so as to freeze as columns of transparent ice. Rather than freeze within the pores of the moist earthy substance the water prefers to rise out of the moist earth, particle by particle; and to push before it upwards the load of columnar ice, gravel etc. above it, and to draw up after it the water held hanging below it through capillary tension or pull: or what may be called a less than atmospheric pressure. Thus the remarkable phenomenon showed itself of water passing from a region of less than atmospheric pressure in the wet pores of the earth, into a place and condition in the base of the columns of ice where it was subject to more than atmospheric pressure; and to elastic stresses unequal in different directions, from being loaded with the mass of ice and gravel above it: and this action went on rather than that the water should freeze within the pores of the moist earthy bottom on which the columns stood, and which was above the water surface level of the pond. The columns were arranged in several tiers one tier below another*, the lower ones having been later formed than those above them, and having pushed the older ones up. From day to day during the frost the earth remained unfrozen, while a thick slab of columnar ice, made up of successive tiers of columns, formed itself by water coming up from the pond and insinuating itself forcibly under the bases of the ice columns so as to freeze there, pushing them up, not by hydraulic pressure, but on principles which, while seeming not to have been noticed previously to their having been suggested by the author at the Cambridge meeting, appear to involve considerations of scientific interest, and to afford scope for further experimental and theoretical researches.

* [See reproduction of author's sketch on p. 269.]

43. LATER INVESTIGATIONS ON THE PLASTICITY OF GLACIER ICE.

Speculations, partly new, On Plasticity of Ice and of Hot Iron and Steel, etc.

[MS. signed J. T. 11th Feb. 1888.]

NOTWITHSTANDING the results published by Dr J. F. Main* (*Royal Soc. Proceedings* May 5, 1887, pp. 329—30), and other statements, to like effect, that ice can manifest some sort of viscous or slow plastic yielding to stress even when decidedly below the freezing point, I think the *principle* of my own hypothesis or theory of The Plasticity of Ice published in the Paper on Crystallisation and Liquefaction as influenced by Stresses in the Crystals† *holds as good as ever* if we merely substitute the word *Fluifaction*‡ for *Liquefaction* in the title and in the general tenor of the paper.

It is quite as easy to imagine gaseous water substance permeating between contiguous crystalline particles which are not in continuous crystalline arrangement, as to imagine liquid water substance permeating these interstices of discontinuity.

In iron and steel at high temperatures as in casehardening and in cementation? (making of blistered steel from iron bars) it is quite certain that atoms movable, diffusible, do permeate among parts that have crystalline characters, and that rearrangements of crystallisation do take place. This seems closely allied to the kinds of interpenetration imagined in my theory of Plasticity of

---

* [In reply to a request from Lord Rayleigh, Prof. Thomson sent the following opinion on the paper to the Royal Society:—

"I have read the paper with great interest, I consider the Experimental Researches to be of a very important kind:—of a kind which I have long wished that some person would enter on and conduct with care and discrimination.

"I think it highly desirable that Dr Main's paper should be published *in full* in the *Proceedings*."

The experimental investigation reported in Dr Main's paper was followed up by an important memoir of date June 11, 1888, by James C. McConnel, M.A., Fellow of Clare College, Cambridge, and Dudley A. Kidd (*Roy. Soc. Proc.* Vol. XLIV. pp. 331—367) regarding which correspondence is printed *infra*.]

† [No. 36 *supra*, p. 236.]

‡ *Postscript, 27th June* 1889, *as to nomenclature only*.

Solidify, Solidification suggest Fluidify, Fluidification. Solidify and Fluidify correspond well in form for two mutually reverse processes.

Ice and other bodies exemplified in my Experiments on Plasticity of Salt in its own brine.

Note as to Plasticity of Ice at temperatures considerably below the Freezing Point.   Date of Note is Sat. 22 June, 1889 : but the ideas noted have been mainly or wholly devised by me long before now in connection with my noted theory on the subject of a few years ago :—

The movable diffusing atoms may just as well be gaseous as liquid in the interstices between crystals of the ice.   Crystals of ice quite below the freezing point evaporate and may so give material for condensing again somewhat as in hoar frost.   Solid camphor evaporates without passing from solid into liquid state and thence to gaseous.

Correspondence with the late Mr J. C. McConnel, Clare College, Cambridge.

HOTEL BUOL, DAVOS, SWITZERLAND.
May 26th, 1889.

MY DEAR SIR,

I can hardly hope that you will have any spare copies of papers written nearly thirty years ago, but if you have I should be greatly obliged if you would send me one or two.   The winter before last a friend and I made some experiments on the plasticity of ice, in which we found amongst other things that a single crystal showed no signs of continuous yielding.   Lord Rayleigh informs me that you predicted this in a paper in the *R. S. Proc.* for 1860.   There is also another paper quoted by Everett (*Deschanel*, p. 313) which appeared in the *R. S. Proc.* for 1861 on the Effect of Stress on Melting and Solution.   I am very anxious to read both papers.   But living out here as I do, I have very seldom the chance of referring to a good library, so it would be a great boon to me if you could send me copies.   I send you by this post a copy of my paper.

Yours sincerely,

JAMES C. M^{c}CONNEL.

Memo. of letter to Mr McConnel of date 11th June 1889.

...I tell him that the theory or speculation offered by me in the Paper on Crystallization and Liquefaction as influenced by

T.   18

Stresses tending to change of form in the Crystals* I have ever since believed to be true; and that some time ago I wrote out a statement of further extensions of that theory towards explaining plasticity in ice at temperatures below the freezing point, and various other cases of plasticity probably including that of red hot and white hot iron, and the welding of iron. I say also to him:— "Your experiments which, among other conclusions seem to show an absence of plasticity in a continuous crystal, I regard as very important."

J. T.

HOTEL ENDERLIN, PONTRESINA.
June 16th, 1889.

MY DEAR SIR,

Your very kind letter has been forwarded to me here where I am staying for a short visit. I shall probably be back in Davos within a fortnight. I am extremely obliged for your kind offer to take so much trouble on my account and if you can let me have printed copies of the two papers I should be very grateful. But I can hardly ask you to have a MS. copy made. Indeed it is probable that my doctor will allow me to make a short visit to England—mid. August to mid. September—, and in passing through London I can read the two papers at the British Museum Library and copy such parts as are to me specially important. So I beg that you will not think of going to Glasgow on my account and that you will not spend much time in the search for the printed copies.

I am curious to learn your explanation of the plasticity of ice at temperatures well below the freezing point. My own views are as follows. Since a bar of ice before and after stretching consists of irregular crystals perfectly fitting each other, and since a single crystal will not stretch, it is evident that there must be accommodation at the interfaces. One crystal must grow at the expense of the next.

It remains for experiment to decide or rather discover which crystal grows, i.e. how the motion of the molecules from crystal to crystal depends on the direction of the optic axes and the stress. I made an observation yesterday on this matter. In one of the glacier grottoes I saw a projecting tongue of ice which had, as the

* [No. 36, *supra*, p. 236.]

layers of stratification showed, originally been horizontal but had
bent down under its own weight into this position.   I cut away
some pieces of ice about the point marked with a
cross, where there had been great longitudinal
pressure and long continued yielding, and took
them into the sunshine.   As usual the little discs
appeared in the ice, which lie at right angles to
the optic axis, and enabled me to see that the majority of the
crystals had their optic axes at right angles to the stress.   This
suggests that under pressure the crystals whose axes are end on,
give up their material to others.

Buchanan supposes that at any temperature glacier ice contains
liquid in the shape of strong brine, which diminishes in quantity
and increases in strength as the temperature falls.   But if this
lies in thin films separating the crystals, would a bar be able to
support a tension of several atmospheres ?

I am yours very truly,

JAMES C. M^cCONNEL.

KNOCKDOLIAN, COLMONELL, AYRSHIRE.

September 5th, 1889.

DEAR SIR,

Your paper on " Crystallisation and Liquefaction as influenced by Stresses Tending to Change of Form in the Crystals "
has been forwarded to me here.   It was extremely kind of you to
have a manuscript copy made.   On my way through London I
paid a visit to the library at the British Museum and read your
paper.   But it will be most useful to me to have the full text at
hand, as it appears to me to have a most important bearing on
the plasticity of ice.   I have not yet attempted to follow out the
new thoughts suggested by it; for I am taking complete holiday,
spending a few weeks at home before starting again for another
year in Switzerland.   I shall leave for Davos in about a fortnight.

I was very sorry to hear of the failure of your eyesight and
that you have been compelled to resign your Professorship.   But
I hope that your eyes may, as you say, do much good service yet.
With many thanks I am

Yours very truly,

JAMES C. M^cCONNEL.

18—2

CONTINUITY OF STATES IN MATTER

44. THE TRANSITION FROM THE LIQUID TO THE GASEOUS STATE IN MATTER. (*Unpublished notes.*)

[As is well known, the idea of connecting the liquid with the gaseous state, by a continuous inflexional curve of transition, at temperatures below Andrews' critical point, originated with Prof. James Thomson. Very little had been published beyond the mere suggestion until the matter was taken up in detail in 1873 by van der Waals; thus it has been decided to include the following manuscript notes for historical reasons. The next stage in the evolution of the subject is represented adequately by Clerk Maxwell's important review of van der Waals' famous dissertation in *Nature*, Vol. x. Oct. 15, 1874, pp. 477—480, reprinted in Maxwell's *Collected Works*, Vol. II. p. 407, to which reference should be made. The models of surfaces, referred to on p. 282, as illustrating the continuous transition for specified kinds of matter, still exist, and have been reproduced here (p. 277). They represent the germ of the more effective types of thermodynamic surfaces introduced by Willard Gibbs in 1873; cf. Gibbs' *Collected Works*, Vol. I. pp. 1—32, 33—54, especially notes on pp. 34, 35, and Maxwell's *Theory of Heat*, ed. 4, 1875, pp. 195—207.]

[THE first of the small models of the various states of carbonic acid, as represented by points determined by volume temperature and pressure as coordinates, here reproduced, is marked as "Model cut out between June 6 and June 9, 1862"; the other one, with a horn to represent the unstable region, is marked as "Cut 9th May 1869." A larger copy of the first one, made exactly to scale from Dr Andrews' experiments, was exhibited at the South Kensington Loan Exhibition of Scientific Apparatus in 1876. The model was exhibited by Dr Andrews at a Royal Institution lecture in 1871: see his *Scientific Papers*, p. 340. Afterwards Clerk Maxwell went much further by constructing to scale for carbonic acid the Thermodynamic Surface then recently introduced by Willard Gibbs; see his *Theory of Heat*, ed. 4, 1875, pp. 195—207; cf. also a letter in Stokes' *Scientific Correspondence*, Vol. II. p. 34.]

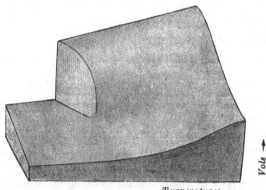

Temperatures →

" Model cut out between June 6 and June 9, 1862 : see MS. notes of that time."

A query (?) is marked on the model as to whether the plane face (which is a section through the boiling-line in No. 2) should be undercut as in No. 2 into a curved surface.

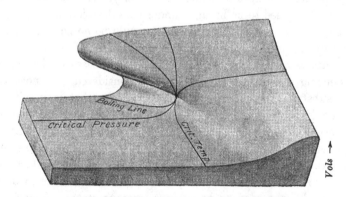

Temperatures →

" Cut 9th May 1869 : see also 1862 papers."

The third dimension is *pressure*: the critical lines are plane sections.

SOLID MODELS GIVING CURVES OF CHANGE OF STATE.

45. CONSIDERATIONS ON THE ABRUPT CHANGE AT BOILING OR CONDENSING IN REFERENCE TO THE CONTINUITY OF THE FLUID STATE OF MATTER.

[From the *Proceedings of the Royal Society*, Nov. 16, 1871.]

WHEN we find a substance capable of existing in two fluid states different in density and other properties while the temperature and pressure are the same in both, and when we find also that an introduction or abstraction of heat without change of temperature or of pressure will effect the change from the one state to the other, and also find that the change either way is perfectly *reversible*, we speak of the one state as being an ordinary gaseous, and the other as being an ordinary liquid state of the same matter; and the ordinary transition from the one to the other we would designate by the terms boiling or condensing, or occasionally by other terms nearly equivalent, such as evaporation, gasification, liquefaction from the gaseous state, etc. Cases of gasification from liquids or of condensation from gases, when any chemical alteration accompanies the abrupt change of density, are not among the subjects proposed to be brought under consideration in the present paper. In such cases I presume there would be no perfect reversibility in the process; and if so, this would of itself be a criterion sufficing to separate them from the proper cases of boiling or condensing at present intended to be considered. If, now, the fluid substance in the rarer of the two states (that is, in what is commonly called the gaseous state) be still further rarefied, by increase of temperature or diminution of pressure, or be changed considerably in other ways by alterations of temperature and pressure jointly, without its receiving any abrupt collapse in volume, it will still, in ordinary language and ordinary mode of thought, be regarded as being in a gaseous state. Remarks of quite a corresponding kind may be made in describing various conditions of the fluid (as to temperature, pressure, and volume), which would in ordinary language be regarded as belonging to the liquid state.

Dr Andrews (*Phil. Trans.* 1869, p. 575) has shown that the ordinary gaseous and ordinary liquid states are only widely separated forms of the same condition of matter, and may be made to pass into one another by a course of continuous physical changes presenting nowhere any interruption or breach of continuity. If we denote geometrically all possible points of pressure and temperature jointly, by points spread continuously in a plane surface, each point in the plane being referred to two axes of rectangular coordinates, so that one of its ordinates shall represent the temperature and the other the pressure denoted by that point, and if we mark all the successive boiling- or condensing-points of temperature and pressure as a continuous line on this plane, this line, which may be called *the boiling-line*, will be a separating boundary between the regions of the plane corresponding to the ordinary liquid state and those corresponding to the ordinary gaseous state. But, by consideration of Dr Andrews's experimental results, we may see that this separating boundary comes to an end at a point of pressure and temperature which, in conformity with his language, may be called the *critical point* of pressure and temperature jointly; and we may see that, from any ordinary liquid state to any ordinary gaseous state, the transition may be effected gradually by an infinite variety of courses passing round outside the extreme end of the boiling-line.

Now it will be my chief object in the present paper to state and support a view which has occurred to me, according to which it appears probable that, although there be a practical breach of continuity in crossing the line of boiling-points from liquid to gas or from gas to liquid, there may exist, in the nature of things, a theoretical continuity across this breach having some real and true significance. This theoretical continuity, from the ordinary liquid state to the ordinary gaseous state, must be supposed to be such as to have its various courses passing through conditions of pressure, temperature, and volume in unstable equilibrium for any fluid matter theoretically conceived as homogeneously distributed while passing through the intermediate conditions. Such courses of transition, passing through unstable conditions, must be regarded as being impossible to be brought about throughout entire masses of fluids dealt with in any physical operations. Whether in an extremely thin lamina of gradual transition from a liquid to its own gas, in which it is to be noticed the substance would not be homogeneously distributed, conditions may exist in a stable state having some kind of correspondence with the unstable conditions here theoretically conceived, will be a question suggested at the close of this paper in connexion with some allied considerations.

It is first to be observed that the ordinary liquid state does not necessarily cease abruptly at the line of boiling-points, as it is well known that liquids may, with due precautions, be heated considerably beyond the boiling temperature for the pressure to which they are exposed. This condition is commonly manifested in the boiling of water in a glass vessel by a lamp placed below, when the temperature of the internal parts of the water, or, in other words, of the parts not exposed to contact with gaseous matter, rises considerably above the boiling-point for the pressure, and the water boils with bumping*. At this stage it becomes desirable to refer to Dr Andrews's diagram of curves showing his principal results for carbonic acid, and to consider carefully some of the remarkable features presented by those curves. In doing so, we have first, in the case of the two curves for 13°·1 and 21°·5, which pass through the boiling interruption of continuity, to guard against being led, by the gradually bending transition from the curve representing obviously the liquid state into the line seen rapidly ascending towards the curve representing obviously the gaseous state, to suppose that this curved transition is in any way indicative of a gradual transition from the liquid towards the gaseous state. Dr Andrews has clearly pointed out, in describing those experimental curves, that the slight bend at about the commencement of the rapid ascent from the liquid state is to be ascribed to a trace of air unavoidably present in the carbonic acid; and that if the carbonic acid had been absolutely pure, the ascent from the liquid to the gaseous state would doubtless have been quite abrupt, and would have shown itself in his diagram by a vertical straight line, when we regard the coordinate axes for pressure and volumes as being horizontal and vertical respectively. Now in the diagram here submitted the continuous curves (that

---

* It has even been found by Dufour (*Bibliothèque Universelle, Archives*, year 1861, Vol. xii. "Recherches sur l'ébullition des Liquides ") that globules of water floating immersed in oil, so as neither to be in contact with any solid nor with any gaseous body, may, under atmospheric pressure, be raised to various temperatures far above the ordinary boiling-point, and occasionally to so high a temperature as 178° C., without boiling.

On this subject reference may also be made to the important researches of Donny, " Sur la cohésion des Liquides et sur leur adhérence aux Corps solides," *Ann. de Chimie*, year 1846, 3rd Ser. Vol. xvi. p. 167.—July 28, 1871.

is to say, those which are not dotted) are obtained from Dr Andrews's
diagram, with the slight alteration of substituting, in accordance
with the explanations just given, an abrupt meeting instead of the
curved transition between the curve for the liquid state and the
upright line which shows the boiling stage.   Looking to either
of the given curves which pass through boiling, and, for instance,
selecting the curve for 13°·1, we perceive, from what has been said
as to the conditions to which boiling by bumping is due, that for

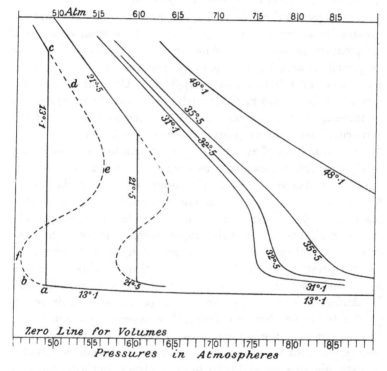

the temperature pertaining to this curve the liquid state does not
necessarily end at the boiling pressure for this temperature, and
that thus in the diagram the curve showing volumes for the liquid
state must not cease at the foot of the upright line* which marks
the boiling stage of pressure, but must extend continuously, for
some distance at least, into lower pressures in some such way as

* [The position of the upright straight line is determined by its enclosing
equal areas with the dotted curve on its two sides, as was first remarked by Maxwell in 1875 in a lecture to the Chemical Society, *Nature*, Vol. XI. p. 358, *Collected
Papers*, Vol. II. p. 425.]

is shown by the dotted continuation from *a* to *b*. But now the question arises, Does this curve necessarily end at any particular point *b*? We know that the extent of this curve in the direction from *a* towards or past *b*, along which the liquid volume will continue to be represented before the explosive or bumping change to gas occurs, is very variable under different circumstances, being much affected by the presence of other fluids, even in small quantities, as impurities in the fluid experimented on, and by the nature of the surface of the containing vessel, etc.

The consideration of the subject may be facilitated, and aid towards the attainment of clear views of the mutual relations of temperature, pressure, and volume in a given mass of a fluid may be gained, by actually making, or by conceiving there to be made, for carbonic acid, from the data supplied in Dr Andrews's experimental results, a solid model consisting of a curved surface referred to three axes of rectangular coordinates, and formed so that the three coordinates of each point in the curved surface shall represent, for any given mass of carbonic acid, a temperature, a pressure, and a volume which can coexist in that mass. It is to be noticed here that in his diagram of curves the results for each of the several temperatures experimented on are combined in the form of a plane curved line referred to two axes of rectangular coordinates, one of each pair of ordinates representing a pressure, and the other representing the volume corresponding to that pressure at the temperature to which the curve belongs. Now to form a model such as I am here recommending, and have myself made, Dr Andrews's curved lines are to be placed with their planes parallel to one another, and separated by intervals proportional to the differences of the temperatures to which the curves severally belong, and with the origins of coordinates of the curves situated in a straight line perpendicular to their planes, and with the axes of coordinates of all of them parallel in pairs to one another, and then the curved surface is to be formed so as to pass through those curved lines smoothly or evenly*. The curved surface so obtained exhibits in a very obvious way the remarkable phenomena

---

* For the practical execution of this, it is well to commence with a rectangular block of wood, and then carefully to pare it down, applying, from time to time, the various curves as templets to it, and proceeding according to the general methods followed in a shipbuilder's modelling-room in cutting out small models of ships according to curves laid down on paper as cross sections of the required model at various places in its length.

of the voluminal conditions at and near the critical point of
temperature and pressure, in comparison with the voluminal
conditions throughout other parts of the range of gradually varying
temperatures and pressures to which it extends, and even throughout a far wider range into which it can in imagination be conceived
to be extended. It helps to afford a clear view of the nature and
meaning of the continuity of the liquid and gaseous states of
matter. It does so by its own obvious continuity throughout its
expanse round the end of the range of points of pressure and
temperature where an abrupt change of volume can occur by
boiling or condensing. On the curved surface in the model
Dr Andrews's curves for the temperatures 13°·1, 21°·5, 31°·1,
32°·5, 35°·5, and 48°·1 Centigrade, which afford the data for its
construction, may with advantage be all shown drawn in their
proper places. The model admits of easily exhibiting in due
relation to one another a second set of curves, in which each would
be for a constant pressure, and in each of which the coordinates
would represent temperatures and corresponding volumes. It
may be used in various ways for affording quantitative relations
interpolated among those more immediately given by the
experiments.

We may now, aided by the conception of this model, return
to the consideration of continuity or discontinuity in the curves
in crossing the boiling stage. Let us suppose an indefinite
number of curves, each for one constant temperature, to be drawn
on the model, the several temperatures differing in succession by
very small intervals, and the curves consequently being sections
of the curved surface by numerous planes closely spaced parallel
to one another and to the plane containing the pair of coordinate
axes for pressure and volume. Now we can see that, as we pass
from curve to curve in approaching towards the critical point from
the higher temperatures, the tangent to the curve at the steepest
point or point of inflection is rotating, so that its inclination to
the plane of the coordinate axes for pressure and temperature,
which we may regard as horizontal, increases till, at the critical
point, it becomes a right angle. Then it appears very natural to
suppose that, in proceeding onwards past the critical point to
curves successively for lower and lower temperatures, the tangent
at the point of inflection would continue its rotation, and the
angle of its inclination, which before was acute, would now become

obtuse. It seems much more natural to make such a supposition
as this than to suppose that in passing the critical point from
higher into lower temperatures the curved line, or the curved
surface to which it belongs, should break itself asunder, and should
come to have a part of its conceivable continuous course absolutely
deficient. It thus seems natural to suppose that in some sense
there is continuity in each of the successive curves by courses
such as those drawn in the accompanying diagram as dotted curves
uniting continuously the curves for the ordinary gaseous state with
those for the ordinary liquid state.

The physical conditions corresponding to the extension of the
curve from a to some point b we have seen are perfectly attainable
in practice. Some extension of the gaseous curve into points of
temperature and pressure below what I have called the boiling-
or condensing-line (as, for instance, some extension such as from c
to d in the figure) I think we need not despair of practically
realizing in physical operations. As a likely mode in which to
bring steam continuing gaseous to points of pressure and temperature at which it would collapse to liquid water if it had any
particle of liquid water present along with it, or if other circumstances were present capable of affording some apparently *requisite
conditions for enabling it to make a beginning of the change of
state**, I would suggest the admitting speedily of dry steam
nearly at its condensing temperature for its pressure (or, to use a
common expression, *nearly saturated*) into a vessel with a piston
or plunger, all kept hotter than the steam, and then allowing the
steam to expand till by its expansion it would be cooled below
its condensing-point for its pressure; and yet I would suppose
that if this were done with very careful precautions the steam
might not condense, on account of the cooled steam being surrounded
entirely with a thin film of superheated steam close to the superheated

---

* The principle that "the particles of a substance when existing all in one
state only, and in continuous contact with one another, or in contact only under
special circumstances with other substances, experience *a difficulty of making a
beginning of their change of state*, whether from liquid to solid, or from liquid to
gaseous, or probably also from solid to liquid," was proposed by me, and, so far as
I am aware, was first announced in a paper by me in the *Proceedings of the Royal
Society* for November 24, 1859 (Vol. x. p. 158) [see *supra*, p. 228], and in a paper
submitted to the British Association in the same year.

In the present paper, at the place to which this note is annexed, I adduce the
like further supposition that *a difficulty of making a beginning of change of state
from gaseous to liquid* may also probably exist.

containing vessel.   The fact of its not condensing might perhaps
best be ascertained by observations on its volume and pressure.
Such an experiment as that sketched out here would not be
easily made, and unless it were conducted with very great precautions, there could be no reasonable expectation of success in
its attempt; and perhaps it might not be possible so completely
to avoid the presence of dust or other dense particles in the steam
as to make it prove successful.   I mention it, however, as appearing
to be founded on correct principles, and as tending to suggest
desirable courses for experimental researches.   The overhanging
part of the curve from e to f seems to represent a state in which
there would be some kind of unstable equilibrium; and so,
although the curve there appears to have some important theoretical
significance, yet the states represented by its various points would
be unattainable throughout any ordinary mass of the fluid.   It
seems to represent conditions of coexistent temperature, pressure,
and volume in which, if all parts of a mass of fluid were placed,
it would be in equilibrium, but out of which it would be led to
rush, partly into the rarer state of gas, and partly into the denser
state of liquid, by the slightest inequality of temperature or of
density in any part relatively to other parts.   I might proceed to
state, in support of these views, several considerations founded on
the ordinary statical theory of capillary or superficial phenomena
of liquids, which is dependent on the supposition of an attraction
acting very intensely for very small distances, and causing intense
pressure in liquids over and above the pressure applied by the
containing vessel and measurable by any pressure-gauge.   That
statical theory has fitted remarkably well to many observed
phenomena, and has sometimes even led to the forecasting of new
results in advance of experiment.   Hence, although dynamic or
kinetic theories of the constitution and pressure of fluids now
seem likely to supersede any statical theory, yet phenomena may
still be discussed according to the principles of the statical theory;
and there may be considerable likelihood that conditions
explained or rendered probable under the statical theory would
have some corresponding explanation or confirmation under any
true theory by which the statical might come to be superseded.
With a view to brevity, however, and to the avoidance of putting
forward speculations perhaps partly rash, though, I think, not
devoid of real significance, I shall not at present enter on details

of these considerations, but shall leave them with merely the slight suggestion now offered, and with the suggestion mentioned in an earlier part of the present paper, of the question whether in an extremely thin lamina of gradual transition from a liquid to its own gas, at their visible face of demarcation, conditions may not exist in a stable state having a correspondence with the unstable conditions here theoretically conceived.

46. SPECULATIONS ON THE CONTINUITY OF THE FLUID STATE OF MATTER, AND ON RELATIONS BETWEEN THE GASEOUS, THE LIQUID, AND THE SOLID STATES.

[From the *Report of the British Association*, Edinburgh, Section A, 1871, p. 30.]

THROUGH the recent discovery of Dr Andrews on the relations between different states of fluid matter, a difficulty in the application of our old ordinary language has arisen. He has shown the existence of continuity between what is ordinarily called the liquid state and what is ordinarily called the gaseous state of matter. He has shown that the ordinary gaseous and ordinary liquid states are only widely separated forms of the same condition of matter, and may be made to pass into one another by a course of continuous physical changes presenting nowhere any interruption or breach of continuity. If, now, there be no distinction between the liquid and gaseous states, is there any meaning still to be attributed to those two old names, or ought they to be abandoned, and the single name *the fluid state* to be substituted for them both? The answer must be that in speaking of the whole continuous state we have now to call it simply *the fluid state*; but that there are two regions or parts of it, meeting one another sharply in one way, and merging gradually into one another in a different way, to which the names *liquid* and *gas* are still to be applied. We can have a substance existing in two fluid states different in density and other properties, while the temperature and pressure are the same in both: and we may then find that an introduction or abstraction of heat without change of temperature or of pressure will effect the change from the one state to the other, and that the change either way is perfectly reversible. When we thus have two different states present together in contact with one another, we have a perfectly obvious distinction, and we can properly continue to call one of them a liquid state and the other a gaseous state of the same matter. The same two names may also reasonably be applied to regions or parts of the fluid state extending away on both sides of the sharp or definite boundary, wherever the merging of the one into the other is little or not at all apparent. If we denote geometrically all possible points of temperature and pressure jointly, by points spread continuously in a plane surface, each point in the plane being referred to two axes of rectangular coordinates, so that one of its ordinates shall represent the pressure and the other the temperature denoted by that point, and if we mark all the successive boiling- or condensing-points of pressure and temperature as a continuous line on this plane, this line, which may be called *the boiling-line*, will be a separating boundary between the regions of the plane corresponding to the ordinary liquid state and those corresponding to the ordinary gaseous state. But by consideration of Dr Andrews's experimental results (*Phil. Trans.* 1869), we may see that this separating boundary comes to an end at a point of temperature and pressure which, in conformity with his language, may be called the *critical point* of pressure and temperature jointly; and we may see that from any ordinary liquid state to any ordinary gaseous state the transition may be gradually effected by an infinite variety of courses passing round the extreme end of the boiling-line.

Fig. 1 is a diagram to illustrate these considerations and some allied considerations to which they lead in reference to transitions between the three states, the gaseous, the liquid, and the solid. This figure is intended only as a sketch to illustrate principles, and is not drawn according to measurements for any particular substance, though the main features of the curves shown in it are meant to relate in a general way to the substance of water, steam, and ice. AX and AY are the axes of coordinates for pressures and temperatures respectively; A, the origin, being taken as the zero for pressures and as the zero for temperatures on the Centigrade scale. The curve L represents the boiling-line. This terminates towards one direction in the *critical point* E; it passes in the other direction to T, the point of pressure and temperature where solidification sets in. This point T is to be noticed as a remarkable point of pressure and temperature, as being the point at which alone the substance, pure from admixture with other substances, can exist in three states, solid, liquid, and gaseous, together in contact with one another. In making this statement, however, the author wishes to submit it subject to some reserve in respect to conditions not as yet known with perfect certainty. He observes that we might not be quite safe in assuming that the melting point of ice solidified from the gaseous state is the same as the melting point of ice frozen from the liquid state, and in making other suppositions, such as that the same quantity of heat would become latent in the melting of equal quantities of ice formed in these two ways. Such considerations as these into which we are forced if we attempt to sketch out the course of the

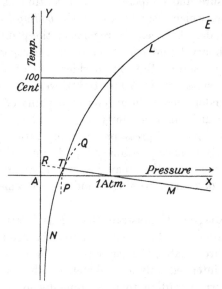

Fig. 1.

boiling-line, and to examine along with it the corresponding boundary-lines between liquid and solid and between gas and solid, may be useful in suggesting questions for experimental and theoretical investigation which may have been generally overlooked before. Proceeding, however, upon assumptions such as usually are tacitly made, of identity in the thermal and dynamic conditions of pure ice solidified in different ways, the author points out that we must suppose the three curves (namely, the line between gas and liquid, the line between liquid and solid, and the line between gas and solid) to meet in one point, shown at T in the figure. This point of pressure and temperature for any substance may then be called *the triple point* for that substance. In the figure the line TM represents the line between liquid and solid. It is drawn showing in an exaggerated degree the lowering of the freezing temperature of water by pressure, the exaggeration being necessary in order to allow small changes of temperature to be perceptible in the diagram. The line TN represents the line between the gaseous and the solid states of water substance. The two curves TL and TN, one between gas and liquid and the other between gas and solid, have been constructed for water substance through a great range of temperatures and pressures by Regnault, from his experiments on the pressure of saturated aqueous gas at various temperatures above and below 0° Centigrade[*]. He has represented and discussed his results above and below the temperature at which the water freezes (which in strictness is not 0° C., but is the freezing temperature of water in contact with no atmosphere except its own gas), as if one continuous curve could extend for both. As brought out experimentally, indeed, they present so little appearance of any discontinuity that the distinctness of the two curves from one another might readily escape notice in the consideration of the experimental results. Prof. Thomson points out, however, that the range from temperatures below to temperatures above freezing comprises what ought to be regarded as two essentially distinct curves meeting one another in the point T; and he further suggests that continuations of these curves, sketched in as dotted lines TP and TQ, may have some theoretical or practical significance not yet fully discovered[†]. He thinks it likely that out of the three curves at least the one, MT, between liquid and solid may have a practically attainable extension past T, as shown by the dotted continuation TR. Various known experiments seem to render this supposition tenable, whether the condition supposed may have been actually realized in experiments hitherto or not. He thinks, too, that there is much reason to suppose that the curve LT between gas and liquid has a practically attainable extension past T, as shown by the dotted continuation TP.

[*] *Mémoires de l'Académie des Sciences*, 1847, pl. viii.
[†] [Cf. *infra*, p. 298.]

In reference to the continuity of the liquid and gaseous states, Prof. Thomson showed a model in which Dr Andrews's curves for carbonic acid are combined in a curved surface, obtained from them, which is referred to three axes of rectangular coordinates, and is formed so that the three coordinates of each point in the curved surface shall represent, for any given mass of carbonic acid, a pressure, a temperature, and a volume, which can coexist in that mass. This curved surface shows in a clear light the abrupt change or breach of continuity at boiling or condensing, and the gradual transition round the extreme end of the boiling-line. Using this model and a diagram of curves represented here in fig. 2 [*supra*, p. 281], the author explained a view which had occurred to him, according to which it appears probable that although there be a practical breach of continuity in crossing the line of boiling-points from liquid to gas, or from gas to liquid, there may exist, in the nature of things, a theoretical continuity across this breach, having some real and true significance. The general character of this view may readily be seen by a glance at fig. 2 [*supra*, p. 281] in which Dr Andrews's curves are shown by continuous lines (not dotted), and curved reflex junctions are shown by dotted lines connecting those of Dr Andrews's curves which are abruptly interrupted at their boiling- or condensing-points of pressure. It is to be understood that each curve relates to one constant temperature, and that pressures are represented by the horizontal ordinates, and corresponding volumes of one mass of carbonic acid constant throughout all the curves are represented by the vertical ordinates. The author points out that, by experiments of Donny, Dufour, and others*, we have already proof that a continuation of the curve for the liquid state past the boiling stage for some distance, as shown dotted in fig. 2, from *a* to some point *b* towards *f*, would correspond to states already attained. He thinks we need not despair of practically realizing the physical conditions corresponding to some extension of the gaseous curve such as from *c* to *d* in the figure. The overhanging part of the curve from *e* to *f* he thinks may represent a state in which there would be some kind of unstable equilibrium; and so, although the curve there appears to have some important theoretical significance, yet the states represented by its various points would be unattainable

---

* Donny, *Ann. de Chimie*, 1846, 3rd series, Vol. xvi. p. 167; Dufour, *Bibliotheque Universelle, Archives*, 1861, Vol. xii.

throughout any ordinary mass of the fluid. It seems to represent conditions of coexistent temperature, pressure, and volume, in which, if all parts of a mass of fluid were placed, it would be in equilibrium, but out of which it would be led to rush, partly into the rarer state of gas, and partly into the denser state of liquid, by the slightest inequality of temperature or of density in any part relatively to other parts.

47. SPECULATIONS ON THE CONTINUITY OF THE FLUID STATE OF MATTER, AND ON TRANSITIONS BETWEEN THE GASEOUS, THE LIQUID, AND THE SOLID STATES.

[From the *Proceedings of the Belfast Natural History and Philosophical Society*, November 29, 1871.]

THE author commenced by referring to two recent communications made by him to the Royal Society of London* and the British Association†, in which he had advanced some new views as to relations among the states of matter commonly named, distinctively, the Gaseous, the Liquid, and the Solid states. The paper, of which an abstract is given here, was arranged for the purpose of submitting to this society an account of the chief new views comprised in those two previous papers, together with such additional explanations and illustrations as might tend to render the subject generally intelligible; and farther, in this new paper some later theoretical considerations, constituting an additional development of one part of the subject, were, for the first time, brought forward‡.

The question may be asked at the outset, *What is to be understood by the three states—the Gaseous, the Liquid, and the Solid?* These names are familiar to all of us, and we are all accustomed to associate each with some idea, more or less definite, of a character or condition of matter. We have all been long accus-

---

* *Royal Society Proceedings*, No. 130, 1871 [*supra*, p. 278].

† *British Association Transactions*, Edinburgh Meeting 1871, p. 30 [*supra*, p. 286].

‡ See p. 32, where there are introduced considerations showing the way in which the curves between gas and liquid, and between gas and solid, must cross one another. [These are here omitted for brevity. They are further developed in Paper No. 48 *infra*, p. 297.]

tomed to conceive of matter as existing in one or other of three
states spoken of as the gaseous, the liquid, and the solid states.
The transition from each of these to any other has usually been
regarded as abrupt; or at least this may be said, if we except the
as yet imperfectly understood phenomena of gradual hardening of
some substances such as melted glass, or melted pure iron. We
have been accustomed to notice, as among the usual properties of
liquids kept at any of various temperatures, their powerfully
resisting compression, and gradually, but very slightly, expanding
on the diminution of the pressure to which they are exposed, till
they attain a certain volume dependent on their temperature,
beyond which volume they will not admit of farther gradual
expansion with gradual diminution of pressure, but at which, with
no change of pressure, an abrupt augmentation of volume supervenes, accompanied by the sudden attainment of a greatly altered
set of characters and properties commonly recognised as specially
pertaining to the gaseous state of matter. Among these altered
properties, one of the most remarkable is that of unlimited farther
gradual expansibility by farther gradual reduction of pressure.
The abrupt change of volume and of other properties thus noticed,
has ordinarily been understood as an absolute and complete
separation between all liquid states and all gaseous states of the
same matter. It is now found, however, that this distinction is
incomplete; and that it is insufficient to separate any one state,
or set of states, from any other state or set of states; or to form
a separation or distinction between what in common language
are called liquids, and what in common language are called gases.

If we consider water and steam present together in mutual
contact in a boiler or in a glass flask heated by a flame or otherwise,
we observe a very decided distinction between the liquid water,
and the steam, or gaseous water-substance, enclosed together in
the vessel. The two are identical in substance, they have both
the same pressure (one atmosphere if the vessel is open), and they
have both the same temperature (100° Centigrade; the ordinary
boiling temperature, if the vessel is open). Yet while both of
them are alike in substance, and in pressure, and in temperature,
they are very different in density, or in volume for a given mass;
and very different too in respect to their possession of latent
heat. Here we have a sharp definite distinction of two states
with a perfectly abrupt change from the one to the other, called

evaporation or condensation, which is effected without any change whatever of either pressure or temperature.

Having here a perfectly abrupt distinction, we can properly speak of the two states under different names, *liquid* and *gaseous*.

But now, to view the matter otherwise, let us commence with this liquid water and steam in mutual contact; and, whatever be their existing temperature and consequent pressure, density, and latent heat, let us for distinction call the existing condition of each its original condition; and let us take some of the liquid water away from the steam, and treat it separately. We can cool this water, and so bring it to a new condition. We can then at pleasure either relieve it from so much pressure, or we can compress it more; and in either case we effect another alteration of its state or condition. When we have it under more than its original pressure, we can heat it to more than its original temperature, and thus we bring it into yet another new condition or state. We can make as many such changes as we please, in varied ways, without meeting any abrupt change at all; or without finding any change sufficient to lead us to suspect any important alteration to have occurred requiring a new name for a new state. We only say that the liquid water has been cooled or compressed, or relieved of pressure, or compressed and warmed, or otherwise changed, by separate or combined changes of temperature and pressure; but we consider the fluid dealt with as being still liquid water in continuously variable conditions. The steam, too, which was originally in contact with the water; and so had the same temperature and pressure as the water had; is, if taken away from the water and treated separately, capable of undergoing gradually varied changes of state or condition quite corresponding to those already described which the liquid water is capable of undergoing. Thus, for instance, the steam or gaseous water-substance may without any sudden or abrupt change be raised in temperature, or lowered in pressure, or altered in both these ways conjointly, or it may be raised in temperature and raised in pressure also; and, from each state so attained to every other, transitions may be effected in an infinite variety of perfectly gradual ways. Thus we have seen (1st), that between water and steam in contact with one another, there is an abrupt transition; (2nd), that the water-substance, taken initially in its liquid condition when in contact with the steam, admits of receiving an infinite variety of perfectly gradual changes of state; (3rd), that the water-substance, taken initially in its gaseous condition when in contact with the liquid water, admits of receiving an infinite variety of perfectly gradual changes of state. Now, farther, through the researches and discoveries of Dr Andrews in recent years, we are led to conclude that these varied states of the water, and varied states of the steam, in their wide divergence from the initial conditions of the water and steam in mutual contact; which were alike with one another in pressure and temperature, but quite distinct in other respects; come to meet again: and in coming to this new meeting, very different from their previous abrupt meeting, they merge perfectly gradually into one another. He has shown by experiments on carbonic acid and several other substances, that the ordinary gaseous and the ordinary liquid states are only widely separated forms of the same condition of matter, and may be made to pass into one another by a course of continuous physical changes presenting nowhere any interruption or breach of continuity.

These experiments of Dr Andrews are to be found described in detail in his original memoir in the *Transactions of the Royal Society for* 1869, and they have been explained by himself to the Belfast Natural History and Philosophical Society. The carbonic acid or other fluid experimented on was confined in the upper part of a small glass tube, hermetically sealed at the upper end, the cavity for containing the fluid being made variable in capacity at pleasure by an arrangement for forcing in a mercurial column from below more or less by means of a hand screw. He made experiments on the effects of changes of pressure and of temperature on various fluids; but carbonic acid was the one which he selected for his most extensive and elaborate experiments. He had means provided for making observations on the co-existing pressures, temperatures, and volumes of the carbonic acid in successive experiments, with all requisite accuracy for the establishment of very remarkable conclusions. In this apparatus, when gaseous carbonic acid commencing with a light pressure is enclosed and maintained at any ordinary temperature of the atmosphere, at, for instance, 10° Centigrade or 50° Fahr., it shows liquefaction on being compressed. The liquid carbonic acid may be seen lying in the bottom of the cavity with gaseous carbonic acid above it, there being a visible surface of demarcation between the two.

But, otherwise, if this same carbonic acid, still enclosed in the cavity, be taken at first at 10° Centigrade as before, and at any pressure which allows it to remain gaseous, and if it be warmed to any temperature above 31° Cent. (that is to say, above 88° Fahr.), it may, at this higher temperature, be subjected to any pressure, however great, without showing any liquefaction. While subject to heavy pressure, it may be cooled to its original temperature of 10° Cent., and yet no appearance of liquefaction will have set in, though it be now in a state even more dense than when it was in the previous example compressed at 10° to the liquid state from the gaseous. It may now, while maintained at 10° Cent., be relieved of pressure, till at last it will begin to boil; thereby showing that it is now liquid and must have really passed from the gaseous to the liquid state, through elevated temperatures and pressures, by a series of gradual changes presenting nowhere any interruption or breach of continuity. Through this abrupt process of boiling, it reverts to the gaseous state; yet it has nowhere passed through the contrary abrupt process of condensing. Again, the reverse of this entire cycle of operations may be effected. We may this time commence at 10° Centigrade, with the cavity containing gaseous carbonic acid at top, and liquid carbonic acid below. Compress so as to diminish the volume, and condensation from gas to liquid will set in, and with continued diminution of volume will go on, without any change of temperature or pressure, till no part remains gaseous. Retaining the same temperature of 10°, press more heavily and then raise the temperature. Thus the critical temperature of 31° Cent. may be surpassed without boiling setting in. Now enlarge the cavity maintaining any temperature above 31° Cent., and the confined fluid, which we may have hitherto regarded as liquid, enlarges with the cavity and abates in pressure without boiling or showing any appearance of evaporation. Continue this process till the pressure in abating reverts to its original amount. We may now lower the temperature, retaining this regained original pressure, and on arriving at 10° Cent., while diminishing the volume, we shall find the fluid beginning to show liquefaction setting in, so that we may be sure it is now gaseous. Thus we have had a gradual transition from the liquid state to the gaseous state made manifest without any abrupt change such as that of boiling or evaporation having occurred.

We may next consider a series of changes during which gaseous and liquid carbonic acid are kept constantly present together in the cavity. Suppose that we commence at an ordinary atmospheric temperature of 10° Centigrade, and that we have already compressed gaseous carbonic acid till a portion of it has condensed to the liquid state. We are now to proceed by raising the temperature and raising the pressure so as to keep the fluid continually partly liquid and partly gaseous; and so we are to keep continually passing through a succession of boiling or condensing points of pressure and temperature jointly. To do this we may proceed thus:—

(1) Maintaining the volume constant, heat the heterogeneous fluid, but not so much as to evaporate the whole. This causes a part of the liquid to evaporate, and expands the remainder in spite of the increased pressure of the gas above. This makes the remaining liquid less dense and the gas more dense than before; and so the two have approached nearer to one another in density.

(2) Keeping temperature constant, diminish the cavity, compressing its contents so as to condense to the liquid state what had evaporated in the previous process. This does not alter the pressure, but leaves the original quantities in the liquid and gaseous states, and nearer to one another in density than at first. Go on in the same way by heating and compressing and taking care so to modulate the two processes as to maintain always a portion of the fluid ·gaseous and a portion liquid. The two operations of heating, and diminishing the volume, might equally well be carried on simultaneously; but perhaps the changes may be more clearly understood by conceiving the two processes as kept distinct, and, considering their effects, separately. In this way, as the temperature and pressure are augmented, the gaseous part is always increasing in density, and the liquid part is diminishing in density, till at last the two come to have the same density with one another, and then they are perfectly alike in every respect, all distinction between them having vanished.

At this stage the temperature is 31° Cent., and the pressure is about 75 atmospheres. Above this temperature of 31° no change of pressure can cause gasification or liquefaction; and above this pressure of about 75 atmospheres, no change of temperature can cause gasification or liquefaction.

[Then follows as in No. 46, *supra*.]

48. On Relations between the Gaseous, the Liquid, and the Solid States of Matter.

[From the *Report of the British Association*, Section A, Brighton, 1872, pp. 24—30.]

THE object of this paper is to submit some new theoretical considerations which constitute a further development of one portion of the views offered, at last year's Meeting of the Association, by the author, in his paper entitled "Speculations on the Continuity of the Fluid State of Matter, and on Relations between the Gaseous, the Liquid, and the Solid States." He has now to make reference to the abstract of that paper printed in the *Transactions* for last year at page 30*; and, in particular, to the diagram of three curves shown sketched in fig. 1 of that abstract.

In respect to these curves several essential features had been, at the time of last year's Meeting, clearly discerned, and were pointed out and reasoned on by the author in his paper then read. His attempt to sketch out the curves, however, in such a way as that they should be in agreement with the known conditions then taken into consideration, soon forced on his attention the question whether the two curves, one of which is that between gas and liquid, and the other is that between gas and solid, ought to be drawn crossing as represented here in fig. 1 *a*, or as in fig. 1 *b*; and his object at present is to give a demonstration, subsequently developed, showing that they must cross as in fig. 1 *a*; or, in other words, as in the diagram which he gave in the abstract of his last year's paper.

It is to be understood that AX and AY are the axes of co-ordinates for pressures and temperatures respectively; A, the origin, being taken as the zero for pressures, and as the zero for temperatures on the Centigrade scale; and, for simplicity in expression and in thought, the diagram may be taken as relating to the particular substance of water, steam, and ice, rather than to substances in general. The curve ETP is the *boiling-line*, or the line which has its successive points such that for any one of them the two coordinates represent a pressure and temperature for a boiling-point, or a pressure and temperature which the water and steam can have when in mutual contact. It may also be

* [*Supra*, p. 286.]

called, for brevity, the *steam-with-water line*. In like manner the
curve *NTQ* is the *steam-with-ice line*; and the curve *MTR* is the
water-with-ice line. The full meaning of these diagrams may
become more distinctly intelligible to the reader if he will advert
to the explanations given in the paper already referred to in last
year's *Transactions*, as to fig. 1 in that paper,—explanations
which, though now useful, need not be wholly repeated here, as
the present paper is meant to be read in connexion with that
previous one.

If we now look to fig. 1 *a* and suppose that we have water and
steam in mutual contact, the pressure and temperature must be

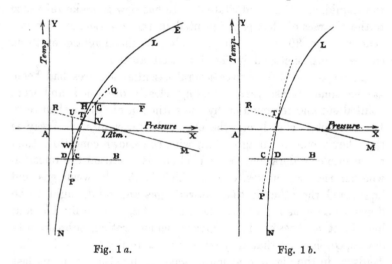

Fig. 1 *a*.   Fig. 1 *b*.

represented by the coordinates of some point of the steam-with-
water curve *LTP*. Let us now suppose that we lower the
temperature gradually while keeping water and steam in mutual
contact: the point whose coordinates show the successively coexistent temperatures and pressures will pass downwards along
the steam-with-water curve *LTP*. Let us suppose this operation
continued so far as to bring this point into that part of the curve
which belongs to temperatures below that of the triple point *T**.
This supposed extension of the steam-with-water curve into
temperatures below that of the triple point, where freezing would
certainly set in if any ice were present, is to be conceived of as a

---

* The meaning of the " Triple Point " is explained in the paper already referred
to in last year's *Transactions*, p. 32 [p. 289].

curve corresponding to states of equilibrium between the steam
and water. It is well known that water can, in various circumstances, be reduced in temperature below its freezing point without
its freezing; and this the author attributes to a difficulty of making
a beginning of change of state*. It is also known that the presence
of a gaseous atmosphere, of common air with aqueous vapour in
contact with water, does not necessarily introduce any condition
which will give liberty to the water-substance to make a beginning
of change of state into ice, either from the liquid or the gaseous
part, or from both at their face of contact. Thus there can
scarcely be a doubt but that the steam-with-water curve, LT, has
a practically attainable extension past T; and valid reasoning,
the author thinks, may certainly be founded on the supposition
of this curve as one of equilibrium between steam and water,
whether or not, in various modes of experimenting, it might be
easy or difficult or unmanageable to practically exclude all conditions which would give liberty to make a beginning of the
formation of ice. We may then see that, supposing steam and
water to be present together in a condition of temperature and
pressure represented by any point such as C in fig. 1 a, there is
perfect freedom for the transition either way between water and
steam. That is to say, while the water and steam are maintained
at the temperature and pressure of the point C, the water is
perfectly free to change to steam, and the steam is perfectly free
to change to water. Let, for brevity, the temperature and pressure
of the point C be denoted by t_1 and p_1 respectively.

Now, to aid our conception in a process of theoretical reasoning, let us imagine an apparatus possessing certain qualities in
theoretic perfection, thus:—(see fig. 2).

Let there be a cylinder, standing upright, closed at bottom,
open at top, and with a piston which works without leakage and
without friction.

Let the weight of the piston, together with the atmospheric
load on it, be balanced by a counterpoise B; or else let the whole
apparatus be conceived to be enclosed in a large external vessel
from which the air has been extracted, and then the counterpoise

---

* In papers by the author (*Proceedings of Royal Society*, Nov. 24, 1859, p. 158
[*supra*, p. 222]; and *British Association Report*, Transactions of Sections, 1859,
p. 25), the principle of attributing such phenomena to *a difficulty of making a
beginning of change of state* was, so far as he is aware, first announced.

B must just balance the weight of the piston. Let weights A, A be laid on the piston, which will give exactly the pressure p_1 to the fluid enclosed in the cavity of the cylinder; and let this enclosed fluid be supposed to be water-substance taken at first in the state of steam with water, as shown in the figure, where S is steam and W is water.

Let the entire cylinder and its contents be maintained at the temperature t_1 (a temperature below that of the triple point) by immersion in a bath at that temperature, t_1, as shown in the figure.

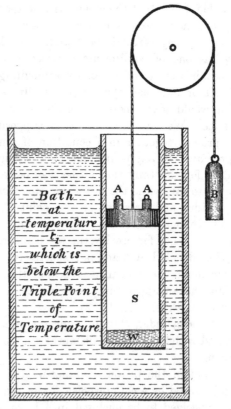

Fig. 2.

Now apply an infinitely small extra weight on the piston, so that the internal pressure becomes $p_1 + \delta$, where δ is infinitely small. This causes the steam to go perfectly gently down to water.

Now insert a particle of ice. Brisk action or agitation instantly sets in. Thus:—

(1) The water with ice cannot repose without both coming to the temperature which, for the pressure $p_1 + \delta$, or we may here as well say for the pressure p_1, belongs to water with ice; that is to say, in reference to fig. 1a, the water with ice cannot repose without both coming to the temperature of the point U on the water-with-ice line in that figure.

(2) The water at this raised temperature, or at any of the intermediate temperatures between this and the temperature t_1 of the surrounding bath, is in a state tending to ebullition into steam, a state in which boiling will ensue if a beginning be made at all, or if due facility to begin be afforded in any way.

(3) Conduction of heat, or conduction with convection, is briskly going on, conveying heat out to the bath, since the temperature inside is at some parts warmer than the bath, and is nowhere cooler.

Now, either ebullition ensues, or it does not.

First. Suppose it not to take place:—

Parts of the water are warmed by the freezing-process. They briskly transmit heat out to the bath, the freezing goes briskly on, and the same process of transmission of heat from a higher to a lower temperature goes briskly forward. This continues till all the enclosed fluid has become ice.

Now it is obvious that if there is a brisk action, with rapid conduction of heat, when steam, or water-substance partly steam and partly water, is allowed to pass into the state of ice while the pressure is $p_1 + \delta$ and the surrounding temperature is t_1, there could be no return or reversal to the old condition of steam, or of steam with water, caused or allowed by merely an infinitely small abatement of pressure from $p_1 + \delta$ to p_1. To cause the ice to evaporate, or to get it to remain in equilibrium with steam, which we know experimentally it can do at a low enough pressure, a finite (not infinitely small) abatement of pressure is necessary.

Thus has been proved what was wanted, provided we be right in supposing ebullition not to take place.

But now:—

Second. Suppose ebullition to ensue on the introduction of the ice—a complicated interaction of water, steam, and ice, involving brisk agitation, must set in. At any face of contact of water and ice, the temperature must be that of the point U in fig. 1a; at any face of contact of steam and ice the temperature must become that which belongs to the pressure p_1 on the steam-with-ice line, and which is shown at the point W in fig. 1a on the supposition of the curves crossing as represented in that figure; and at any face of contact of steam with water the temperature must be that of the point C. As yet we need not assume that we know whether the point W for pressure p_1 on the steam-with-ice line is at a higher temperature than that of C, as is represented in fig. 1a, or at a lower temperature than that of C, as it would be if the curves crossed as in fig. 1b; but clearly we know that the temperature of U is higher than that of C, which is the same as that of the bath; and we can also see that any steam in contact with water and surrounded with the bath at temperature t_1 while the pressure is p_1 will be ready to condense to water, or will actually so condense if the pressure be increased by the infinitely small augmentation δ, just as did the steam originally supposed to occupy part of the cavity. Thus we must have an action going briskly on, involving rapid conduction of heat, an action involving the continual conversion of water-substance from the fluid state (gaseous or liquid) to ice, and which goes on till no steam remains to condense to water at a face of contact with water, and till no water remains to be frozen at a face of contact with ice. As this process goes on with briskness or agitation, involving rapid conduction of heat, we can see that, as in the previously supposed case, the process is irreversible by an infinitely small abatement of pressure; and we can see that to get steam to remain in repose in contact with ice at the temperature t_1 of the surrounding bath, we must have the pressure abated by a finite amount, so as to be decidedly less than the pressure p_1 belonging to steam with water at the fixed temperature of the bath: that is to say, for a temperature below the *triple point* the pressure of steam with ice is less than the pressure of steam with water.

Hence, referring to fig. 1a, we see that in the steam-with-ice curve the point D, having the same temperature t_1 as the point C of the steam-with-water curve has, must, while situated in the isothermal line BD passing through C, be away from C at the side where the pressure is less than at C; or it must lie between C and the coordinate axis YA produced past A.

This may be regarded as very nearly establishing that the

curves cross one another, as drawn in fig. 1a. It shows that they do not, as in fig. 1b. Up to the present stage, however, the reasoning does not exclude the suppositions :—1st, that the curves might meet tangentially in the triple point T, and pass on without crossing ; 2nd, that they might cross in the triple point, meeting each other there tangentially ; 3rd, that the steam-with-ice line might absolutely stop short in the triple point.

The first and second of these remaining suppositions, depending, as they do, on supposed tangential meeting instead of meeting or crossing angularly, the author thinks very unlikely. One reason is that the condensed water-substance in contact with the steam makes a perfectly sudden change in its character in changing from water to ice or from ice to water; and he therefore thinks that in the curve which represents steam with water above the triple point, and steam with ice below it, we should expect to find a sudden change of direction at the point where this great physical change suddenly takes place.

Another reason against the first of these suppositions will be given in what follows almost immediately, by a proof that after meeting in the triple point in rising from lower temperatures, they cannot go on further without crossing. The third supposition, namely, that the steam-with-ice line might stop short in the triple point, the author thinks very unlikely to be the truth ; but he is not aware of any experimental proof to offer against it.

Now, that the curves, after meeting in the triple point in rising from lower temperatures, cannot go on further without crossing, will be proved if it be shown that on the supposition of the steam-with-ice curve not stopping short on rising to the triple point, it must, on passing that point, have its course on the side of the steam-with-water curve remote from the coordinate axis YA ; or, in other words, if it be shown that, for any temperature t_2 above the triple point, the pressure of steam with water is less than the pressure of steam with ice.

This can easily be done by a demonstration quite like the one already given for a temperature below that of the triple point ; and a brief sketch of it will here suffice.

Let us imagine that we have a cavity of variable dimensions, such as a cylinder with a piston which can be loaded so as to apply any desired pressure to fluid substance enclosed within. Let this vessel contain steam with ice at a temperature t_2, which is above that of the triple point; and let the cylinder be immersed in a bath maintained constantly at the temperature t_2. Let the pressure of the steam with ice for this temperature be called p_2.

Now increase the pressure by an infinitely small amount δ, making it $p_2 + \delta$. While this is kept applied to the steam, the steam is by it kept going down to the state of ice; and thus we can conceive of the whole or any desired part being converted quite gently to ice*. Next, while maintaining the pressure p_2 or $p_2 + \delta$ in the steam, if any remains, or in the water next to be introduced, introduce a particle of water. Instantly the ice begins to melt, and falls in temperature, at the place of contact with water, to the temperature of water with ice for the applied pressure p_2 or $p_2 + \delta$; that is, to the point V in the figure. But the surrounding bath is warmer than this, and so a decided difference of temperature is maintained, involving a rapid conduction of heat from the warmer bath to the colder melting ice and the cold water in contiguity to that ice. There can be no repose till all the water-substance originally enclosed as steam with ice has become water; because, while the steam can pass gently to ice under the pressure p_2, on the supposition that some particle of ice is kept present, and will be forced down by the infinitely small excess of pressure δ, the ice must briskly rush to the state of water. But we know we can have steam present in repose with water at the maintained temperature t_2 if we make the pressure small enough. An infinitely small abatement of pressure will not counteract or reverse the change which has been briskly taking place; and so the pressure must be made decidedly lower than either $p_2 + \delta$ or p_2 to allow of the water resting in equilibrium in contact with steam at the temperature t_2.

That is to say, referring to fig. 1a, on any isothermal line, such as FG, the point H, where it is cut by the steam-with-water line, must be nearer to the axis YA than is the point G, where it is cut by the steam-with-ice line.

This, then, closes the course of reasoning entered on hitherto in these pages, and establishes (the author thinks with very little if any room left for doubt) that the two curves do not cross as in fig. 1b; and that in meeting at the triple point, they do not meet

---

* The fact that the ice being rigid would oppose a mechanical obstruction to the complete pressing of the steam down to ice by a piston, may be noticed in passing, but it does not introduce any theoretical difficulty into the reasoning.

and pass tangentially without crossing, but that they must cross as in fig. 1*a*.

The conclusion here arrived at the author thinks may admit of experimental verification; and he thinks it opens a desirable field for further and more perfect experimental researches than have hitherto been made on the coexisting pressures and temperatures of steam and other gaseous substances, each in contact with its own substance, either in the liquid or in the solid state, at temperatures ranging above and below the triple point for each substance. Without its being necessary to make experiments on substances in the conditions represented by the dotted extensions of the curves past the triple point, he thinks that very accurate experiments might show, for steam, an obtuse re-entrant angle or corner at *T*, in the line *LTN*, which appears not to be one curve, but two distinct curves meeting in *T*, and crossing each other at that point.

Through an examination which the author has made of the experimentally derived curve given by Regnault* for what is shown as *LTN* here in fig. 1*a*, he finds that the curve seems to show a slightly perceptible feature of the kind here anticipated— a slight re-entrant angle, or at least a slightly flattened place, or place of diminished curvature at the triple point; but this feature does not appear sufficiently marked to admit of its being relied upon as a decisive experimental confirmation of the theoretical view here submitted.

The author also submitted to the Meeting the following additional considerations on the subject.

It can easily be shown that the *perpetual motion* would be theoretically attainable unless (1) the pressure of steam with ice for a temperature t_1, which is below the triple point, were less than the pressure of steam with water for the same temperature t_1; and also (2) unless the pressure of steam with water for a temperature t_2, taken above the triple point, were less than the pressure of steam with ice for the same temperature t_2.

To prove the first of these, we have to observe that at t_1, which is below the triple point, in pressing steam down into water, we give mechanical work to the substance (call this *a*). Then when we insert ice, there is a finite difference of temperatures, with conduction of heat out to the bath; now by making this heat

---

* *Mémoires de l'Académie des Sciences*, 1847, plate viii.

pass, not by conduction, but through a thermodynamic engine (an air-engine for instance), we can obtain work, which let us call b. During this freezing, too, we get back from the water-substance a little work, owing to the expansion of the water in freezing under the pressure p_1 (call this c). Next allow the volume to increase while arranging that the ice shall be evaporating into steam under the temperature of the bath t_1; we obtain mechanical work, which call d.

Now if, in this expanding process of ice to steam, the pressure were as great as p_1, which was the pressure during the compressing to water, we would get back on the whole from the piston all the work we gave to it; that is, the two portions c and d of work got back would together be as much as we gave, namely a; and we would have made a clear gain of the work b obtained from the thermodynamic engine.

A like proof could be given in respect to the second case— that in which the temperature is above the triple point.

A slight extension of this reasoning will prove that the curves, in crossing at the triple point, cannot cross tangentially.

This can be seen obviously from the consideration that the work obtainable by the thermodynamic engine is proportional to the difference of the temperatures between which the heat is transmitted; and that the difference between the work given to the piston of the cavity in compressing steam to water, and that obtained back again during the evaporation of the ice to steam, and then pressing the steam when the evaporation is complete a little down till it attains again its original pressure and volume, will be proportional, very approximately, to the difference of the pressures existing during the compression of steam to water, and the expansion of ice to steam, which latter pressure let us now call p_1'. Also let us call the temperature of the triple point t_0.

Thus it is obvious that we must have, as long as we keep very near the triple point,

$$p_1 - p_1' \propto t_0 - t_1.$$

And this shows that the crossing of the curves must be angular, not tangential.

The author further suggested that the reasoning here adduced may be followed up by a quantitative calculation founded on experimental data, most if not all of which are already available, by which calculation the difference of the pressures of steam with water and steam with ice for any given temperature very near the triple point may be found with a very close approximation to the truth.

49. A QUANTITATIVE INVESTIGATION OF CERTAIN RELATIONS BETWEEN THE GASEOUS, THE LIQUID, AND THE SOLID STATES OF WATER-SUBSTANCE.

[From the *Proceedings of the Royal Society*, No. 148, 1873.]

IN two communications made by me to the British Association at its Meetings at Edinburgh in 1871*, and at Brighton in 1872†, and printed as abstracts in the *Transactions of the Sections* for those years, considerations were adduced on relations between the gaseous, the liquid, and the solid states of matter. The new subject of the present paper constitutes a further development of some of those previous considerations; and a brief sketch of these is necessary here as an introduction for rendering intelligible what is to follow.

Taking into consideration any substance which we can have in the three states, gaseous, liquid, and solid, we may observe that, when any two of these states are present in contact together, the pressure and temperature are dependent each on the other, so that when one is given the other is fixed. Then, if we denote geometrically all possible points of temperature and pressure jointly by points spread continuously in a plane surface, each point in the plane being referred to two axes of rectangular coordinates, so that one of its ordinates shall represent the temperature and the other the pressure denoted by that point, we may notice that there will be three curves—one expressing the relation between temperature and pressure for gas with liquid, another expressing that for gas with solid, and another expressing that for liquid with solid. These three curves, it appears, must all meet or cross each other in one point of pressure and temperature jointly, which may be called the triple point‡.

---

* [*Supra*, p. 286.]   † [*Supra*, p. 302.]

‡ In making this statement, that it appears that the three curves must all cross each other in one point, I would wish to offer it here (as I previously did in the 1871 British-Association paper) subject to some reserve in respect of conditions not

The curve between gas and liquid, which may be called the *boiling-line*, will be a separating boundary between the regions of the plane corresponding to the ordinary liquid and those corresponding to the ordinary gaseous state. But by consideration of Dr Andrews's experimental results (*Phil. Trans.* 1869), we may see that this separating boundary comes to an end at a point of temperature and pressure which, in conformity with his language, may be called the *critical point* of pressure and temperature jointly; and we may see that, from any liquid state to any gaseous state, the transition may be gradually effected by an infinite variety of courses passing round the extreme end of the boiling-line*.

The accompanying figure [fig. 1a, p. 298] serves to illustrate these considerations in reference to transitions between the three states, the gaseous, the liquid, and the solid. The figure is intended only as a sketch to illustrate principles, and is not drawn according to measurements for any particular substance, though

yet known with perfect clearness and certainty. I have to suggest that we might not be quite safe in assuming that, within a cavity containing nothing but pure water-substance partly gaseous, the melting temperature and pressure of ice solidified from the gaseous state would be the same as the melting temperature and pressure of ice frozen from the liquid state, and in making other suppositions, such as that the same quantity of heat would become latent in the melting of equal quantities of ice formed in these two ways, and in neglecting conceivable but, I presume, as yet imperfectly known distinctions of capillary conditions between ice amply wet with water and ice only moistened with the last vestiges of water before the whole liquid may be either evaporated or frozen. It might be a question in like manner whether we can be sure that there can be theoretically a condition of repose in a cavity containing only perfectly pure water-substance in which the three states are present together, each in contact with the other two, so that there would be ice partly wet with water, and partly dry in contact with gaseous water-substance, or steam as it may be called, while the water and steam were also in contact with each other. I offer these remarks by way of caution, as they force themselves into notice when we attempt to sketch out the features of the three curves under consideration, and because they may serve to suggest questions for experimental and theoretical investigation which may have been generally overlooked before. In the present paper, however, I proceed on assumptions, such as are usually tacitly made, of identity in the thermal and dynamic conditions of pure ice solidified in different ways, assumptions which, so far as is known, may be, and probably are, perfectly true; and I proceed on the supposition that there can be theoretically the condition of repose here alluded to, of the solid, liquid, and gaseous states, present together each in contact with the other two—and consequently that the three curves would meet or cross each other in one point, which I have called the *triple point*.

* Mention of this condition has been already made in a former paper by me in the *Proceedings of the Royal Society*, November 16, 1871, p. 2 [*supra*, p. 279].

the main features of the curves shown in it are meant to relate in
a general way to the substance of water, steam, and ice. AX and
AY are the axes of coordinates for temperatures and pressures respectively; A, the origin, being taken as the zero for pressures and
as the zero for temperatures on the Centigrade scale. The curve L
represents the *boiling-line* terminating in the critical point E.
The line TM represents the line between liquid and solid. It is
drawn showing in an exaggerated degree the lowering of the
freezing temperature of water by pressure, the exaggeration being
necessary to allow small changes of temperature to be perceptible
in the diagram. The line TN represents the line between the
gaseous and the solid states of water-substance. The line LTN
appears to have been generally (in the discussion of experimental
results on the pressure of aqueous vapour above and below the
freezing point) regarded as one continuous curve; but it was a
part of my object in the two British-Association papers referred
to, to show that it ought to be considered two distinct curves
(LTP and NTQ) crossing each other in the triple point T.

In the second of the two British-Association papers already
referred to (the one read at the Brighton Meeting, 1872)*, I gave
demonstrations showing that these two curves LT and NT should
meet, as shown in the accompanying figure, with a re-entrant
angle at T, not with a salient angle such as is exemplified in the
vertex of a pointed arch, and offered in conclusion the suggestion
that the reasoning which had been adduced might be followed up
by a quantitative calculation founded on experimental data, by
which calculation the difference of the pressures of steam with
water and steam with ice for any given temperature very near the
triple point may be found with a very close approximation to the
truth.

In the month of last October (October, 1872) I explained to
my brother, Sir William Thomson, the nature of that contemplated
quantitative calculation: I mentioned to him the method which I
had prepared for carrying out the intended investigation, and
inquired of him for some of the experimental data, or data already
deduced by theory from experiments, which I was seeking to
obtain. On his attention being thus turned to the matter, he
noticed that the desired quantitative relation could be obtained
very directly and easily from a simple formula which he had given

in his paper on the Dynamical Theory of Heat, *Transactions of the Royal Society of Edinburgh*, March 17, 1851*, § 21 (3), to express the second law of thermodynamics for a body of uniform temperature throughout, exposed to pressure equal in all directions.

That formula is $$\frac{dp}{dt} = CM;$$

in which p denotes the amount of the pressure, and dp/dt its rate of increase per unit increase of temperature, the volume being kept constant; C denotes Carnot's function; and M denotes the rate of absorption at which heat must be supplied to the substance per unit augmentation of volume, to let it expand without varying in temperature. The body may be either homogeneous throughout, as a continuous solid, or liquid, or gas; or it may be heterogeneous, as a mass of water and aqueous vapour (*i.e.* steam), or ice and water, or ice and aqueous vapour (*i.e.* steam).

Now apply that formula, 1st, to steam with water, and, 2nd, to steam with ice, the temperature of the heterogeneous body in each case being that of the triple point; or we may, for the present purpose, say 0° Centigrade, which is almost exactly the same. It is to be observed that while in the general application of the formula the rate of increase of the pressure with increase of temperature, *when the volume is kept constant*, has been denoted by dp/dt, yet in each of the two particular cases now brought under consideration, it is a matter of indifference whether the volume be kept constant or not; because the pressure of steam in contact either with water or with ice, for any given temperature, is independent of the volume of the whole heterogeneous body; so that the change of pressure for change of temperature is independent of whether there be change of volume or not. As C is a function of the temperature which has the same value for all substances at the same temperature, it has the same value for the two cases now under consideration. Hence, retaining for the first case (that, namely, of steam with water) the same notation as before, but modifying it by the use of an accent where distinction is necessary in the second case (that of steam with ice), and thus using dp/dt to denote the rate of increase of the pressure per unit increase of temperature for steam with water at the triple point (0° Centigrade nearly), and M to denote the rate of absorption at

---

* [Sir Wm Thomson, *Math. and Phys. Papers*, Vol. I, p. 187.]

which heat must be supplied to a body consisting of steam and water at the triple point, per unit augmentation of volume of that whole heterogeneous body, to let it expand without varying in temperature, and using dp'/dt and M' to denote the corresponding rates for steam with ice at the triple point, we have

$$\frac{dp}{dt} \Big/ \frac{dp'}{dt} = \frac{M}{M'}.$$

The latent heat of evaporation of one pound of water at the freezing point (or triple point) into steam at the same temperature, as determined by Regnault, is 606·5 thermic units, the thermic unit being here taken as the heat which would raise the temperature of one pound of water one degree Centigrade; and the latent heat of fusion of ice is about 78 or 79 of the same thermic units. Hence, though M and M' belong each to a cubic foot of steam at the triple point, not to a pound mass of it, still the ratio M/M' is

$$= \frac{606}{79 + 606}.$$

Hence $\qquad \dfrac{dp}{dt} \Big/ \dfrac{dp'}{dt} = \dfrac{606}{79 + 606} = \dfrac{1}{1\cdot13}.$

This shows that for any small descent in temperature from the triple point (where the pressure of steam with ice is the same as that of steam with water), the pressure of steam with ice falls off 1·13 times as much as does the pressure of steam with water.

In submitting the quantitative calculation now given, I have preferred to adopt the method proposed and developed by my brother rather than that which I had myself previously devised, because his method is simpler, and brings out the results more briefly by established principles from existing experimental data. I may say, however, that the method devised by myself was also a true method, and that I have since worked it out to its numerical results, and have found that these are quite in accordance with those brought out by my brother. The two indeed may be regarded as being essentially of the same nature; and I think it unnecessary to occupy space by giving any details of the method I planned and have carried out. Its general character may be sufficiently gathered from the concluding passages of the British-Association 1872 paper, as printed in the *Transactions of the Sections*, Brighton Meeting*.

* [*Supra*, pp. 305, 306.]

In order to discover whether the feature now developed by
theoretical considerations is to be found showing itself in any
degree in the experimental results of Regnault on the pressures of
steam at different temperatures*, I have made careful examinations of his engraved curve (plate viii of his memoir), and of his
empirical formulæ adapted to fit very closely to the results
exhibited in that curve, and of his final Tables of results at the
close of his memoir; and by every mode of scrutiny which I have
brought to bear on the subject (in fact by each of some seven or
eight varied modes) I have met with clear indication of the
existence of the expected feature; and by some of them I have
found that it can readily be brought prominently into notice.
The engraved curve drawn on the copper plate by Regnault
himself is offered by him as the definitive expression of his experiments, as being an expression which satisfies as well as possible
the aggregate of his observations—subject, however, to a very
slight alteration, which he has pointed out as a requisite amendment in the part of the curve immediately below the freezing point,
a part with which the investigations in the present paper are
specially concerned.

After telling (page 581) of the great care with which he had
marked the curve on the copper plate and got it engraved, he
says :—" Je n'ai pas pu éviter cependant quelques petites irrégularités dans les courbes; mais une seule de ces irrégularités me
paraît assez importante pour devoir être signalée. Elle se présente
pour les basses températures comprises entre 0° et − 16°; la courbe
creuse trop vers l'axe des températures, elle laisse, notablement
au-dessus d'elle, toutes les déterminations expérimentales qui ont
été faites entre 0° et − 10°. Ainsi les valeurs, que cette petite
portion de la courbe donne pour les forces élastiques, sont un peu
trop faibles, et j'ai eu soin de les augmenter, de la quantité
convenable, dans les nombres que je donnerai plus loin." Whether
we are now to think that this bend downwards† of the curve
towards the axis of temperatures, involving what Regnault regarded
as a small faulty departure of his drawn curve from his actual
experiments, was introduced merely by a casual want of accuracy

---

* Regnault, " Des Forces Élastiques de la Vapeur d'Eau aux différentes Températures," *Mémoires de l'Académie des Sciences*, 1847.

† In M. Regnault's curve the temperatures are measured horizontally across
the sheet, and pressures are measured upwards.

in drawing, or whether we may suppose that possibly there may have been some experimental observations which attracted the curve downwards, but were afterwards rejected on a supposition of their being untrustworthy, it appears that such a bend is a feature which the curve really ought to possess, and is one which even after being partially smoothed off by way of correction is not obliterated, but still remains clearly discoverable in the final numerical tables of results.

This is best brought to light by means of the empirical formulæ devised and employed by Regnault for the collating of his results. He proceeded evidently under the idea of the curve being continuous in its nature, so that a single formula might represent the pressures of aqueous vapour throughout the whole of his experiments; but before seeking for such a formula he proceeded to calculate several local formulæ of which each should represent very exactly his experiments between limits of temperature not wide apart; and afterwards he worked out several general formulæ, each adapted singly for the whole range of his experiments.

In regard to the one of these general formulæ which he designates as formula (H)*, he says that it represents the aggregate of his determinations of the pressures of the vapour of water, referred to the air-thermometer, and extending between the extreme temperatures of $-33°$ and $+232°$ with such precision that there could not be any hope of attaining to representing them better by any other mode of interpolation, because the differences, he says, between the calculated numbers and the numbers deduced from his graphic constructions are always smaller than the probable errors of observation. Still, for making out his final general Table of pressures of steam for every degree of the air-thermometer from $-30°$ to $+230°$, he used three local formulæ, finding that by them he could get slightly closer agreements with his experimental determinations than by using the single formula (H) for the whole range. Thus between $-32°$ and $0°$ he used his formula designated as (E); from $0°$ to $100°$ he used his formula (D); and between $100°$ and $230°$ he used his

---

* This and other formulæ in M. Regnault's memoir are here referred to only by their letters of reference, because to cite the formulæ themselves with their necessary accompanying explanations, would extend the present paper to too great a length; and any person wishing to scrutinize the formulæ would naturally prefer to have recourse to the original memoir.

formula (H). He points out (page 623) that he might have calculated this Table throughout its entire extent by the single formula (H), and that he would thus have got almost identically the same values by it from 100° down to 40° as those he calculated by the formula (D), but that between $+40°$ and $-20°$ the pressures given by the formula (H) would be slightly too small. This gives indication of the existence of the feature which it is my object at present to bring into view; and an examination of the column of Differences in Regnault's Table on his page 608, adapted for comparing the pressures got from experiments as expressed by his graphic curve with those got from the formula (H), shows distinctly a re-entrant angle, or at least a flattened place, in the curve at or about 0°. Several other like comparisons, by means of his other formulæ, give like indications; but most of these may for brevity be passed over without further mention here. The most decisive indication comes out in the following way. We may observe that for temperatures adjacent to the freezing point and extending both ways from it, Regnault finally adopted as fitting best to his experiments the formula (E) for temperatures descending from 0°, and the formula (D) for temperatures ascending from 0°. He tried (at pages 598, 599 of his memoir) the continuing of the application of his formula (D) beyond the inferior of the two limits 0° and 100°, for which he had specially aimed at adapting it to his experimental determinations; and he found that in calculating by it the pressures which it would give for temperatures below 0°, these pressures come out always slightly in excess of those which were given by his experiments. I have developed this mode of comparison in a more complete manner, and have arrived at remarkable results. The formula (D) may be regarded as the formula for giving the pressure p of steam with water, and (E) as that for giving the pressure p' for steam with ice. The following two Tables show the pressures p and p' for temperatures, in each case, both below and above the freezing point, as calculated from these two formulæ; and they show, also in each case, the consequent differences of pressure for 1° change of temperature at several different temperatures, or, what is the same, the values of dp/dt and dp'/dt for several temperatures slightly above and slightly below the freezing point.

TABLE I. By Formula (D); Steam with Water.

Temperatures	Pressures $= p$	Differences for 1°, which are values of dp/dt for intermediate temperatures	Temperatures to which the values of dp/dt belong
$-3°$	3·703		
		·280	$-2\frac{1}{2}°$
$-2°$	3·983		
		·298	$-1\frac{1}{2}°$
$-1°$	4·281		
		·319	$-\frac{1}{2}°$
$0°$	4·600		
		·340	$+\frac{1}{2}°$
$+1°$	4·940		
		·362	$+1\frac{1}{2}°$
$+2°$	5·302		
		·385	$+2\frac{1}{2}°$
$+3°$	5·687		

TABLE II. By Formula (E); Steam with Ice.

Temperatures	Pressures $= p'$	Differences for 1°, which are values of dp'/dt for intermediate temperatures	Temperatures to which the values of dp'/dt belong
$-3°$	3·644		
		·297	$-2\frac{1}{2}°$
$-2°$	3·941		
		·322	$-1\frac{1}{2}°$
$-1°$	4·263		
		·347	$-\frac{1}{2}°$
$0°$	4·610		
		·375	$+\frac{1}{2}°$
$+1°$	4·985		
		·405	$+1\frac{1}{2}°$
$+2°$	5·390		
		·437	$+2\frac{1}{2}°$
$+3°$	5·827		

From these two Tables we obtain the following values of $\dfrac{dp'}{dt} \Big/ \dfrac{dp}{dt}$ as deduced from Regnault's formulæ (D) and (E).

TABLE III.

Temperatures	Values deduced for $\dfrac{dp'}{dt}\Big/\dfrac{dp}{dt}$
$-2\frac{1}{2}°$	$\dfrac{297}{280} = 1\cdot06$
$-1\frac{1}{2}°$	$\dfrac{322}{298} = 1\cdot08$
$-\frac{1}{2}°$	$\dfrac{347}{319} = 1\cdot09$
$+\frac{1}{2}°$	$\dfrac{375}{340} = 1\cdot10$
$+1\frac{1}{2}°$	$\dfrac{405}{362} = 1\cdot12$
$+2\frac{1}{2}°$	$\dfrac{437}{385} = 1\cdot13$

This gives for $\dfrac{dp'}{dt}\Big/\dfrac{dp}{dt}$ at the freezing point the value of about 1·09 or 1·10; while its value brought out in the earlier part of the present paper by my brother's quantitative calculation was 1·13; and so the feature expected shows itself here in Regnault's results almost in the full extent in which theory shows that it ought to exist.

Regnault gives in the same memoir (page 627 and following pages) another Table, one intended chiefly for meteorological purposes, and in which the pressures are stated from $-10°$ to $+35°$ for every $\frac{1}{10}$ of a degree. In this Table the numbers inserted as representing the pressures below the freezing point are slightly different from the corresponding ones in his general Table already referred to; and he mentions that this slight discrepancy has resulted from the fact that the two Tables were formed at different periods, and were not calculated by the same formula; but he remarks that the differences are insignificant, as they scarcely amount to ·02 millimetre. Here, too, as in the general Table, the feature expected shows itself, though in a diminished degree. By a careful examination of its column of Differences for $\frac{1}{10}$ of a degree, and by making a few small arithmetical adjustments which may be regarded as amendments in the way of interpolation in that column, I find that, according to the experimental results as they are represented in this Table, the value of $\dfrac{dp'}{dt}\Big/\dfrac{dp}{dt}$ at the freezing point would come out to be about 1·05 or 1·06. We have seen by the new calculation, based on theory, in the present paper that it ought to be 1·13; so here the feature is found showing itself in about half the degree in which, according to the new quantitative calculation, it ought to be met with. When we consider that Regnault's reductions of his experimental results in the making out of curves, formulæ, and tables for representing them in the aggregate were, as we have sufficient ground to suppose, carried out under the idea, now proved to be erroneous, of there being, for aqueous vapour, continuity in variations of pressure with variations of temperature past the freezing point, just as past any other point of temperature, and when we further consider that the quantities with which we are here concerned are indeed very small, it is not surprising that there should have been a tendency to smooth off this feature on the supposition that any departures of the experimental observations from the course of a continuous or smooth curve were only slight irregularities due to experimental errors or imperfections.

It may now, in conclusion, be remarked that if from experiments independent of those which have been made, or may be made, directly on the pressure of aqueous vapour at different temperatures near the freezing point, both above and below it, very correct determinations of the values of the quantities C, M, and M' can be made, such determinations will lead to more correct evaluations of dp/dt and dp'/dt for aqueous vapour in contact in the one case with liquid water, and in the other with ice, than we at present possess. Such determinations, we may presume further, would, if very trustworthily arrived at, conduce to the attainment of a more correct estimate of the density of steam at the freezing point (or at the triple point) than we now possess. In fact, in connexion with the subject which has been here under consideration, there are various important quantities so connected that improved determinations of one or more of them may lead to more correct evaluations of others.

50. RELATION BETWEEN GASEOUS AND LIQUID STATES [UN-PUBLISHED NOTES BEARING ON ANDREWS'S EXPERIMENTS].

[*May 21st and 22nd*, 1862.]

IN order to show at least some arbitrary limits to gases and liquids, as distinguished from their substance in a fluid state which can scarcely be called distinctively either a liquid or a gas:

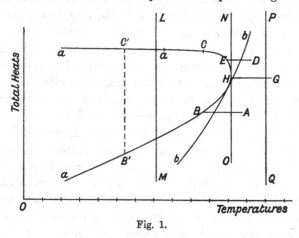

Fig. 1.

Let *aaa* be the curve of total heat of the liquid at its boiling point and of the gas at saturation:

Let *bb* be that one of the curves of total heat for the various temperatures when the pressure is constant, which touches the foregoing curve *aaa*. Then, according to one arbitrary assumption, we might say that the condition of the fluid expressed by the space *A* to the right of the latter curve is one of fluidity which is neither to be called liquid nor gaseous distinctively, because no addition nor abstraction of heat while that pressure is maintained will make the fluid either boil off or condense. Still according to another view we might call the fluid at *A* a liquid, because by simply diminishing the pressure without altering the total heat, either 1st without adding or abstracting heat, or 2nd while keeping total heat the same by making an addition of heat to correct for the heat disappearing as work, we may make it boil off. In this process the evaporation would produce such cold as to keep a portion liquid while another portion gets an increase of total heat at its expense and becomes a gas. On getting to B the fluid divides into two parts. One starts to C and passes towards C'. The other goes towards B' but keeps always as it goes giving off portions up to the curve CC'.

On the other hand, if the fluid began by going from D to E without addition or diminution of heat from an external thermal source, any further diminution of pressure will cause one part to condense while the other will continue on the upper branch of the curve going from E towards C', but always obtaining its accession of total heat (if the form of the curve requires that*) by the deposition of some of itself as liquid.

Consider next the passage from G to H by diminution of pressure and then the separation along the two branches:—a separation into gas and liquid *without boiling* or *condensation*.

Consider and describe what happens when pressure is continuously altered for a constant temperature: by passing along lines LM, NO, PQ. In the line LM we shall arrive at a place where an infinitely small increase of pressure will expel a large finite quantity of heat accompanied by a sudden diminution of volume, after which the expulsion of heat and diminution of volume become very slow. In the line NO I expect (if the curve has not a cusp) that we shall meet with a place of extremely rapid, at one point infinitely rapid, expulsion of heat and rapid falling down of volume (temperature being constant) then by further increase of pressure we get a very slow expulsion of heat and diminution of volume. In PQ nearly the same but gradual.

[*June 6th*, 1862.]

As the forms of the various curves which I have sketched on previous papers on this subject are uncertain and can only be determined by experiment, it is to be observed that the sketching of those curves is as yet useful chiefly to aid in the consideration of experimental observations; and that while *some* of the features of these curves are already indicated approximately by Dr Andrews's experiments, it is necessary to guard against deducing conclusions from assumed forms of any of these curves as *delineated* partly at

---

* The study of this may lead to some conclusions as to the relative forms of the two branches of the curve.

random in advance of experiments*. The sketching of probable features or approximate forms of these curves may serve useful purposes in indicating desirable courses for experimental investigation.

The foregoing being borne in mind as to the uses of such kinds of sketches of curves, I think the following curves will deserve consideration.

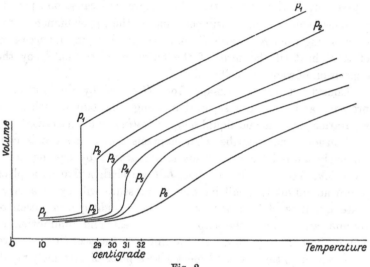

Fig. 2.

These curves mean those of volumes for varying temperatures and each curve being for a constant pressure p_1 or p_2 or p_3 or p_4 or p_5. The chief feature that I mean here to show is the relation between the wave in p_4, p_5 and p_6 as compared with the abrupt rise in p_1, p_2 and p_3.

The experiments of Dr Andrews (whether or not those of Cagniard de la Tour did) show that there is a certain superior limit of temperature above which the fluid carbonic acid does not show

* [The first brief report by Andrews on the influence of very high pressures on the more permanent gases is a short condensed note in *Report of Brit. Assoc.* 1861; he communicated a brief statement of the discovery of the critical point for carbonic acid to Miller's *Chemical Physics* in 1863 : the first "Bakerian Lecture" (*Phil. Trans.*) on the continuity of the liquid and gaseous states was published in 1869— see Introduction thereto, reprinted in Andrews' *Scientific Papers*, p. 297. The marked lowering of the critical point by admixture of a small portion of non-condensable gas was communicated to the Royal Society in 1875, and the second "Bakerian Lecture" in 1876.]

any sudden transition from "liquid" to "gas," or rather does not arrive at any point of total heat beyond which it refuses to take more diffused through it, and sends all additions of heat only to parts of itself, leaving the remainder unchanged.

Now it seems clear that we have a similar superior limit of pressure, so that for pressures above that limit we meet with no discontinuity in the possible volume or in the possible quantity of heat.

[June 9th, 1862.]

Having now cut out in wood* a curved surface to exhibit, and to aid the further consideration of my view of, the relation between liquids and their gases, or rather of the various conditions of pressure, temperature, volume, and quantity of heat† in which a fluid can exist, I shall here note a few points on the subject.

1st. It is only by experiment that the exact form of the curved surface can be found; but some of the facts of the case being already certainly known it is possible already to carve or mould some of the chief features of the curve surface, and these may serve to aid in showing the courses along which further experimental researches might best be directed, and also to aid in understanding the correlation of experimental results; and to aid in forming opinions in advance of experiments as to what is likely to result in intermediate cases between various experimental results which may already or at any time be arrived at.

2nd. It has occurred to me to look for what may be the nature of the discontinuity of the increase of volume with increase of temperature at the boiling point for any given pressure; and I now note for consideration the question whether it may not perhaps be the case that the curves, when the constant pressure for each of them is low enough to admit of boiling or of the existence of the fluid in two different states of density or in two different states of quantity of heat per unit of mass, have a fold over themselves as shown by dotted curved junctions $a_1 a_1$, $a_2 a_2$, $a_3 a_3$.

With reference to this it is to be observed that, in the curve

---

* [See figures from photographs, p. 277.]

† June 14. It is to be observed that this model for giving the relation between pressure, temperature and volume *aids* in considerations of quantities of heat, but that a separate model is intended to give the quantity of heat in terms of pressure and temperature.

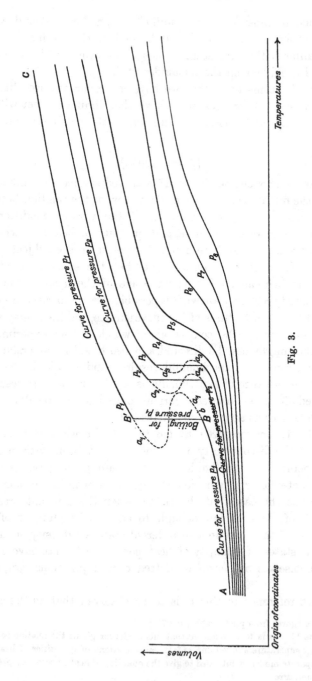

Fig. 3.

for pressure p_1 for instance, while the boiling point for the pressure p_1 is at BB', it is possible to heat the liquid to a higher temperature without making it boil till arrival at some point of temperature b, where it boils with bumping and from which the temperature falls to B when the steam and water are present together. This observed fact, of the liquid being capable of being raised above its boiling point without boiling, makes it reasonable to suppose that the curve AB does not terminate abruptly at B but may be continued; and my present suggestion is that it seems not unlikely that the curve would really be continued by a_1a_1 so as to be continued as $B'C$. In reference to this it seems to me that it will be important to ascertain whether steam can possibly be cooled below its regular condensing point for its pressure without its condensing. I think probably this has never yet been properly examined into ; and has probably been never ascertained. To test the point fairly precautions would require to be taken to have *no water* present along with the steam, in order that the steam might experience a difficulty of making a beginning of its change of state if the occurrence of such difficulty be possible. The internal surface of the enclosing vessel, too, might require special preparation to avoid anything that could set the steam off to change to water.

If the hypothesis here suggested, as to there being something in nature corresponding to the dotted junctions a_1a_1, a_2a_2, a_3a_3, between the curves AB and $B'C$, &c., be correct (though at present it is merely a suggestion for examination) it may further be considered whether or not there may in any circumstances be a possibility of changing from a liquid to its gas by passing along a curve a_1a_1, or a_2a_2, or a_3a_3. If so, it would seem that in a properly prepared vessel containing only a liquid, and that being kept at a constant pressure, we might by continuously adding heat, get 1st, a rising temperature and increasing volume, then 2nd, a falling temperature and increasing volume, then 3rd, a rising temperature and increasing volume, and then have vapour at the boiling temperature for the pressure.

If the dotted curves a_1a_1, &c. have a real existence, I suppose they will represent possible but *unstable* states of the fluid :— states which perhaps men cannot bring into occurrence : unless perhaps in some extreme or special circumstances.

A reason in favour of the supposition of the existence of the dotted junctions is that (so far as I can judge from Dr Andrews's experiments hitherto made) it seems probable that in passing from heavy pressures to light pressures, as from the heavy pressure p_6 to p_5 and p_4 successively, there is a point of steepest slope in each of the curves (that is a point of inflexion) and that in the successive curves the steepness of the slope or the angle of the tangent to the abscissa at that point goes on increasing from curve to curve till it becomes 90°: and then it would seem probable that the angle would go on increasing beyond 90° in which case the curve would stand thus :—

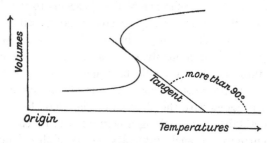

Fig. 4.

PROPOSED EXPERIMENT WITH REFERENCE TO PART OF THE FOREGOING. To take a glass phial or hollow globe such as a Florence flask. Put a thermometer into it, put water in and draw the mouth down to a small aperture which can be easily hermetically sealed. Immerse the flask in a bath at 213° Fahr. or so (a little above boiling temperature). The water will boil off and carry out all air with the steam. Melt the tube so as to close it. Then the flask contains only steam: or steam with a little water which may perhaps have condensed in the tube but which it might be well to avoid allowing to be there, unless in very small quantity. Then immerse the whole in a bath at say 220° Fahr. or at a high enough temperature to make quite sure that not a molecule of water is left in the liquid state. Then remove the flask from the bath, or better allow the bath very gradually to cool, and watch the enclosed thermometer. It will of course sink gradually. Watch particularly to see whether after sinking to a certain extent it rises again suddenly. If such rise be found to occur it will take place at the moment of the commencement of condensation of water from the steam; and if it occur it will prove that the steam had been cooled below its boiling point without condensing, and that on the commencement of precipitation of water or dew the difficulty of making a beginning was got over and the boiling point for the existing pressure was assumed.

[*June* 14, 1862.]

In respect to the curve shown thus on former papers; I now observe that it is simply the projection, on a plane parallel to the

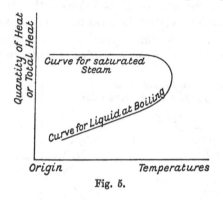

Fig. 5.

reference line of temperature and the reference line of total heat, of the abrupt edge of the curve surface which gives the quantity of heat as a function of the temperature and pressure;—that abrupt edge being the line where the sudden transition takes place from gas to liquid or liquid to gas.

It will be a matter of great interest to discover the relations of pressure, temperature and volume, and of pressure, temperature and total heat for a given substance (carbonic acid for instance), through so wide a range of pressures and temperatures as to introduce the change from either liquid or gas to solid. There will be in each of the two general curve surfaces which represent by their ordinates the volume and the total heat as functions of the temperature and pressure, *two* abrupt places;—one for the change from gas to liquid, and the other for the change from liquid to solid;—but as, at ordinary atmospheric pressure, the change occurs at once from gas to solid, without liquefaction, it is clear that these two abrupt places will merge into one as the pressure diminishes from very high pressures down towards one atmosphere.

In reference to the idea of there being perhaps a possibility of cooling steam or other vapours below their ordinary saturation points of temperature; it strikes me that perhaps a film of clean oil washed completely round the inside of the hermetically sealed glass globe or flask containing the liquid or vapour might give the requisite kind of surface for withholding whatever may tend to give freedom to change of state from gas to liquid. My impression at any rate is that I have often seen on a piece of glass or other substance dew deposited at some parts, while none gets a beginning of forming itself at others;—the difference being associated essentially with differences of cleanness, &c. of the glass or other surface. As to the deposition of hoar frost on scratched lines on a window in preference to other parts there is no doubt.

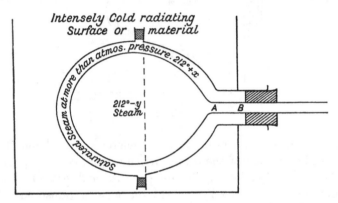

Cork may be far away from flask so as to give a long hot tube *AB*.

Fig. 6.

Brief, *hasty* memorandum of idea which has occurred to me for devising a method for use in case it be found that no kind of internal surface whatever enclosing the steam can be obtained free from sufficing to give by its contact, perfect freedom to the steam to begin to change its state to water. The method consists in keeping the internal surface slightly above the boiling point of temperature, and making the internal part of the steam cooler by radiation. If any such method is to succeed, I think it will be necessary to avoid having any thermometer in the middle of the flask, because the glass of the thermometer, I presume, would be cooled by the radiation so as to be quite as cool as, or cooler than, the surrounding steam cooled by the same radiation. It strikes me (but I have not yet properly considered this) that the *steam itself might be used like an air thermometer with due precautions by its expansion or contraction* which might give indications in the hot tube AB; perhaps by acting on a piston of mercury in that tube.

This page and foregoing are mostly taken from an old scribbled temporary pencil note of some years ago which was not meant to be kept, except till I should note the subject better. J. T., 29 March 1868.

Notes and Queries.—On Gases, Liquids, Fluids.

[*Primary date* 10*th May*, 1869.]

Various conversations with Dr Andrews in recent weeks, and at least one new experimental result of his (showing a "*critical point*" in mixed air and carbonic acid) together also with my having recently attained to a somewhat better understanding and clearer views of some principles or theoretical reasonings as to molecular attraction in capillarity, have led me in the last few weeks to give much consideration to the nature and relations and distinctions among the Gaseous and Liquid states of fluids, and the transitions between those states.

I believe I have arrived at some new and improved views; and further developments of old views; and I now intend to note down some views which I have come to hold as being either true or as being at least suppositions or hypotheses worthy of further study and research.

I proceed on the hypothesis of the substance dealt with, whether called Gas, Vapour, Liquid or Fluid; or at one part of itself Gas or Vapour, and at another part Liquid; being of the nature of a compressible fluid in every part down to the smallest parts that are to come into account; or in other words I shall not go on any supposition of gases being made up of separate particles repelling one another and leaving greater or less void spaces between them according as the gases are more expanded or more reduced in volume. I shall treat the matter as if from every point in the fluid to every other, no matter how near they be, down to the smallest distances that require consideration for the matters in question, the fluid is continuous as a fluid:—and as if the fluid everywhere possesses an expansive tendency, so that pressure must everywhere be received by the fluid on one side of a dividing surface (or as I call it *interface*) from the fluid, or solid, on the other side, to prevent the fluid from expanding indefinitely, or to balance its expansive force. I proceed on the basis of supposing that it is either true; or that it closely enough, for present purposes, resembles the truth; to suppose that contiguous parts of the fluid press on one another by forces applied at their faces of mutual contact, distributed over those faces, and that molecular attractions, very powerful in the case of liquids, act between all portions of fluid that are very near one another: the attractive force from any one material point to any small portion or mass of adjacent fluid having its points of application distributed everywhere in the substance of that attracted mass.

In a gaseous fluid at its condensing point of pressure and temperature it may be said that *in some sense* a state of instability is arrived at: because in that condition further reduction of the space confining the whole mass does not cause compression of every part of the fluid, but the fluid separates into two parts in different states of density and of total heat per unit of mass. The one part undergoes no change whatever;—meets with no compression;—and the diminution of the whole volume influences the other part alone by giving it diminished volume or increased density; but in fact neither the one part nor the other of the separated parts undergoes any change except in this that a further reduction of the whole volume causes portions of the gaseous part to change to the denser state of the liquid part. The kind of instability which, apparently at least, is arrived at, or in *some sense* is actually arrived at, is of this nature :—that if we conceive *without the transformation* of any part of the gaseous fluid to liquid, any further reduction of volume without change of temperature to be effected, the gaseous fluid will be in a state in which it cannot practically remain; at least in which so far as at present known to me, or I suppose to anybody, it cannot be got to remain, and certainly unless with very special arrangements will not remain ; but out of which it will pass by one part going down in volume to the liquid state and the other reverting to its original density and pressure, the original temperature being understood to be maintained, say, by immersion in a bath of constant temperature.

Thus *practically* we may be right or very nearly right in representing the case of the gas compressed to its condensing point of temperature and pressure, as having just arrived at the verge of instability so that it cannot practically (by ordinary means at least) be further compressed at its constant temperature because a state of complete practical instability would then be entered on and a part would collapse to liquid while another would revert to the gaseous state at the verge of condensing.

Now it appears to me that this practical instability really does not hold good in principle in a gas at its condensing point if we consider the gas as totally free from contact with any solid or liquid: this supposition is practically impossible, but I am not sure that it would be impossible to devise some means of actually getting the gas brought to a temperature and pressure below the condensing point by some method that might be analogous to methods used by Donny and others to get perhaps water above its boiling point without boiling, or below its freezing point without freezing*.   In principle I think the gas without presence of water or of any solid or liquid dense substance is not in an unstable state when cooled and pressed to below its ordinary condensing point.

In fact I think that in a gaseous fluid at its condensing point of temperature and pressure *we have not the condition* that any further compression would make it collapse to the liquid state: or we have *not* the condition that if the adjacent parts be brought nearer together, their attraction will overcome the expansive force and cause collapse: but for causing the condensation *it is necessary* to have dense matter in close proximity: such dense matter may be either the solid containing vessel, or some of the fluid in its liquid state.   I think it probable however (and I hope to find it to be the case) that a perfectly dry interior of a containing vessel with perhaps some specially prepared face, may give less freedom for the beginning of the change of state from gaseous to liquid than would a portion of the liquid itself in contact with the gas.

Now see fig. 7.   The shaded part represents (according to my idea) a bar of continuous fluid normal to and passing through the surface or rather the lamina of demarcation:—one end of the bar being gaseous and the other liquid, and the transition along the bar from gas to liquid being perfectly gradual by a gradually increasing pressure and density.   The lamina of demarcation extends from *aa* to *bb*.   Now at any such depth as *cc* the fluid (as I think) is subject to far more pressure and has far more

---

* *Qy., might this be effected in steam by allowing rapid expansion in a super-heated vessel as in a steam engine cylinder completely jacketed?   Does not the expansion of the steam while performing work cause it to pass partly into the liquid state? and might not perfect jacketing prevent the beginning of liquefaction?*

density than those due at the condensing point to the temperature: and yet it does not collapse: but on the contrary is quite
ready to rise in volume to the gaseous state on any retirement
of the insistent bar of gas; such retirement being however un-
accompanied by any reduction of the gaseous pressure.

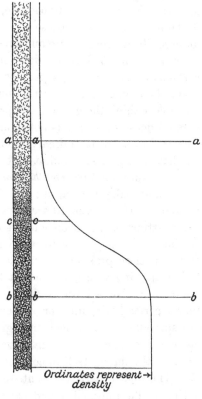

Fig. 7.

Memo written at Largs, Dec. 31, 1869.—Be careful to recollect that throughout
this diagram, the pressure as contemplated in Dr Andrews's Experiments and in
my wooden curve surfaces * would be CONSTANT. It is the *abutting push* that varies
here in descending from the gas into the liquid.

Now further we can see how raising temperature and pressure
jointly we arrive (under the present theory or hypothesis) at a
critical point:—that which Dr Andrews has found experimentally:—

* [See figures from photographs, p. 277.]

By raising the temperature of the whole mass of fluid, and increasing the pressure at the same time so as to retain the same portion of the mass as before still gaseous; and the same part liquid as before; the consequence of applying the heat must be that the liquid part will expand in volume and get less density; so that it will have less molecular attraction and will attract less than before the fluid in the lamina of transition per unit of mass of that fluid. This would allow that fluid to expand and disperse or go off were there not a pressure applied from without sufficient to give increased density to the whole gas in spite of the expanding tendency of the heat. Now by having denser gas than before at about the gaseous side of the lamina of transition we can attain to the requisite pressure in the middle of the lamina in spite of the diminished density of the liquid :—the liquid less dense than before attracting the gas more dense than before being able to give to the intervening fluid the requisite pressure and density, even though these require to be more than before. Thus the gaseous part will be more pressed, and instead of being allowed to expand will actually become more dense than before; and will have increased molecular attraction; and, by our carrying forward far enough the augmentation of temperature and of corresponding pressure, we *must* ultimately (under the theory I am giving) arrive at a stage of temperature and corresponding pressure at which the gaseous part and the liquid part shall have the same density, and then the molecular attractive cause of difference of density in the two parts will no longer exist, but the distinction will have faded away.

It appears to me to be very important now to get what I may call the prevalent prejudice eradicated from people's minds, under which people usually imagine or tacitly assume the pressure applied to a gas by its confining vessel as being the pressure in the gas. The so-called " Law of Mariotte" is not a law of nature in any sense. First because Natterer's experiments show that when a density such as that of liquids is attained by air, the volume is several times as great as it should be if Mariotte's law held good for the relation between the applied pressure and the volume.

Secondly we must specially guard against taking Mariotte's law even if amended or modified to adapt it so that it amended, or the new formula substituted for it, would express the relation between density and *applied pressure*;—against taking that altered or

amended formula as showing the expansive force in terms of the
density (or what is the same in terms of the volume). The
expansive force which is the real pressure is altogether different
from the applied pressure; and the one does not vary proportionally to the other during the compression of air or oxygen or
hydrogen or other commonly called perfect gases, nor of steam,
carbonic acid or other condensible gases.

When Natterer applied 2700 atmospheres to a gas:—he may
so far as can now be known have increased its molecular attraction
millions of times: but we do not at present know, for the gas to
begin with, what the amount of the pressure due to molecular
attraction was, over and above the applied pressure:—so we cannot
at present express the molecular attraction (I think) in atmospheres
either in the gas in its large state at atmospheric applied pressure,
or when under the applied pressure of 2700 atmospheres.

Curve for constant temperature is [not*] impossible as here
drawn below. Because from n up to m we would have at constant
temperature an increase of pressure giving an increase of volume,
which I think obviously cannot be †.

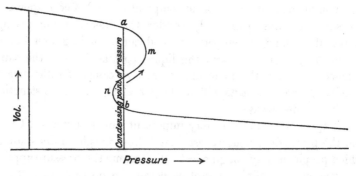

Fig. 8.

* A little later I see that it is not impossible in principle, only that equilib.
would be unstable; equilib. of gas alone would be stable from a to m, ... fluid
alone or in the single state unstable . . . . liquid alone stable. [Author's MS. here
partly torn.]

† [The considerations developed in these MS. memoranda are recapitulated in
Clerk Maxwell's *Theory of Heat* at the end of Chapter VII. See also Prof. Thomson's
paper, No. 46, *supra* p. 278, from *Proc. Roy. Soc.* 1871. Maxwell, and Clausius
later, afterwards pointed out that the line in the figure marked condensing point of
pressure must cut off equal areas on its two sides.]

Supposition.

I assume as admitted, or proceed hypothetically on supposition, that the pressure due to molecular attraction, or added by it to that applied by the containing vessel, is as square of the density at the place.   Water at ordinary boiling point is about 1728 times as dense as its steam; $1728^2 = 3,000,000$; so that the pressure due to molecular attraction in the water would be about three million times as much as that due to molecular attraction in the steam.

DYNAMICS AND ELASTICITY

51. ON THE STRENGTH OF MATERIALS, AS INFLUENCED BY THE
EXISTENCE OR NON-EXISTENCE OF CERTAIN MUTUAL STRAINS[*]
AMONG THE PARTICLES COMPOSING THEM.

[From the *Cambridge and Dublin Mathematical Journal*, November 1848,
pp. 252—258. This paper is also reprinted, with a few changes made
in it with the author's concurrence, in Article "Elasticity," by Sir William
Thomson, in *Encyclopaedia Britannica*, 9th Edition, Vol. VII. pp. 798—
800.]

MY principal object in the following paper is to show that
the absolute strength of any material composed of a substance
possessing ductility (and few substances, if any, are entirely devoid
of this property), may vary to a great extent, according to the
state of tension or relaxation in which the particles have been
made to exist when the material as a whole is subject to no
external strain.

Let, for instance, a cylindrical bar of malleable iron, or a piece
of iron wire, be made red hot, and then be allowed to cool. Its
particles may now be regarded as being all completely relaxed.
Let next the one end of the bar be fixed, and the other be made
to revolve by torsion, till the particles at the circumference of the
bar are strained to the utmost extent of which they can admit
without undergoing a permanent alteration in their mutual connexion[†]. In this condition, equal elements of the cross section

---

[*] [Note added November 1877, in *Encyc. Brit.* reprint by Lord Kelvin.
More nearly what is now called *stress* than what is now called *strain* is meant by
"strain" in this article, which was written before Rankine's introduction of the
word stress, and distinct definition of the word strain.]

[†] I here assume the existence of a definite "elastic limit," or a limit within
which if two particles of a substance be displaced, they will return to their original
relative positions when the disturbing force is removed. The opposite conclusion,
to which Mr Hodgkinson seems to have been led by some interesting experimental
results, will be considered at a more advanced part of this paper.

of the bar afford resistances proportional to the distances of the elements from the centre of the bar; since the particles are displaced from their positions of relaxation through spaces which are proportional to the distances of the particles from the centre. The couple which the bar now resists, and which is equal to the sum of the couples due to the resistances of all the elements of the section, is that which is commonly assumed as the measure of the strength of the bar. For future reference, this couple may be denoted by L, and the angle through which it has twisted the loose end of the bar by Θ.

The twisting of the bar may, however, be carried still farther, and during the progress of this process the outer particles will yield in virtue of their ductility, those towards the interior assuming successively the condition of greatest tension; until, when the twisting has been sufficiently continued, all the particles in the section, except those quite close to the centre, will have been brought to afford their utmost resistance. Hence, if we suppose* that no change in the hardness of the substance composing the material has resulted from the sliding of its particles past one another; and that, therefore, all small elements of the section of the bar afford the same resistance, no matter what their distances from the centre may be, it is easy to prove that the total resistance of the bar is now $\frac{4}{3}$ of what it was in the former case; or, according to the notation already adopted, it is now $\frac{4}{3}L$†.

* [Note added October 1877, *Encyc. Brit.*

This supposition may be true for some solids; it is certainly not true for solids generally. A piece of copper or of iron taken in a soft and unstrained condition certainly becomes "harder" when strained beyond its first limits of elasticity, that is to say, its limits of elasticity become wider; and a similar result will probably be found in ductile metals generally. Thus the resistance of the outer elements will be greater than those of the inner elements in the case described in the text, until the torsion has been pushed so far as to bring about the greatest hardness in all the elements at any considerable distance from the axis. It may be that before this condition has been attained the hardening of the outer elements will have been overdone, and they may have begun to lose strength, and to have become friable and fissured. The principle set forth in the text is not, however, vitiated by the incorrectness of a supposition introduced merely for the sake of numerical illustration.]

† To prove this, let r be the radius of the bar, η the utmost force of a unit of area of the section to resist a strain tending to make the particles slide past one another; or to resist a shearing strain, as it is commonly called. Also let the section of the bar be supposed to be divided into an infinite number of concentric

If, after this, all external strain be removed from the bar, it will assume a position of equilibrium, in which the outer particles will be strained in the direction opposite to that in which it was twisted and the inner ones in the same direction as that of the twisting, the two sets of opposite couples thus produced among the particles of the bar, balancing one another. It is easy to show that the line of separation between the particles strained in the one direction, and those in the other, is a circle whose radius is $\frac{3}{4}$ of the radius of the bar. The particles in this line are evidently subject to no strain* when no external couple is applied. The bar with its new molecular arrangement may now be subjected, *as often as we please*†, to the couple $\frac{4}{3}L$, without

annular elements; the radius of any one of these being denoted by x, and its area by $2\pi x\,dx$.

Now, when only the particles at the circumference are strained to the utmost; and when, therefore, the forces on equal areas of the various elements are proportional to the distances of the elements from the centre, we have $\eta\,(x/r)$ for the force of a unit of area at the distance x from the centre. Hence the total tangential force of the element is

$$= 2\pi x\,dx \,.\, \eta\,(x/r),$$

and the couple due to the same element is

$$= x \,.\, 2\pi x\,dx \,.\, \eta\,(x/r) = 2\pi\eta\,(1/r) \,.\, x^3 dx:$$

and therefore the total couple, which has been denoted above by L, is

$$= 2\pi\eta\,\frac{1}{r}\int_0^r x^3 dx,$$

that is, $L = \tfrac{1}{2}\pi\eta r^3$ ....(a).

Next, when the bar has been twisted so much that all the particles in its section afford their utmost resistance, we have the total tangential force of the element $= 2\pi x\,dx \,.\, \eta$, and the couple due to the same element

$$= x \,.\, 2\pi x\,dx \,.\, \eta = 2\pi\eta \,.\, x^2 dx.$$

Hence the total couple due to the entire section is

$$= 2\pi\eta\int_0^r x^2 dx = \tfrac{2}{3}\pi\eta r^3.$$

But this quantity is $\frac{4}{3}$ of the value of L in formula (a). That is, the couple which the bar resists in this case is $\frac{4}{3}L$, or $\frac{4}{3}$ of that which it resisted in the former case.

* Or at least they are subject to *no strain of torsion* either in the one direction or in the other; though they may perhaps be subject to a strain of compression or extension in the direction of the length of the bar. This, however, does not fall to be considered in the present investigation.

† This statement, if not strictly, is at least extremely nearly true: since from the experiments made by Mr Fairbairn and Mr Hodgkinson on cast iron (see various *Reports of the British Association*), we may conclude that the metals are i nfluenced only in an extremely slight degree by time. Were the bars composed of some substance, such as sealing wax, or hard pitch, possessing a sensible amount of viscosity, the statement in the text would not hold good.

undergoing any farther alteration; and therefore its ultimate strength to resist torsion, in the *direction of the couple L*, has been considerably increased. Its strength to resist torsion in the opposite direction has, however, by the same process, been much diminished: for, as soon as its free extremity has been made to revolve backwards through an angle* of $\frac{2}{3}\Theta$ from the position of equilibrium, the particles at the circumference will have suffered the utmost displacement of which they can admit without undergoing permanent alteration. Now it is easy to prove that the couple required to produce a certain angle of torsion is the same in the new state of the bar as in the old†. Hence the ultimate strength of the bar when twisted backwards, is represented by a couple amounting to only $\frac{2}{3}L$. But, as we have seen, it is $\frac{4}{3}L$ when the wire is twisted forwards. That is, then, *The wire in its new state has twice as much strength to resist torsion in the one direction as it has to resist it in the other.*

Principles quite similar to the foregoing, operate in regard to beams subjected to cross strain. As, however, my chief object at present is to point out the existence of such principles, to indicate the mode in which they are to be applied, and to show their great practical importance in the determination of the strength of materials, I need not enter fully into their application in the case of cross strain. The investigation in this case closely resembles that in the case of torsion, but is more complicated on account of the different ultimate resistances afforded by any material to tension and to compression, and on account of the numerous varieties in the form of section of beams which for different purposes it is

---

* [Note added October 1877, in *Encyc. Brit.*

This assumes that the limits of elasticity in a substance which has already been strained beyond its limits of elasticity are equal on the two sides of the shape which it has when in equilibrium without disturbing force—a supposition which may be true or may not be true. Experiment is urgently needed to test it; for its truth or falseness is a matter of much importance in the theory of elasticity.]

† To prove this, let the bar be supposed to be divided into an infinite number of elementary concentric tubes (like the so-called annual rings of growth in trees). To twist each of these tubes through a certain angle, the same couple will be required whether the tube is already subject to the action of a couple of any moderate amount in either direction, or not. Hence, to twist them all, or what is the same thing, to twist the whole bar, through a certain angle, the same couple will be required whether the various elementary tubes be or be not relaxed, when the bar as a whole is free from external strain.

found advisable to adopt. I shall therefore merely make a few remarks on this subject.

If a bent bar of wrought iron, or other ductile material, be straightened, its particles will thus be put into such a state, that its strength to resist cross strain, in the direction towards which it has been straightened, will be very much greater than its strength to resist it in the opposite direction, each of these two resistances being entirely different from that which the same bar would afford, were its particles all relaxed when the entire bar is free from external strain. The actual ratios of these various resistances depend on the comparative ultimate resistances afforded by the substance to compression and extension; and also, in a very material degree, on the form of the section of the bar. I may however state that in general the variations in the strength of a bar to resist cross strain, which are occasioned by variations in its molecular arrangement, are much greater even than those which have already been pointed out as occurring in the strength of bars subjected to torsion.

What has been already stated is quite sufficient to account for many very discordant and perplexing results which have been arrived at by different experimenters on the strength of materials. It scarcely ever occurs that a material is presented to us, either for experiment or for application to a practical use, in which the particles are free from great mutual strains. Processes have already been pointed out by which we may at pleasure produce certain peculiar strains of this kind. These or other processes producing somewhat similar strains are used in the manufacture of almost all materials. Thus, for instance, when malleable iron has received its final conformation by the process termed cold swaging, that is by hammering it till it is cold, the outer particles exist in a state of extreme compression, and the internal ones in a state of extreme tension. The same seems to be the case in cast iron when it is taken from the mould in which it has been cast. The outer portions have cooled first, and have therefore contracted while the inner ones still continued expanded by heat. The inner ones then contract as they subsequently cool, and thus they as it were pull the outer ones together. That is, in the end, the outer ones are in a state of compression and the inner ones in the opposite condition.

The foregoing principles may serve to explain the true cause of an important fact observed by Mr Eaton Hodgkinson in his valuable researches in regard to the strength of cast iron (*Report of the British Association* for 1837, p. 362)*. He found that, contrary to what had been previously supposed, a strain, however small in comparison to that which would occasion rupture, was sufficient to produce a set in the beams on which he experimented. Now this is just what should be expected in accordance with the principles which I have brought forward: for if, from some of the causes already pointed out, various parts of a beam previously to the application of an external force have been strained to the utmost, when, by the application of such force, however small, they are still farther displaced from their positions of relaxation, they must necessarily undergo a permanent alteration in their connexion with one another, an alteration permitted by the ductility of the material; or, in other words, the beam as a whole must take a set.

In accordance with this explanation of the fact observed by Mr Hodgkinson, I do not think we are to conclude with him that "the maxim of loading bodies within the elastic limit has no foundation in nature." It appears to me that the defect of elasticity, which he has shown to occur even with very slight strains, exists only when the strain is applied for the first time; or, in other words, that if a beam has already been acted on by a considerable strain, it may again be subjected to any smaller strain in the same direction without its taking a set. It will readily be seen, however, from Mr Hodgkinson's experiments, that the term "elastic limit," as commonly employed, is entirely vague, and must tend to lead to erroneous results.

The considerations adduced seem to me to show clearly that there really exist *two elastic limits* for any material, between which the displacements or deflexions, or what may in general be termed the changes of form, must be confined, if we wish to avoid giving the material a set; or, in the case of variable strains, if we wish to avoid giving it a continuous succession of sets which would gradually bring about its destruction: that these two elastic limits are usually situated one on the one side, and the other on

* For farther information regarding Mr Hodgkinson's views and experiments, see his communications in the *Transactions of the Sections of the British Association* for the years 1843 (p. 23) and 1844 (p. 25), and a work by him entitled *Experimental Researches on the Strength and other Properties of Cast Iron*. 8vo. 1846.

the opposite side of the position which the material assumes when subject to no external strain, though they may be both on the same side of this position of relaxation *, and that they may therefore with propriety be called the *superior* and the *inferior limit* of the change of form of the material for the particular arrangement which has been given to its particles; that these two limits are not *fixed* for any given material; but that if the change of form be continued beyond either limit, two new limits will, by means of an alteration in the arrangement of the particles of the material, be given to it in place of those which it previously possessed: and lastly, that the processes employed in the manufacture of materials are usually such as to place the two limits in close contiguity with one another, thus causing the material to take in the first instance a set from any strain however slight, while the interval which may afterwards exist between the two limits, and also as was before stated, the actual position assumed by each of them, is determined by the peculiar strains which are subsequently applied to the material.

The introduction of new, though necessary, elements into the consideration of the strength of materials may, on the one hand, seem annoying from rendering the investigations more complicated. On the other hand, their introduction will really have the effect of obviating difficulties, by removing erroneous modes of viewing the subject, and preventing contradictory or incongruous results from being obtained by theory and experiment. In all investigations, in fact, in which we desire to attain, or to approach nearly, to truth, we must take facts as they actually are, not as we might be tempted to wish them to be, for enabling us to dispense with examining processes which are somewhat concealed and intricate, but are not the less influential from their hidden character.

* Thus, if the section of a beam be of some such form as that shown in either of the accompanying figures, the one rib or the two ribs, as the case may be, being very weak in comparison to the thick part of the beam, it may readily occur that the two elastic limits of deflexion may be situated both on the same side of the position assumed by the beam when free from external force. For if the beam has been supported at its extremities and loaded at its middle till the rib AB has yielded by its ductility so as to make all its particles exert their utmost tension, and if the load be now gradually removed, the particles at B may come to be compressed to the utmost before the load has been entirely removed.

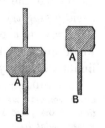

52.   ON THE ELASTICITY AND STRENGTH OF SPIRAL SPRINGS, AND OF BARS SUBJECTED TO TORSION.

[From the *Cambridge and Dublin Mathematical Journal*, No. XVII. November 1848, pp. 258—266.]

A SPIRAL Spring of the most usual kind consists of a long bar or wire, generally of a circular section, coiled up into the form of the thread of a screw.

For the purpose of attaining precision in speaking of such springs, the following definitions and preliminary explanations will be useful. The curve in which the centres of all the sections of the bar are situated may be called the *spiral axis* of the spring. This lies in the surface of a cylinder, the axis of which may be called the *longitudinal axis* of the spring. The angle which the spiral axis makes with a plane perpendicular to the longitudinal axis may be called the *inclination of the coil* or *of the spiral*. Each end of the bar is bent in such a way that the force applied to elongate or compress the spring may act in the longitudinal axis*. In what follows, unless the contrary be specified, the spring may be supposed to be suspended by one end, the force applied being one of tension produced by a weight hung at the lower end.

The Elasticity and Strength of Spiral Springs have not, so far as I am aware, been hitherto subjected to scientific investigation; and erroneous ideas are very prevalent on the subject, which are not unfrequently manifested in practice by the adoption of forms very different from those which would afford the greatest advantages. Having had occasion to construct some spiral springs which, in their elasticity, strength, and dimensions, should fulfil certain definite conditions, I was led to seek for principles to

---

* In fact, even if, in the construction of the spring, this matter has not been attended to, and the ends of the bar forming the points of application of the opposite forces have not been placed in the longitudinal axis; immediately on a tensile force being applied, the spring will, if it be of considerable length, adjust itself so that the force will act in that axis, except in the neighbourhood of the two ends. At those parts the spring will be weaker than towards the middle; and, if the force be sufficient to induce a permanent alteration on its form, this change will commence by the ends assuming forms of greater resistance, the longitudinal axis approaching more nearly to the line of action of the forces.

guide me in determining the forms and dimensions best adapted
to accomplish the ends desired.

With this view, the first matter to be considered was the
exact nature of the strains which act
on the various parts of a spiral spring;
and the whole subject became at once
simple, as soon as I perceived that the
only strain which produces a sensible
effect is one of torsion, acting alike on
every part of the coil.   To render this
clear, let us consider any section P of
the bar, made by a plane passing
through the longitudinal axis AB.
Now, when a weight w is suspended
at B, the forces transferred from one
side to the other of this section consist of a couple whose force is w, and
whose arm is the distance PC from
the centre of the section at P to the
line AB, together with a force w

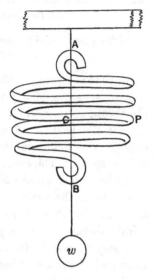

parallel to AB, tending to make the one side of the section
slide upon the other in the direction of the longitudinal axis.
The effect of this force must be extremely small, in fact
evanescent, compared to that of the couple; and the force may
therefore be neglected, especially when the coefficients for the
elasticity and strength of the material are determined in the way
which will be hereafter pointed out.   The slight deviation of the
above-mentioned section of the bar from a circle due to the
inclination of the spiral, may also be neglected in the theory, as
the minute influence which it may have will also be, in a great
degree, corrected for by the mode of determining the coefficients.

Let now r be the radius of the bar; l the total length coiled
up; a the radius of the coil, or the distance from the longitudinal to the spiral axis; w any weight which may be hung on
the spring; e the elongation corresponding to that weight;
and θ the angle through which a bar composed of the same
substance as the spring, having its length and radius each
unity, is twisted when subjected to a unit couple.   This quantity
θ may be called the coefficient of deflexion by torsion for the
substance of which the bar is composed.

Let the spiral part of the bar be supposed to be straightened out, and let the bar be supposed to be placed with one end fixed at AD; and with a couple applied at the other extremity by means of a weight, equal to w, suspended at B, and a force in the line EF equal, parallel, and opposite to that of w; the arm of this couple being equal to a the radius of the coil.

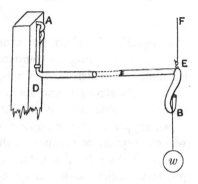

Now it can be shown that, in bars subjected to torsion, the angle of torsion is proportional to the length of the bar, to the couple applied, and inversely to the fourth power of the radius of the bar. Hence the angle of torsion in the present case, or the angle described by EB, in passing from its natural position to one of equilibrium with the couple wa is $= \theta \dfrac{lwa}{r^4}$ *.

Hence, the space moved over by a point at a distance a from the centre is $= \theta \dfrac{lwa^2}{r^4}$. Now a little consideration will show that this space must be equal to e, the elongation of the spiral spring

---

* As these principles regarding torsion have been laid down in various works on Mechanics and Engineering, I here take them for granted. The methods of deducing them have, however, been insufficient, as it has been, tactily at least, assumed that all the elements in the section of a bar are free from strain when the bar as a whole is free from external strain; since it is assumed that when the bar is twisted to any extent less than that which would strain its circumference to the utmost, any equal elements of its section undergo strains proportional to the distances of the elements from the centre. It seldom occurs, however, that the real condition of a bar is in accordance with this assumption; for I have shown in the preceding paper that the various particles of materials usually exist under great strains in opposite directions, which are in equilibrium with one another. The conclusions which had previously been derived are, however, in themselves correct; and in a note to the paper just referred to, I have supplied the step which was wanting in the proof by showing that the angle of torsion of a bar, or in other words, its stiffness is not influenced by the presence or absence of internal opposing strains among its particles; although the case is very different with regard to ultimate strength of the bar, which is materially altered by changes in those strains. From this it follows that the coefficient for the stiffness or deflexion of a bar composed of a given substance has but one value, while that for its ultimate strength may have various values. Of this more will be said in what follows.

due to the application of the weight w. Hence we have

$$e = \theta \frac{lwa^2}{r^4} \quad \dots\dots\dots\dots\dots\dots\dots(1).$$

This equation involves the conditions of the *elasticity*, or the *stiffness* of a spring as compared with its dimensions, and the substance of which it is composed. We will next proceed to those of its strength and its power; or, in other words, we will enter on considerations connected with the greatest weight which it can support, and the space through which it can be elongated without rupture or permanent alteration.

In addition to the notation given above:—Let W be the greatest weight which the spring, if its particles are all relaxed when it is not loaded, can bear without taking a set; E the greatest elongation, that namely which corresponds to W; μ the utmost couple producing torsion which can be resisted by a bar whose radius is unity, composed of the same substances as the spring, and having its particles at various distances from its centre free from mutual opposing strains when it, as a whole, is subject to no strain. In the preceding paper, "On the Strength of Materials," I showed that the utmost couple which can be resisted by the bar will vary with the internal arrangement of the particles, its greatest value being $\frac{4}{3}$ and its least $\frac{2}{3}$ of its mean value, which, in the bar whose radius is unity, has just been denoted by μ. Any value of this couple, different from the mean one, adapted to a particular arrangement of the particles, may be denoted by μ', and the utmost weight and elongation in a spring having a similar arrangement may likewise be denoted by W' and E'. It will readily be seen that μ is to be regarded as the coefficient for any given substance, of the utmost strength of a cylindrical bar composed of it to resist torsion, when the particles of the bar have been so arranged that they may be all relaxed when the bar is free from external strain. It must, however, be remarked that μ would still represent the strength of the bar, even though there were *some* mutual opposing strains among the particles, provided that the particles at the circumference be relaxed when the bar is free from external strain; and that none of the internal particles exist under so great displacements from their positions of relaxation, as to occasion their being strained to the utmost sooner than the particles at the circumference, during the twisting of the bar.

If now, in the formula $L = \frac{1}{2}\pi\eta r^3$, for the strength of a bar when it is of its mean amount, which was proved in the paper before referred to, we take $r = 1$, L will become what we have denoted by μ. Hence $\mu = \frac{1}{2}\pi\eta$, and the formula becomes L or the utmost couple $= \mu r^3$.

Again, since W is the utmost weight, and a the arm at which this acts, the utmost couple may also be expressed by Wa. Hence $Wa = \mu r^3$, and

$$W = \frac{\mu r^3}{a} \quad \ldots\ldots\ldots\ldots\ldots\ldots\ldots\ldots(2).$$

The greatest elongation of which the spring can admit will be found by substituting in (1), W for w and E for e.

To justify us in making this substitution it must, however, be here remarked that, in ordinarily formed spiral springs, the elongations continue proportional to the weights added, even up to the very greatest that can be resisted. This fact I have myself observed by an experiment conducted with considerable care, and in which any deviation from the foregoing relation which may have existed was less than the inaccuracies of observation*.

Now, by making the substitution above indicated, of W for w, and E for e, in (1) we obtain

$$E = \theta \frac{l W a^2}{r^4} \quad \ldots\ldots\ldots\ldots\ldots\ldots\ldots(3),$$

which, by (2), may be put in the following form,

$$E = \theta\mu \frac{la}{r} \quad \ldots\ldots\ldots\ldots\ldots\ldots\ldots(4).$$

The equations (1) and (2) together with (3) or (4) involve the various circumstances connected with the elasticity and strength of ordinary spiral springs. For enabling us to determine, by means of these equations, the actual amounts of any of the

---

* From this two distinct conclusions may be inferred : 1st, that the angle of torsion of a bar continues proportional to the applied couple as long as the arrangement of the particles remains unaltered ; and, 2nd, that the alteration in the form of the spring by the increase of its angle of inclination and the consequent diminution of the radius of the coil does not produce a sensible effect. For, were there any deviations from the relation stated in the former proposition, this must in the experiment have been exactly counteracted by an effect of the change of form of the spring ; which, it is clear, would have been a coincidence very unlikely to occur.

variable quantities concerned, when a sufficient number of the variable quantities have been already fixed upon in accordance with the purposes to be effected; the constant coefficients θ and μ for the substance must be determined by experiment.

Without however knowing the actual amounts, we may, by interpreting the equations, arrive at many useful conclusions for the comparison of the properties of springs constructed of the same substance, but having various dimensions.   From (1) we see that:

1st.   If r the radius of the bar, and a that of the coil, be fixed; the elongation produced by any weight w will be proportional to l the length rolled up to form the coil.

2nd.   If a bar or wire of a certain length and radius be given to form a spring; the elongation produced by a certain weight w will be proportional to the square of the radius which we may adopt for the coil.

3rd.   If the radius of the bar be fixed, and the length of the spring when closed so that the coils may touch one another, or what is the same, the number of coils be also fixed; l must be proportional to a; and therefore the elongation due to a weight w will be proportional to the third power of the radius which we may adopt for the coil.

4th.   If the length of the bar and the radius of the coil be fixed, the elongation due to a weight w, will be inversely proportional to the fourth power of the radius of the bar which we may adopt.

5th.   With a given weight of metal and a given radius of the coil; the elongation, due to a weight w, will be proportional to l^3, or inversely to r^6, since l must be proportional to $\dfrac{1}{r^2}$.

From (4) we see that the ultimate elongation is:

1st.   Proportional to the length of the bar, if the radius of the bar and that of the coil be fixed.

2nd.   Proportional to the radius of the coil, if the length and the radius of the bar be fixed.

3rd.   Inversely proportional to the radius of the bar, if the length of the bar and the radius of the coil be fixed.

From (2) we perceive that the absolute strength of the spring, being independent of the length, is proportional to the third power of the radius of the bar, if the radius of the coil be fixed; and that it is inversely proportional to the radius adopted for the coil, if the radius of the bar be fixed.

By combining (2) and (4) we arrive at the interesting conclusion that the *"resilience"* of a spiral spring, that is the total quantity of work which can be stored up in it, is independent of the form or proportions of the spring, and is simply proportional to the quantity of metal contained in the coil. For, since the weights producing any elongations are proportional to those elongations, it follows that the resilience is $= \frac{1}{2} WE$.

Hence, by (2) and (4), we find that the resilience is $= \frac{1}{2}\theta\mu^2 l r^2$, which, since θ and μ are constant, is proportional to the volume of the coil or to the weight of metal composing it.

Many other relations might be deduced in similar ways, but those already pointed out will suffice, as others will be readily perceived when they may be wanted, by properly interrogating the formulas.

For determining the values of μ and θ for iron wire such as is commonly used for making spiral springs (called Charcoal Spring Wire); a spring constructed of this material was subjected to careful measurement and experiment, and the following data were obtained:

$$\text{Dimensions of the spring} \begin{cases} r = \cdot 0923 \text{ inches}, \\ l = 215 \cdot 6 \text{ inches}, \\ a = 1 \cdot 315 \text{ inches}. \end{cases}$$

When the spring was successively loaded with weights, each four pounds heavier than the one before, it was found that 56 pounds was the weight which just commenced to produce a permanent elongation, and the elongation corresponding to this weight was observed to be 16·9 inches. By the method which had been employed for bending the wire into the spiral form, and for separating the coils so that they might not press on one another when the spring was unloaded, the wire had been put into the condition which it would have received by having been twisted beyond the original elastic limit, that condition namely, in which nearly all the particles in its section would come to be strained to the utmost at the same time. Hence, according to the notation which has been adopted, this ultimate weight and elongation must be denoted by W' and E'; which, by the principles given in a former paper, already referred to, will be such that

W' shall be equal to, or rather less than, $\frac{4}{5}W$,

and E' equal to, or rather less than, $\frac{4}{5}E$*.

By taking W', μ', and E', for W, μ, and E in equations (2) and (3), we get

$$W' = \frac{\mu' r^3}{a}, \text{ and } E' = \theta \frac{l W' a^2}{r^4};$$

from which, by the foregoing experimental data, we obtain

$$\theta = 0\cdot000,000,059† \text{ and } \mu' = 94,000.$$

But μ' is equal to, or rather less than, $\frac{4}{5}$.   Hence

$$\mu = 70,000, \text{ or rather more.}$$

The values here assigned to θ and μ will, I think, be found quite sufficiently accurate for all practical purposes; especially as the metal, in other cases, cannot be assumed to be of exactly the same quality as that used in the present one. The experiment from which these values have been deduced is the only one I have as yet been able to make; I hope, however, as soon as circumstances may permit, to carry out a consecutive series of comparative experiments with springs of various dimensions and forms and of several different kinds of metal, and also to make corresponding experiments by subjecting to direct torsion, bars similar to those forming the springs. In this way would be found the amount by which the coefficients may vary in accordance with different circumstances; and, in applying the formulas afterwards to any particular case, we should be able to choose such values of the coefficients as might appear most suitable; or at least we should know what degree of dependence ought to be placed in results obtained by applying the formulas to such cases. If a spring having the radius of the bar very small compared to

---

* If we could be sure that the twisting of the wire, before the experiment, had been sufficiently great, it might be stated with certainty that $W' = \frac{4}{5}W$, and $E' = \frac{4}{5}E$. The uncertainty in this respect, though it cannot much affect the resulting value of μ, is the greatest to which the present determination of the coefficient μ is subject.

† It may here be observed that for the determination of θ the extreme weight and elongation W' and E' should not have been used, unless it had been found that these would lead to the same results as any smaller weight and elongation w and e. It was however found, as has been already stated, that the elongations were throughout proportional to the weights applied; and therefore it is a matter of indifference what weight and corresponding elongation we adopt for the determination of θ.

that of the coil, and having the angle of its spiral as small as possible, were used, there can be no doubt but that the coefficients so obtained would agree very closely with those deduced from experiment by direct torsion.

Before concluding, I shall now merely remark that, according to the principles already given, spiral springs should always be made of bars of circular section, since such bars have a much greater resilience* when subjected to torsion than others whose section is different. Thus we see that the rectangular section which has been frequently adopted in large spiral springs for railway carriage buffers is very disadvantageous. It has probably been derived from the idea that the bar is subjected to a transverse strain; an idea which, however erroneous, is the one that usually presents itself to persons considering in a cursory manner the action of spiral springs.

53. On the Principles of Estimating Safety and Danger in Structures, in respect to their Sufficiency in Strength.

[Lecture delivered as introductory to the course of Civil Engineering and Mechanics in the University of Glasgow, in Session 1873–4.]

It is a frequent remark with parents and teachers, that it augurs ill of a boy when he tends too much to ask the *cui bono* question—to ask what is the use of his tasks in Latin and Greek, or of his studies in Euclid or Algebra. To a certain extent that remark holds good. It is often better for a youth to trust to others that a course of learning may likely be valuable to him in the future, than to let opportunities for learning slip past through his being over-careful against the trouble of learning things unless their utility to himself has been first clearly and certainly shown. At the period of life, however, at which you whom I now address have arrived, and when you have entered on a course of Engineering, intended, as it is, especially for practical applications, I think the *cui bono* question is often a very proper one, and I think that, as much as possible, you should aim at having in view at least

* The meaning of the term "*resilience*" was before stated to be the quantity of work which can be stored up in the material, by giving to it its utmost change of form.

some practical uses for most of the subjects which occupy your attention or engage your labours.

Holding this view, I have selected, for an introductory lecture, a subject which I think may with advantage be brought before your minds at this early stage, as showing or suggesting some of the important practical uses of the detailed studies in strength and elasticity of materials, and arrangement of structures, to which your attention will subsequently be directed. The lecture may be briefly entitled—

ON THE PRINCIPLES OF ESTIMATING SAFETY AND DANGER IN STRUCTURES, IN RESPECT TO THEIR SUFFICIENCY IN STRENGTH.

The sources from which dangers in the employment of structures may arise are almost infinitely numerous and varied, and no brief discussion of them is possible. Even with structures perfectly sufficient in strength for their legitimate uses, dangers may arise in the management of them through casualties, misunderstandings, negligences, and many other faults or misfortunes of people. It is not of such things that I propose to treat on the present occasion. I do not intend to refer at present to safety or danger as involved in the methods of organizing and conducting active, busy, and complicated operations or processes, such as the traffic of railways and steamboats, or the industrial work in factories. In short it is on points respecting safety and danger as involved in the adaptation of matter rather than in the management of men that on the present occasion I propose to treat.

I wish, specially, to turn attention to the means which can be used, and either are, or ought to be used, for arriving at our opinions or judgments on the sufficiency or insufficiency of some of the most important, and some of the most common structures on which the lives of the public generally, or of many of its members, must necessarily be often staked. As examples of the class of structures to which I allude, I may cite steam-engine boilers, iron and timber bridges, floors, cranes, ships, water reservoirs, and numerous small things involving more or less of danger, as, for instance, slating of roofs, chimney tops on houses, and gasaliers with their counterpoise weights, which, like the sword of Damocles, hang threateningly over our heads from day to day.

The modes of judging of the sufficiency of structures are mainly of two kinds:

1st, By calculating, or guessing what the capabilities of the structure are likely to be, on the supposition of the material being good, and free from important flaws; and trusting to the care of the maker, and to such inspection as may be practicable, for avoiding original defects, or subsequent dangerous deterioration. And, 2nd, By testing or proving the sufficiency originally, and at suitable times subsequently. My main object is to advise, and to assign reasons for the recommendation, that the latter method, the one by actual testing, should be used generally in cases in which it is practicable, in preference to the former alone.

It must, however, be understood that it is indeed only in a small proportion out of the whole number of cases occurring in practice, that any very definite or reliable testing can be carried out. We cannot, for instance, proceed to test a ship as to its capability of resisting in all its multitudinous parts, the varied and unknown violences to which it may probably sometime be subjected by winds and waves; and, perhaps, by sandbanks and rocks in addition. For railway carriage wheels, axles, springs, &c., in like manner, we can arrange no very accurate or precise tests; for this reason, among others, that, where shocks and tremors are concerned, we have no sufficient knowledge of what stresses or forces the material may be subjected to in its several parts; and that such violences as those just now mentioned often bring about their destructive effects by gradual accumulations of injury, as, for instance, by the introduction of local over-straining, or incipient cracking, which will gradually increase.

The method of testing besides, even in the cases to which it is applicable, cannot supersede the calculating, or, in many cases, merely guessing what may probably be required for sufficiency, because the structures must first be made before they can be tested; and in the original designing of them the best principles of calculation, or of scientific estimation, would still have their most legitimate scope. Even when well designed and carefully constructed, however, or presumed to be so, in many cases the structures cannot or ought not to be regarded as either perfectly or in a high degree safe until after they have been actually tested, and actually found to have some very wide margin of excess of strength beyond their utmost known or possible requirements.

What then are we to understand by the words "safety" and "danger" in reference to such questions as are now proposed to be considered? Safety and danger in structures as to their sufficiency or insufficiency of strength, when they are to be subjected to any definite or more or less indefinite forces or influences, do not depend merely on the sufficiency or insufficiency of the structures themselves, but they depend essentially also on our knowledge or ignorance of whether the structure be sufficient in strength or not so. Danger or risk consists in uncertainty; perfect safety as to strength consists in knowledge of sufficiency. A structure may possess tenfold the requisite strength, but our using it may be a very hazardous proceeding, if we do not know that it is strong enough in its dimensions and general arrangement, or if we have reason to doubt that it *may* have unascertained dangerous faults. Let us test it, however, with a threefold or a fivefold load, and if this does not break it nor destructively alter it, we may now have great confidence that it is very safe for being trusted with the smaller intended load for its ordinary use.

A rope or chain either new, or worn and damaged, hanging in the shaft of a mine, or from the top of a building, may reasonably be declared to be unsafe for bearing a man suspended to it in a bucket or basket. Put on a fivefold load, and if this does not break it, it becomes quite safe; and yet its own strength has not been augmented: the change is that our knowledge of its capabilities has been extended and rendered more definite.

Further, in respect to our ordinary risks in the use of chains, let us suppose that we buy a chain at a store, and we have no ground to doubt the respectability of the people from whom we make the purchase. We do not know whether the chain has been proved or not, nor whether the iron in it is of very good quality or not. The salesman *says* it is good, or says the makers have sold it as of good quality. He says it is of such size as has often been sold and used for lifting stones of a ton weight. We pay the price of a good chain, and have not time to set about testing or proving the strength of the chain we have bought; but we proceed to lift weights of one ton by it, though we are aware that we can have no certainty that the iron in the chain is all good, nor that every link is properly welded. In using it thus we can have no certainty that the one ton stone may not come down with a crash; and we could not possibly stand below the hanging stone and regard the chain as being safe for its duty. If, however, we first lift and let down many times repeatedly three tons at a time, and then use the same chain for one ton, we can with great confidence stand under the hanging stone and regard the chain as being very safe.

If a plank is placed across a deep opening to serve as a temporary passage for men—as, for instance, is constantly done during the progress of building operations—and if the plank appears old, knotty, or otherwise weak, a workman thinking to cross it will consider it dangerous. Let him, however, see a load much heavier than himself repeatedly passed over it from end to end, and he will then regard the plank as safe for his own weight. The change here again through which safety has been attained, or risk obviated, is not that the plank has become stronger, but that knowledge has been attained of its sufficiency.

Thus we may clearly observe that while the sufficiency or insufficiency of a structure for bearing a certain load is a property inherent absolutely in the structure itself, and does not depend on our knowledge or uncertainty respecting it; yet the safety or unsafety of the structure for our employment of it depends on whether we know of its being sufficient.

In submitting to consideration the two very simple cases I have adduced—the cases of the chain and of the plank—I wish at the outset to cast aside a notion which is most extensively prevalent, and which is extremely influential in the practical determination of the degrees of mildness or severity to which tests should be carried. It is that structures—as, for instance, boilers or bridges—should be *tenderly* dealt with in testing them; that, in fact, the tests should not, or need not, very much exceed the working loads, because it is alleged that if the limits of elasticity of the material have been exceeded, if the material of the structure has taken a permanent *set* or alteration of form, the material is injured in its strength, and thus the structure may be left in a less safe or more dangerous condition than before testing. With the notion thus, and in various similar statements constantly put forward, I totally disagree; and, briefly to reply to such statements, I would say that in cases where the structure from the nature of its material, or from any peculiarity in the combination of its parts, can be supposed to be liable to receive cumulative injury, or alterations of form from successive applications of the same test load, I would have the severest desirable test repeated several

times, or many times successively. Then, if the structure bears
this without showing signs of failure, it must be *very safe indeed*
in regularly working with a small proportional part of that test
load. Because, if the structure was weakened by the first application of the load, and had been previously too weak, much more
will it now be incapable of resisting a second application of that
same damaging load, and it will either break, or show some
permanent damage; and even if it should not absolutely suffer
rupture at this stage, yet a number of repetitions of the loading
would result in visible damage, or absolute fracture. If no such
result ensues, then, as I said before, the structure must be very
safe for its comparatively small ordinary working load.

That a piece of material, especially of a ductile or malleable
substance, such as wrought iron or copper, is not necessarily
injured, or " *crippled,*" to use a word which has often been applied
in this respect, will become very evident when we consider a few
well known ordinary cases. For an ordinary cylindrical boiler the
plates of the cylindrical part have been all bent to their cylindrical form by being passed through rollers when cold, and have
thus received a *set* or permanent alteration of form; yet they are
never considered as being injured or seriously altered in their
capabilities by this usage.

Again, a bellhanger brings soft copper wire into a house to use
it as material for bell wire; but, before proceeding to adjust it
into its place for permanent service, he attaches one end to any
suitable fixed object, and pulls strongly at the other end. The
wire yields to his pull more and more as he pulls more and more
strongly, till it is permanently stretched by perhaps a fifth part
or a quarter of its original length. The wire cannot have been,
at the beginning, perfectly alike in strength at all parts. On the
pull being applied, the weakest part must yield first; that part
must become smaller in cross section, because of the stretching in
length; yet that part does not break, but becomes actually
stronger than before, and pulls out the other parts which had not
commenced to yield in a ductile manner at first. The wire has
become much stronger, even with its diminished size of cross
section, than it was originally; for it now bears, and can henceforth
bear, without any new ductile stretching, a pull very much
greater than that to which it formerly yielded. This hardening
and strengthening process, performed on the spot where the wire

is to be used, rather than in the manufactory or workshop, serves, in the case of the bellhanger, the useful purpose of producing for him a wire straightened from such crumples and bends as would exist in, and would be difficult to be cleared away from, a coil of copper wire hardened by stretching previously to being brought to the place where the wire is wanted for use.

A gasfitter twists a considerable length of block tin or lead pipe, say, through 90° of torsion. The twist is distributed (or rather distributes itself) over the whole length; but if any part twisted were thereby weakened, the further twisting applied to the pipe as a whole would take place at the first damaged part, and would not be spread along the whole length.

The like applies to the twisting of wire, and to the bending of it; and numberless other instances might readily be adduced.

Now, to return to the practical testing of structures, which I would recommend for attainment of safety, I have to add that if the structure be one liable to gradual deterioration from wear and tear, corrosion, accidental injuries, or the like, as is the case, for instance, with steam boilers, I would, at reasonable intervals of time, repeat the process of testing for safety in its full original severity; and if at any time the structure should fail under a well devised test, I would consider that it was good to have it decisively put out of use at that stage. In instituting the tests, I would prefer to undergo the danger of loss of the property, or supposed value of the structure itself, rather than the danger both to property and to life which might be consequent on the continued use of the unproved and dangerous structure.

I have here spoken of such a degree of severity as may be suitable in a well devised process of testing for safety; and I have recommended the repetition, at reasonable intervals, of the same process in its full original severity. There is, however, a distinction of practical importance which may be made between a test for safety and an original test for another purpose. It may sometimes be wise to make a test (called, it may be, a contract test, or a valuation test) which may be made heavier than what is considered necessary or suitable for repetition after wear. In a boiler, for instance, we may want to ascertain whether or not the new boiler sold is of such good material and workmanship as to stand some test agreed on, which may be more than enough to insure present safety, but may be agreed on and paid for with a view to provision of a sufficient margin of extra strength at first as an allowance for subsequent deterioration by use—through rusting, for instance, and through the injurious effects of unequal heating and cooling in different parts, or of occasional overheating of parts exposed to the fire.   Thus, an original contract or valuation test may often, with good reason, be made more severe than the safety tests which may wisely be subsequently applied from time to time.

The principle of using in some cases a valuation test as distinct from an ordinary safety test may . apply practically to chain-cable testing, anchor testing, boiler testing, and bridge testing, and the list might readily be extended to numerous other cases.   The object here is to find if the thing comes up to a certain standard of goodness; even though in some cases—as, for instance, in anchors and anchor chain-cables—we cannot specify any particular load as the working load.

Another practical case, somewhat different from any I have hitherto adduced, may now be noticed.   Hitherto I have spoken of cases in which the object is to test things intended for use, and, therefore, to be careful that the test shall not injure them unless they be of bad material, or otherwise defective, so as not to possess the desired and expected strength.   But, on the other hand, there are frequent cases in which it is both advisable and customary to test to destruction samples selected at random from among things supplied in large numbers intended to be alike. Such testing serves an important use in checking against the employment, whether knowingly or not, of materials of bad quality.

In such considerations as those with which we are to-day engaged, the term *working load* is used to denote the greatest or most severe load to which a structure is intended to be subjected when serving the purposes for which it is designed.   *Ultimate load* means the load which would just suffice to break or destroyingly damage the structure.   *Proof load* means the load applied, or intended to be applied, in testing and proving its strength. This last term, *proof load*, I must however mention, has been sometimes used to denote the unknown greatest load against the continued application or frequent repetition of which the structure may be supposed to be *safe*, or, as it is sometimes called, *proof*, as in the expressions *water-proof*, *fire-proof*, &c.   The use of the term *proof load* in this sense I regard as not to be recommended; and I shall use it only to signify the load in respect to which we have, or intend to have, *proof* that it can be resisted by the structure. The other meaning, which I reject, I would express in such terms as the *estimated extreme safe load*; and I would not call that the proof load.

FACTORS OF SAFETY are of several kinds:—Thus, (1) A factor of safety may be taken as the ratio in which the estimated breaking or destructively damaging load exceeds the working load. (2) Another factor of safety, having a different meaning, may be the ratio in which the estimated breaking or destructively damaging load exceeds the proof load. (3) Another factor of safety, and that which most truly may be called a factor of safety, is the ratio in which the proof load exceeds the working load.

The first of these factors is that to which the designer usually gives his most special attention, because the destroying load is the one as to which, for any material, we possess the most precise information. He considers what is to be the working load for which his structure must be designed. He then multiplies this the number of times that he thinks the probable ultimate load ought reasonably to exceed this working load; and he then does his best to design the structure so that it would just break or fail at that ultimate load arrived at from the working load by his factor of safety.

The ultimate load in any structure intended to be preserved for use, must, of course, be only judged of by calculations, or estimates from experiments on other like structures or on pieces of the same kind of material tested up to their point of breakage, or destructive alteration of form or of cohesion.

In respect to boilers, where the straining force is applied by steam or water pressure, we may use the terms *working pressure*, *ultimate pressure*, and *proof pressure*, which have an obvious correspondence with the terms *working load, ultimate load,* and *proof load,* already explained.

Now, in respect to boilers there is a prevailing opinion or frequently adopted rule, to the effect that:—

1st, The ultimate strength of a good boiler should be at least eight times that called into play under its working pressure.

That, 2nd, The proof pressure need not be more than twice the working pressure, and that often less than twice is quite enough.

And that, 3rd, The proof pressure ought not to be made more than one-fourth of that corresponding to the ultimate strength of the boiler.

The reason for keeping the proof so low as only twice, or less than twice, the working pressure is fear of damaging the boiler. That is, if the plain truth is to be spoken, fear that the ultimate strength may probably be nothing like eight times the strength called into play at the working pressure. If the fear were for slight leakage alone, and for the consequent trouble and cost of amending the defective parts, there would be some shadow of justification for yielding to that fear; but what is usually said is, that a severer test than about double the working load in the case of a boiler whose calculated ultimate strength is eight times the working load would tend *to danger of bursting* by its, perhaps, overstraining and bulging some parts of the structure, and that thus what would be meant for attaining safety against accidents and loss of life would tend much more to the very opposite result.

In this view, when properly considered, there is to be found a decided logical fallacy. It arises from a confusion between, on the one hand, risk of injuring the usefulness of the boiler for driving the engine and making money to its owner; and, on the other hand, risk of loss of life or limb to people whose lives or limbs are staked on the sufficiency of the boiler, and risk of loss of property to those whose goods may also be staked on its sufficiency.

If there were any validity in the argument, it would go against all proving beyond the working load; for in any new or in any old boiler subjected to testing, some one part—a stay bar perhaps, or an angle iron, or a gusset piece—may have an unseen flaw, or a crack, or an unascertained quality of brittleness, and may be just on the point of rupture under the very mild proof of double its working load. Then, if the view I am combating were valid, this defective part would be rendered more dangerous by the mild proving. I say, however, in answer, that the proof ought rather to have been made much more severe still, and then the defective part would have been broken, or destructively altered, and so would have been rendered powerless for causing future mischief by its delusive estimated, but practically non-existent, strength relatively to the duty to be imposed on it.

I say without hesitation that the greater the pressure that a boiler has actually borne without bursting or leaking, the greater will be its safety for the future; and this the more so if the severe test has been repeatedly applied without the occurrence of rupture or destruction to any part. On the whole, my opinion is that people, while liking to be left fondly to cherish the idea that *perhaps* or *probably* they have a wide margin of extra strength for the sake of security; and while willing to pay for the material required for such extra strengths, are habitually quite too niggardly of *making sure* that they have a good margin at all. Thus the material applied in giving strength that people will not dare intentionally to call into play is really little better than waste material in so far as any working within limits of safety is concerned; and beyond limits of safety people ought not willingly to work their structures, or to allow them to be worked, when very serious consequences would accompany the failure of the structures in strength. Such extra material, no doubt, is a proper means for diminishing the risk in unproved or insufficiently proved structures; and it may often be properly introduced on that ground, as also for affording greater durability against corrosion or wear.

Cases no doubt may occur in which it may be better not to prove a structure beyond the working load. Without waiting to discuss such cases, I may suggest a bridge hastily constructed by an army for its use in retreating before a pursuing enemy, or an existing bridge of doubtful sufficiency which it is about to use in the same circumstances; a worn rope let down a precipice for escape of an unfortunate party of people surprised by a rising tide; the boiler of a merchant screw steamer pursued in time of war by a privateer; or, on the other hand, the boiler of a privateer pursued by a steam ship of war. In all these cases there is danger both ways; and then it may be wise to choose that which seems the less.

Further, there are few engineers who have had much practical concern with steam engines, and who cannot recall to mind cases in which it has been preferred to continue working a boiler found to be in very bad condition from age and corrosion, rather than endanger the stoppage of some very important work, by venturing to prove the boiler beyond its ordinary working load, and thereby using means either to break it, or to remove all grounds of fear in its use. On a tract of fen land, for instance, drained by steam power, the engine boiler was found to be very far gone, a new one was ordered with all speed, the old one was kept working at forty pounds to the square inch, but was not proved at all beyond this. To have proved it even to forty-five pounds would have been deemed to involve too much danger of breaking the boiler, causing the stoppage of the engine, and the flooding of the low-lying lands. This story is not an exact report of any one single occurrence, but it is a fair representation of numerous cases often occurring in the exigencies of the varied operations of men.

I have now been speaking of structures which would admit of being proved beyond a definite working load which they are required to resist. On the other hand, however, it is to be observed that many structures, or most structures, indeed, from their very nature do not admit of being proved experimentally, but must be made and set to work on trust in the sufficiency of the design and of the execution. A ship, as I have already pointed out, is an instance of this kind; and here, for due safety, we have to rely on a large excess of calculated or estimated strength in the various parts beyond that required in their ordinary use, and in many of them beyond what may ever happen to be called into exertion. We have also to submit in such cases to have the clear knowledge that at best the structure is not sufficient for resisting all the influences to which it may possibly be exposed when in use, and that danger in its use cannot be avoided.

Boilers, on the other hand, are a structure most peculiarly well suited for perfect experimental testing. If two boilers executed by different makers, but under the same specification and the same terms of contract, and intended to work at 40 lbs. to the square inch, have their dimensions such that their calculated or estimated bursting pressure will be 320 lbs. to the square inch, or eight times the working pressure, and if one be proved to 80 lbs. and the other to 120 lbs., neither being found to fail,—surely the one which has resisted 120 lbs. will be more trustworthy, under the various chances of possibly augmented pressure, through negligence or other causes, than the one which is only known to have resisted 80 lbs.

Though offering here for consideration a comparison of two cases, in one of which the test is carried to twice, and in the other to three times the working pressure; yet I do not wish to

pronounce upon any one particular margin for safety as that alone
which may be proper in boilers in general or in bridges in general.
The decision as to what may be a suitable and proper margin to
require, or to provide, ought to depend on many circumstances in
individual cases; but I do wish to say that it is a great mistake to
assert a *calculated* strength of eight times the working load, while
at the same time admitting that three times or two and a half
times would be likely to cause damage.

The subject I have entered on, as to physical properties of
matter, and as to practices of men—as to what these practices are,
and as to what perhaps they ought to be, is a very wide one. I
might go on to say much respecting other structures, large and
small, besides those which I have already spoken of at some
length. I might proceed to discuss questions of safety and danger
in bridges, in water reservoirs, and in numerous small structures
in very common use; and I might point out how, by testing and
proving, safety may be attained, or danger abated. I might point
out how, in some cases, multiplicity of parts tends to danger, as in
the numerous links of a chain, where any one being faulty destroys
the utility of all the rest; and, on the other hand, how, through
multiplicity of parts, safety, almost perfect, is attainable without
testing in a wire cable. I cannot, however, attempt to exhaust
the subject, nor would I wish to exhaust your patience in this
introductory lecture. So I will now close, and leave any further
discussion of this and kindred subjects for future occasions.

54. COMPARISONS OF SIMILAR STRUCTURES AS TO ELASTICITY, STRENGTH, AND STABILITY.

[From *Transactions of the Institution of Engineers and Shipbuilders in Scotland.* 21st December, 1875.]

IN the brief considerations which I propose now to offer to
your attention, I do not know that there is anything to be
regarded in the light of a new discovery, or of an entirely new
kind of investigation. I think, however, that I am able to bring
together some easily intelligible and easily recollected principles,
which may often be of great practical use; and that I can offer

very simple and easy considerations for their establishment and application.

The cases of similar structures which I propose to discuss are of two kinds, very distinct, and which stand remarkably in contrast each with the other.

I. The one relates to comparisons of similar structures in respect to their elasticity and strength for resisting bending, or damage, or breakage by similarly applied systems of forces.

II. The other relates to comparisons of similar structures as to their stability, when that is mainly or essentially due to their gravity*, or, as we may say, to the downward force which they receive from gravitation.

For the first of the two kinds of cases just referred to, a comprehensive but simple and easily intelligible principle, which it is proposed now to establish and illustrate, may be stated as follows :—

General Principle.—Similar structures, if strained similarly within limits of elasticity from their forms when free from applied forces, must have their systems of applied forces, similar in arrangement and of amounts, at homologous places, proportional to the squares of their homologous linear dimensions.

To establish this we have only to build up, in imagination, both structures out of similar small elements or blocks, alike

---

* The word *gravity* is here preferred to *weight*, although the word *weight* has been, and still is, very commonly used for conveying the meaning here intended to be expressed. The common word *weight* has unfortunately two meanings in frequent use, and this ambiguity is a troublesome source of perplexity in mechanics and in other branches of Natural Philosophy. Firstly, the word " weight " very commonly signifies *quantity of matter*, as when we speak of a hundred-weight of iron, or of four pounds weight of mercury, or when, in an Act of Parliament (18th and 19th of Victoria, chapter 72, July 30, 1855) for the special purpose of establishing standard weights and measures, it is enacted that a certain piece of platinum referred to as a " weight of platinum," deposited in the Office of the Exchequer, shall be denominated the IMPERIAL STANDARD POUND AVOIRDUPOIS, and shall be deemed to be the only standard of weight from which all other weights, and other measures having reference to weight, shall be derived, computed, and ascertained. And secondly, the same word " weight " is often used to signify the downward force which a piece of matter exerts on whatever supports or suspends it ; as when we speak of the *weight* of a piece of matter hung to a spring-balance as being the *force* which draws out the spring. The word MASS is employed to express distinctly, in scientific language, one of the two meanings of the word " weight "—that, namely, in which it signifies *quantity of matter* ; but for the other meaning, in which it signifies a *force*, we have no established name as yet ; it may, however, very well be called the *gravity* of the piece of matter.

strained, with the same intensity and direction of stress in each
new pair of homologous elements built into the pair of objects.
(See figures below.) To aid our conception, we may first imagine
both structures to be free from forces applied to them from
without, and we may imagine the two to be similarly divided
into a great number of small similar elements, or blocks, similarly
situated. Thus we may notice that since the similar elementary

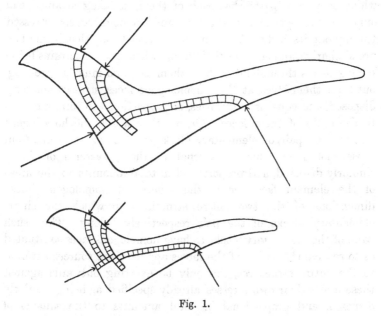

Fig. 1.

blocks, unstrained, would fit together in both cases, so as to
produce two similar unstrained structures, the blocks, or elements,
if alike stressed in homologous pairs in the two cases, so as to be
similarly deformed from their free condition, would also go together
so as to produce two similar structures similarly strained from
their free condition*.

* A case liable to occur in practice, but which is very commonly, and sometimes erroneously, neglected or left unnoticed in investigations on strength and
elasticity of materials and structures, is here, for simplicity, left out of consideration
—the case, namely, in which a piece of material, or a structure, is subject to
stresses in its substance, arising from mutual action of its parts when free from
externally applied forces. The principles discussed in the present paper are, however, easily applied to such cases, but, for avoidance of any extra complication,
the detailed discussion of such cases is not here entered on. It may suffice to say
that, so long as the utmost stresses do not strain any part of the substance of

Now, if we confine attention for a moment to any pair of homologous elementary blocks in the two structures, considering them by themselves, and regarding the immediately adjoining elementary blocks which meet them face to face as being objects external to them, though not external to the entire structures, we may observe, from the nature of the case, since the elementary blocks are extremely small in comparison with the structures of which they form parts, that each of them, in being strained, can only be homogeneously strained and homogeneously stressed throughout its body. We can then obviously see that if the two are similarly strained from their forms when free, the strains being homogeneous throughout each of them, the two must be exerting, out from themselves, at their homologous faces, stresses similarly disposed, and equal in intensities, or proportional in total amounts to the areas of those faces. Or, in other words, the homologous faces of the pair of elementary blocks must be exerting out from those blocks, on objects external to them, systems of forces similarly disposed, and proportional in total amounts to the areas of the element faces, or to the squares of homologous linear dimensions of the two entire structures to which the single elementary blocks of the pair respectively belong. Thus such pairs of the elementary blocks as have homologous faces so situated as to receive the action of the forces applied from sources external to the entire structures, can only be exerting outwards against these forces their own stresses already specified as being similarly disposed, and proportional in total amounts to the squares of homologous linear dimensions of the structures. Conversely we see that the forces applied from without, subject to which the two similar structures will be, at all homologous places, similarly

either structure beyond limits of elasticity, and provided that the ordinarily existing condition of elasticity (known as Hooke's law), and which may be briefly described as *stress proportional to strain*, or more precisely, as *unital stress proportional to unital strain*, applies truly to the material of the structures, the conclusions arrived at here as to the similarly applied forces and the similarity of the strains and stresses produced by them, will be in no degree affected by the antecedent existence or non-existence of internal stresses in the structures when free from forces applied from without. In respect to this subject, reference may be made to a paper by the author, published in the *Cambridge and Dublin Mathematical Journal*, November, 1848, " On the Strength of Materials, as influenced by the Existence or Non-Existence of certain Mutual Strains among the Particles composing them." [See *supra*, p. 334.]

strained and stressed, must be similarly disposed and of amounts at homologous places proportional to the squares of homologous linear dimensions of the two structures.

From what has been now shown, we find that similar structures of different dimensions must not be similarly loaded, that is to say, must not be loaded with similar masses, if they are to be stressed with equal severity, and are thus to undergo similar deformations, or are to continue similar, each to other, after being loaded. We see this because loads similar to the structures would, if actually resisted, apply at homologous places forces proportional to the cubes of homologous linear dimensions; but the similarly deformed structures would exert on external objects homologous forces proportional only to the squares of the homologous linear dimensions; and so, if the load in the smaller

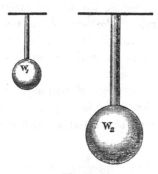

Fig. 2.

structure were just such as to be balanced by the elastic forces exerted by that structure, a similar load, if attempted to be applied to the larger structure, would be too much for the elastic forces exerted by that larger structure when similarly deformed so as to be alike stressed, and so would not, in actual fact, be resisted by the larger structure deformed similarly to the deformation of the smaller.

The simple case of two similar wires with two similar weights suspended (as in the accompanying figure), will illustrate this general statement. The weights, and consequently the forces which they apply, are as the cubes of the diameters of the wires; but the cross-sectional areas of the wires are only as the squares, and so the unital stresses in the wires (that is to say, the stress-rates per unit of area) are in the ratio of the linear dimensions, and are thus more severe in the larger structure than in the smaller. This easy case is here selected for illustration of the general principle applicable alike to simple and to complex cases, because, from its extreme simplicity, its conditions can be perceived without any elaborate reasoning; while, by the agreement of the results obviously exhibited in it with those announced in the general statement, an easy illustration and verification of that statement is afforded.

Numberless other examples or illustrations might be offered, but a few will suffice.

We may consider the case of the comparison of two similar girders or horizontal beams, rectangular in their cross sections (like flooring joists), and each loaded only with its own weight; and we may suppose the similar forms and the dimensions to be such that in neither case do the utmost stresses anywhere introduced exceed the limit of elasticity of the material; and that the beams do not bend, in either case, to any such considerable curvature as to vitiate importantly the application of the ordinary formula which will be used.

For the elastic deflection of rectangular beams of uniform cross section, and loaded uniformly along their length, while placed horizontally and supported at both ends, we may use the well-known formula,—

$$\delta = \frac{5}{32} \frac{pl^3}{Ebd^3};\text{ where}$$

l denotes the length between supports,
b the breadth of the beam,
d its depth,
E the modulus of elasticity,
p the load applied, and
δ the deflection or depression at middle due to the load.

Now, putting u to denote any chosen homologous linear dimension in the larger and smaller beam, we see that $p \propto u^3$, $l \propto u$, $b \propto u$, and $d \propto u$; and hence from the formula we get

$$\delta \propto \frac{u^3\,u^3}{u\,u^3}, \text{ or } \delta \propto u^2.$$

That is, the deflections of similar rectangular beams supported as has been stated, and loaded with their own weights, would increase proportionally to the squares of the linear dimensions. But, obviously, for the production of equally severe stresses, the deflections ought to be proportional only to the linear dimensions, and consequently the larger beam will be more severely stressed by the load of its own weight than the smaller one will be in like manner by the load of its own weight.

Again, as another example, let us suppose that a tank of boiler plate, or of cast iron plates, for holding water, is arranged, for simplicity, to stand on three columns placed triangularly on the ground thus .·., so that the tank is supported simply at three points of itself, and that it has been found of sufficient strength for bearing the load of water which fills it; and let us suppose that it is judged to be of good design, neither too strong nor too weak. Let it be proposed to construct another tank like this one, of exactly the same proportions in all respects, but having all its linear dimensions exactly double of the corresponding linear dimensions of the one already successfully in use. The question arises, What inference is to be drawn, as to whether the proposed larger tank would be of suitable strength, or needlessly too strong or too weak, for bearing to be filled with water, from the information that the similar, but smaller, tank is of just suitable and proper strength for what it has to bear? The answer, which in this case would not be attainable in any such easy ways as were available in the two very simple cases previously adduced, is obtainable at once from the general principles which have been demonstrated. Thus we see, in comparing the loads of water pressure that can be borne on homologous areas in the smaller and the larger tank, that in order to introduce equal intensities of stress at corresponding places in the metal of both tanks, and therefore to produce similar deformations in both, the loads of water pressure on homologous areas must be proportional to the squares of the linear dimensions; or, in the case supposed, must be four times as much in the larger tank as in the smaller, but that the actual loads would be eight times as much in the larger as in the smaller tank. Hence the larger tank would be insufficient in strength for its load, or would have a less margin of superabundant strength for safety than the smaller one has.

Now, from the particular cases which have been adduced, and from the general principles which have been demonstrated, we may see that it would be wrong to judge from the sufficiency of strength of a model beam bridge, or of a water tank or other structure, in circumstances such as have been under consideration, that the larger corresponding structure would be likewise sufficient. In fact, we may notice in general that, in order to have similar structures to take similar deformations under the action of loads of material similar to the structures and similarly disposed, and to have equal margins of superabundant strength for safety, it would be necessary that the specific rigidity and strength of the material should increase directly in proportion to the homologous linear dimensions of the structures. But such is not the case in fact; and so, as we go on enlarging our already large structures (beam bridges or ships, for instance), if we keep to the same proportions, we must come ultimately to dimensions such that the structure will break down by its own gravity.

Referring now again to the general principle which has been laid down and demonstrated in the early part of this paper—viz., That similar structures, if strained similarly within limits of elasticity from their forms when free from applied forces, must have their systems of applied forces similar in arrangement and of amounts, at homologous places, proportional to the squares of homologous linear dimensions—we may see that it affords very tolerably satisfactory means of judging of the relation between large and small similar structures which have to resist the forces of wind, when those structures depend for their sufficiency mainly on their strength, and their gravity does not come into account as having any very important influence among the conditions affecting the results. Thus, if we compare two similar masts carrying similar sails, and being either with or without stay-ropes similar in arrangement, and having their sails alike exposed to the wind— so that we may assume as being at least approximately enough true—that the force of the wind per unit of area is the same at corresponding places in the two structures; or that, in other words, on homologous areas in the two structures, the force of the wind will be similarly applied as to direction, and will, in magnitude, be in the ratio of the squares of homologous linear dimensions, we may judge that the two will be similarly strained, and will be stressed with equal severity at all homologous places, and thus that both will be alike suitable for resisting the force of the same day's wind.

For the same reasons, it would appear that two umbrellas— similar in form at every part, but different in size—would be alike suitable for resisting the same kinds of exposure to the same storminess of weather.

From the considerations already adduced, it is easy to pass on to the consideration of strains beyond elastic limits, and to show that in structures of similar forms, and composed of materials exactly alike in qualities, at all homologous parts, we have a right to expect the introduction of similar *sets*, or permanent changes of form by similar systems of externally applied forces; and to expect that similar deformations past elastic limits should bring the smaller and the larger structure to the point of rupture together.

As an instance of what is comprised in this latter statement, it may be noticed that, contrary to ideas very frequently entertained, and which have sometimes affected decisions in important practical cases, we ought to expect a thick boiler plate or plate for shipbuilding, or a round or square iron bar, large in cross section, to admit of being bent double cold without cracking, just as readily as a thin plate or small similar bar of exactly the same quality of material. People often take the view that in the thicker plate or bar there must be more stretching of the outer part at the bend than in the thinner plate or bar. It ought to be recollected, however, that although in the larger one there be a greater total stretching, yet there is in like proportion a greater length of material to receive this greater stretching, and the deformation of any homologous small elementary blocks in the two will be exactly similar; or equal very small blocks similarly situated in both will be alike deformed, and therefore equally severely dealt with. Now, no doubt we are accustomed to find as a matter of fact that thick bars or thick plates, which men can make with their available means and appliances, do not admit of bending in a ductile or malleable manner without breaking, as well as thin bars or plates of good quality do; but this is to be attributed to the difficulty or impracticability of making large forgings of qualities of material as good as are easily attainable in thin plates or in bars of small cross section.

Passing now to the second class of cases of similar structures referred to at the opening of the present paper—the class, namely, which relates to comparisons of similar structures as to their stability, when that is mainly or essentially due to their gravity (commonly named as their *weight*)—we may see that, if we regard the cases to be dealt with as being such that in them the strength

of the material is amply sufficient to prevent the occurrence of any
such failure of strength at any part of either structure as would
be in any important degree influential on the relative stabilities
of the structures of different sizes, the similar structures will be
brought alike to be just ready to fail in stability (that is, alike
ready to be overturned) by the application on their surfaces of
similar force-systems, the magnitudes of the forces applied on
homologous areas being in the ratio of the cubes of homologous
linear dimensions of the different structures; or, in other words,
the intensities of the forces applied at similarly situated places
being in the ratio of the homologous linear dimensions.

In reference to this, we may first consider the case of two
similar obelisks, or two similar pinnacles, supposing them to be
acted on by wind pressing on all corresponding patches of surface

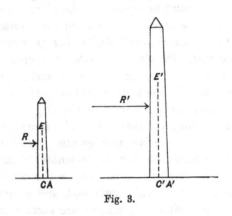

Fig. 3.

of the two objects, similarly in direction, and *with the same intensity
of force*; and we may suppose the material to have sufficient
strength to prevent the introduction of any important influence
on their relative stabilities, from crushing or failing of the material
in either case, at any place where it would become severely pressed
at the edge of the base round which rotation would occur if overturning were effected.

Obviously, the resultants of the forces of the wind, acting as
here supposed, on the two obelisks will be similarly situated, as
shown by R and R' in the figure: and their magnitudes will be
as the squares of homologous linear dimensions; and their arms or
leverages for their overturning moments round A and A' will be
as the linear dimensions. Hence the moments of the overturning

motives will be as the cubes of the linear dimensions. But if the obelisks were both on the point of overturning, the resisting moments would be as the fourth powers of the linear dimensions, because the gravities of the two obelisks are as the cubes of the linear dimensions, and the arms or leverages at which they act round the edges A and A' are CA and $C'A'$ respectively, which obviously are as the linear dimensions. Thus we see that the larger obelisk would resist a force of wind which would blow down the smaller one; and we can readily see, further, that to make both be just on the point of overturning by the action of similar systems of forces applied by wind, the forces per unit of area at corresponding places would be in the ratio of the linear dimensions. If, for instance, one of them is of linear dimensions double of those of the other, it will be able to bear twice as much force of wind per square foot as the smaller one can bear; it will be able to bear eight times as much force on every element of its surface as the smaller one can bear on every corresponding element of its surface; and, in the aggregate, the resultant of all the wind forces which it will be able to bear will be eight times as great as the resultant of all the wind forces which the smaller one can bear; and, further, the moment of the overturning motive, and that of the equal and opposite resisting turning motive, will each be sixteen times as great in the larger structure as in the smaller one.

The consideration of the relations between factory chimneys or other chimneys, similar in form but differing in dimensions, may be aided, though not fully elucidated, by consideration, in connection with them, of the principles just adduced for the supposed cases of obelisks and pinnacles. In the case of large chimneys, the yielding of the material by crushing at the severely compressed edge of the base, or by crushing, splitting, or otherwise failing through deficiency of strength at any part, would be decidedly influential, and in some cases might be largely influential, on the results, and therefore must not be left out of account in any very complete consideration of the subject. In their case too, as also in that of obelisks, standing with their bases at or near the ground, we must bear in mind that the taller of any two structures, comparatively considered, will ordinarily be exposed to a greater force of wind per square foot at any part of its surface than the lower one would be exposed to at the corresponding part of its surface, because the velocity of the wind low down is more abated by friction on the ground and by sheltering obstructions, such as houses and trees, than at higher elevations. Still we may judge with much confidence that the larger of two similar chimneys may be expected to be capable of standing against a greater storm than the smaller could endure; and we may see that it would be an unsafe procedure to judge, from the known sufficiency of a large chimney, that a smaller one, constructed in the same propor- tions as to general configuration, and of like material, would be of sufficient stability.

[This subject was extended and developed by Professor Archibald Barr, D.Sc., in a Paper 'Comparisons of Similar Structures and Machines' read before the Institution of Engineers and Shipbuilders in Scotland, 21st March, 1899. Volume XLII. page 322.]

55. On Metric Units of Force, Energy, and Power, larger than those on the Centimetre-Gram-Second System.

[From the *Report of the British Association*, Glasgow, Section A, 1876, p. 32.]

THE author premises that under the excellent method of Gauss for establishing units of force, a unit of force is taken as being the force which, if applied to a unit of mass for a unit of time, will impart to it a unit of velocity. In the system already adopted by the British Association Committee on Dynamical and Electrical Units (*Brit. Assoc. Report*, part i. 1873, page 222), the Centimetre, the Gram*, and the Second were taken as the units of length, of mass, and of time; and the unit of force thence derived under the method of Gauss was called the *Dyne*.

That force is very small, quite too small for convenient use in all ordinary mechanical or engineering investigations. It is about equal to the gravity of a milligram mass, and that force is so small that it cannot be felt when applied to the hand. That

---

* The spelling Gram, instead of Gramme, for the English word is adopted in the present paper in accordance with the spelling put forward in the Metric Weights and Measures Act, 1864, which legalizes the use of the Metric System in Great Britain and Ireland.

system, designated as the Centimetre-Gram-Second System, re-commended by the Committee of the British Association, and described fully, with many applications, in a book since published by Dr Everett, who was Secretary to the Committee, is well suited for many dynamical and electrical purposes; and it ought certainly to be maintained for use in all cases in which it is convenient. But the object of the present paper is to recommend the employ-ment also of two other systems which are in perfect harmony with it, and to propose names for the units of force under these two systems.

In one of these systems, the Decimetre, the Kilogram, and the Second are the units adopted for length, mass, and time; and thus the system comes to be called the Decimetre-Kilogram-Second System.

In the other, the Metre, the Tonne*, and the Second are adopted as the units of length, mass, and time; and thus the system comes to be called the Metre-Tonne-Second System.

It is to be particularly observed that all the three systems here referred to are framed so as to attain the condition, very important for convenience, that the unit of mass adopted is the mass of a unit volume of water, and that, therefore, for every substance the specific gravity and the density, or mass per unit of volume, are made to be numerically the same.

In the Decimetre-Kilogram-Second System, the unit of force derived by the method of Gauss is 10,000 Dynes, or is about equal to the gravity of 10 Grams. It is impossible, or almost so, to work practically with any such system without having a name for the unit of force. The unit of force in this system is such that a human hair is well suited for bearing it as a pull, with ample allowance of extra strength for safety against breakage; and the author proposes to call it the *Crinal*, from the Latin *crinis* and *crinalis*.

In the Metre-Tonne-Second System the unit of force, likewise derived by the Gaussian method, is 10,000 Crinals, or 100,000,000 Dynes, or is about equal to the gravity of 2 cwt., or of $\frac{1}{10}$ of a ton. This force would be properly borne as a pull by a moderately-sized rope; and the author proposes to call it the *Funal*, from the Latin *funis* and *funalis*.

---

* The Tonne is the mass or quantity of matter contained in a cubic metre of water, and is very nearly the same as the British Ton.

Then we have One Horse-Power, of 33,000 foot-pounds per minute, about equal to 75,000 Decimetre-Crinals per second ; and the Horse-Power is also about equal to ·75 of a Metre-Funal per second.

Also 1 Metre-Funal

$$= 100,000 \text{ Decimetre-Crinals,}$$
$$= 10,000,000,000 \text{ Centimetre-Dynes, or Ergs,}$$
$$= 10^{10} \text{ Ergs.}$$

Also 1 Horse-Power is about

$$= 7,500,000,000 \text{ Centimetre-Dynes per second,}$$

or as the same may be written

$$75 \times 10^8 \text{ Centimetre-Dynes per second.}$$

The number 7,500,000,000, for expressing a Horse-Power under the Centimetre-Gram-Second System, is an exceedingly unmanageable one; and it gives a very decisive indication that the Centimetre and Gram are too small to be suitable as fundamental units of length and of mass for ordinary engineering purposes; and that there is great need for the establishment of systems having larger units, such as those which have been recommended in the present paper, and for which a convenient nomenclature has been offered.

It is to be observed that the provision made by the British Association Committee, in the Report already referred to, of a multiple of the Dyne, such as the Megadyne, or million of Dynes, as a larger unit of force, does not accomplish all that is to be desired, because various important formulas, or convenient methods of statement, will not hold good when any of the units are so derived. Thus, for instance, if the Megadyne be the unit of force, while the Gram and Second are the units of mass and time, the ordinary formulas for giving the so-called "centrifugal force" of a revolving mass,

$$F = \frac{mv^2}{r} \text{ and } F = m\omega^2 r,$$

will not hold good; and, as another instance, we may notice that the proposition that, in respect to a jet of water, the reaction force on the vessel is equal numerically to the momentum generated per second, will not hold good; and numberless other instances might readily be cited, but those given may suffice.

56. On Dimensional Equations, and on some Verbal Expressions in Numerical Science.

[From the *Report of the British Association*, Section A, Dublin, 1878, p. 451.]

In recent years attention has been given, more than before, to relations among standard quantities of variable things, to be taken as units for use in giving numerical expression to various quantities of those things. The quantity of each different variable thing selected as a unit might be, and often has been, arbitrarily chosen, independently of the quantities chosen for units of other things. But great advantages as to convenience and facility are attainable by making a methodical connection among the quantities to be selected for the units of the various things, so that when some of the units are arbitrarily selected, the others will be derived from them in some good systematic way.

The units thus arbitrarily selected are called *fundamental units*; and the others obtainable from them by the systematic method are called *derived units.*

Teaching on this subject is given in the early pages of Professor Clerk Maxwell's *Treatise on Electricity and Magnetism*, and in Professor Everett's *Treatise on the Centimetre-Gramme-Second System of Units.* The subject is important; but much of the nomenclature and notation hitherto used is very confusing and unsatisfactory.

I now wish to propose some amendments, or new modes of expression, which appear to be commendable.

Instead of saying, as is done in Professor Everett's very useful treatise,

"The unit of acceleration varies directly as the unit of length, and inversely as the square of the unit of time;"

I would propose to say

The change-ratio of the unit of acceleration is the product of the change-ratio of the unit of length and the inverse second power of the change-ratio of the unit of time.

The meaning of the new name, *change-ratio*, may be given by the following definition.

DEFINITION.—In respect to changes of any variable, the *change-ratio* is the ratio of the new value to the old, for any change from one value of that variable to another.

Further, in Everett's treatise, the relation already referred to among the units of *length, time,* and *acceleration,* is stated in some other ways, which I will next cite :—

"The dimensions of acceleration are $\dfrac{\text{length}}{(\text{time})^2}$."

"The dimensions of the unit of acceleration are

$$\frac{\text{unit of length}}{(\text{unit of time})^2}\text{ ,}$$

Instead of either of these, I would substitute this—

Change-ratio of unit of acceleration

$$= \frac{\text{change-ratio of unit of length}}{(\text{change-ratio of unit of time})^2}.$$

This expression states clearly and correctly all the truth which is meant to be conveyed by the previous statements.

In order now to be enabled to speak, in language brief and free from ambiguity, of any numerical expression whatever, whether whole or fractional, greater or less than unity, or unity itself, I shall use the word *numeric,* which I recommend for general use, to comprise all the meanings which at present are conveyed in common use, but with much of troublesome ambiguity, by words or phrases such as *number, fraction, number or fraction, number and fraction, number or proper fraction or improper fraction.* I recommend that, as soon as possible, the word number should be restricted to its only proper signification, which is often at present designated in an objectionable way by the two words "*whole number,*" but which is often also expressed, and really properly so, by the single word *number**.

* Thus, for instance, in the public regulations for Post Office Savings Banks, issued by authority of the Postmaster-General (*British Postal Guide*), the intimation is made that "*At these banks deposits of one shilling, or any number of shillings, will be received,*" this being, however, subject to some restrictions, which need not be mentioned here, merely assigning limits to the amounts that will be accepted from any one person. Now the words here quoted would convey a false statement of what the Post Office authorities really mean to announce, if the word "number" were allowed to mean a fractional numerical expression. The announcement is obviously framed on the assumption that the word "number" in it can only mean *legally* what in the present paper is referred to as its only *proper* signification,

Now we have no right to speak of dividing one quantity by another of a dissimilar kind, except, merely for brevity, in the case of dividing the numeric expressing the one quantity by the numeric expressing the other quantity, after we have fixed upon units of the two things. Thus we have no right to speak of *unit of length* divided by *unit of time*, nor to employ, unless perhaps for brevity, and under an implied protest, such a notation as

$$\frac{\text{Unit of length}}{\text{Unit of time}}.$$

Further, we have no right to speak of "*second power of unit of time*," nor of "*square of unit of time*." The name *power* seems admirably well suited (whether by deliberate design entirely, or partly by good chances) for its uses in reference to what are called powers, whether integral or fractional, of any numerics, as for instance

$$x^2, \ x^3, \ x^{\frac{3}{2}}, \ x^{2\cdot13}, \ \&\text{c.}, \ \&\text{c.};$$

and also for its uses in cases such as

$$x^{-2}, \ x^{-3}, \ x^{-\frac{3}{2}}, \ x^{-2\cdot13}, \ \&\text{c.}, \ \&\text{c.};$$

in any of which x may denote any numeric whole or fractional.

Such expressions as the *square of the unit of time* are not to be approved of. That expression, for instance, is too like such combinations of words as *a square second*, or *a square minute*. It is not really quite so unreasonable, however, because there is a good meaning sometimes intended though badly expressed, when the *square of the unit of time* is spoken of, that unit being then regarded as a variable, and the true meaning being usually just what may be distinctly stated as *the second power* of the *change-ratio of the unit of time*. A second is essentially a constant quantity of time; and *the square of a second, a second squared*, and *the second power of a second of time*, are all of them essentially meaningless conjunctions of words.

that, namely, which is commonly designated as "*a whole number*" or "*an integer*." If an intending depositor, understanding the word "number" in the extended sense in which it is very often, and also quite authoritatively used, would offer a deposit of $7\frac{1}{3}s.$, his offer would be refused, as it would amount to 7s. 4d., which is not contemplated in the regulations as an amount to be accepted as a deposit.

More examples to the same effect might be cited from usages in practical business affairs; and also from usages of scientific writers in arithmetic, and in other branches of mathematics, but the one here given may suffice.

It is to be observed that any ratio is a numeric, or may be treated as such to any degree of approximation we please; and so we can have the second power or any other power of a change-ratio; and we are entitled properly to write as a fraction any power of any change-ratio divided by any power of the same or of any other change-ratio. That fraction will be itself a numeric, and may properly form one side of an equation having a numeric for its other side.

It follows, then, under the views already offered, as to legitimate and illegitimate modes of expression and notation, that such notations as

$$1 \frac{\text{yard}}{(\text{minute})^2} = \frac{1}{1200} \frac{\text{foot}}{(\text{second})^2},$$

which is given in Dr Everett's Treatise (page 4), as a very expressive notation now becoming common, are not commendable, and are such as it is desirable promptly to reject. We might quite rightly note that

$$1 \frac{\text{yard}}{\text{foot}} = \frac{1}{1200} \left(\frac{\text{minute}}{\text{second}}\right)^2,$$

but it would be illegitimate to pass from this, by imitation of a real algebraic process, so as to write

$$1 \frac{\text{yard}}{\text{foot}} = \frac{1}{1200} \frac{(\text{minute})^2}{(\text{second})^2},$$

and thence further to make a pseudo-multiplication of both sides by foot, and a pseudo-division of both sides by (minute)^2, and so to bring out the seeming equation above objected to.

The name *dimensions of units* is subject to a distressing ambiguity. It might mean the *greatness* or *smallness* of them; and indeed a dimensional equation is for the very purpose of telling how the *greatness* of some units changes in accordance with changes made in the *greatness* of other units. This is not, however, the idea which is attached to the word *dimensions* in *dimensional equations*. It is mentally associated rather with such notions as the three dimensions in space, length, breadth, and thickness (not to say also with the fanciful notions, so often now put forward, of a fourth dimension in space, or a $2\frac{1}{2}$th, or $4\frac{2}{3}$th, or an infinite number of other alleged dimensions in a dreamland space, not found in our world or conceived in our brain). It has, in fact, to do with change-ratios, or powers of change-ratios of quantitative units, whereby the magnitudes of the various units are mutually connected, and some of them specified by reference to others. There is, I may remark, for instance, in Dr Everett's book, one article on *Dimensions of Units*, and another on *Dimensions of the Earth*. Now the word *dimensions* in these two cases has totally different meanings.

57. ON THE LAW OF INERTIA; THE PRINCIPLE OF CHRONO-METRY; AND THE PRINCIPLE OF ABSOLUTE CLINURAL REST, AND OF ABSOLUTE ROTATION.

[From the *Proceedings of the Royal Society of Edinburgh*, 3rd March 1884, Vol. XII, p. 568.]

THERE is no distinction known to men among states of existence of a body which can give reason for any one state being regarded as a state of absolute rest in space, and any other being regarded as a state of uniform rectilinear motion. Men have no means of knowing, nor even of imagining, any one length rather than any other, as being the distance between the place occupied by the centre of a ball at present, and the place that was occupied by that centre at any past instant; nor of knowing or imagining any one direction, rather than any other, as being the direction of the straight line from the former place to the new place, if the ball is supposed to have been moving in space. The point of space that was occupied by the centre of the ball at any specified past moment is utterly lost to us as soon as that moment is past, or as soon as the centre has moved out of that point, having left no trace recognisable by us of its past place in the universe of space.

There is then an essential difficulty as to our forming a distinct conception either of rest or of rectilinear motion through unmarked space.

We have besides no preliminary knowledge of any principle of chronometry, and for this additional reason we are under an essential preliminary difficulty as to attaching any clear meaning to the words *uniform rectilinear motion* as commonly employed, the uniformity being that of equality of spaces passed over in equal times.

If two balls are altering their distance apart, we cannot suppose that they are both at rest. One, at least, must be in motion.

Men have very good means of knowing in some cases, and of imagining in other cases, the distance between the points of space simultaneously occupied by the centres of two balls; if, at least, we be content to waive the difficulty as to imperfection of our means of ascertaining or specifying, or clearly idealising, simultaneity at distant places. For this we do commonly use signals by sound, by light, by electricity, by connecting wires or bars, and by various other means. The time required in the transmission of the signal involves an imperfection in human powers of ascertaining simultaneity of occurrences in distant places. It seems, however, probably not to involve any difficulty of idealising or imagining the existence of simultaneity. Probably it may not be felt to involve any difficulty comparable to that of attempting to form a distinct notion of identity of place at successive times in unmarked space.

There is, in the nature of things, a real distinction, cognisable by men, between absolute rotation (or absolute clinural motion) and absolute freedom from rotation (or absolute clinural rest)*.

The only motion of a point that men can know of or can deal with is motion relative to one, two, three, or more other points. Three points marked or indicated on one, two, or three bodies, the centres, for instance, of three balls, whether preserving their distances apart unchanging or not, are sufficient for enabling us to construct or to imagine a reference frame of any changeless configuration desired—three rectangular co-ordinate axes, for instance, or three rectangular co-ordinate planes—to which the situations, instantaneous or successive, of points may be referred.

Any arrangement whatever of points, lines, or planes, changeless in mutual configuration, will, for present purposes, be named as a reference frame, or briefly as a *frame*.

The word *motion*, in ordinary usages, has several varied significances.

1. Thus it is often said that a body, or rather some specified point of it, has performed a motion from a point A to a point B, along a straight or curved line of motion AMB. It may be often

---

* The word *clinural* is to be understood as introduced for conveying precisely one out of the various conflicting meanings of the word directional. All straight lines which are mutually parallel are, in this amended mode of nomenclature, said to be in one same *clinure*. In connection with this, it may be convenient here to mention that all parallel planes are, in like manner, said to be in one same *posture*.

said that this same motion has been effected slowly on one
occasion and quickly on another, speed or velocity of the moving
point not being treated as any essential quality or condition of
the motion.

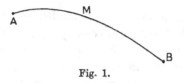

Fig. 1.

2.   Again, it is often said that a point, moving along a curve
GABH, has a certain motion at the instant of its passing *A*, and
that its motion undergoes change during the passage from *A* to *B*,
and that at *B* it has a motion changed from that which it had at
A.   In this sense the motion at *A* is regarded as determined by
the line of motion specified as being the tangent to the curve at

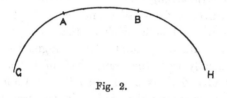

Fig. 2.

A, and the ward, or way, of the motion along its line at the point
of contact *A*, and the velocity of the motion at that point.   The
velocity of the motion is usually understood as meaning a true
time rate of motion—a rate which may be specified, for instance,
as being that of so many feet per second, or the like.   For this
ordinary mode of specifying that which is to be called velocity, it
is necessary that true chronometry should have been previously
attained to, in idea at least, and approximately in fact.

Sometimes it is found convenient to apply, temporarily at
least, the name *velocity* (for want of any other name) to the rate
of travel of a point along a line as referred to the progress of
something else moving relatively to a frame or to a dial.   Thus,
for a point moving along a line, the motion at any point of the
path may sometimes be said to have a velocity of so many feet
per unit of angular space turned by the crank shaft of a steam
engine relatively to the framing of the steam engine.   What is
thus specified might be called a *quasi-velocity*, not a *true velocity*,

as it is customary to regard true velocity as being referred not to the revolution of a steam engine shaft, nor to the revolution of a hand of a badly-going clock, but to the progress of absolutely true time when once the idea of progress of true time has been arrived at.

Before arriving at any principle of absolute chronometry, how-ever, we cannot deal with true velocity at all. We cannot specify a rate of progress of any moving point relatively to progress of true time, or relatively to progress of a clock hand on its dial advancing proportionally to progress of true time. But, without assuming or presupposing any principle of absolute chronometry, we can refer motions of points to an assumed reference frame, jointly with an assumed dial-traveller. The dial-traveller may conveniently be imagined as a clock hand or index travelling continuously along a graduated dial, such as the face of a clock, but without the adaptation of any pendulum or balance wheel, or other chronometric arrangement for regulating the motion of the hand. The traveller, for instance, might be kept moving round its dial by a winch handle, such as that of a grindstone, or of a barrel organ, turned by hand. Or it might be an index pro-jecting out radially from the crank shaft of a steam engine, and revolving round a dial fixed to the adjacent framework of the engine, so as to surround the shaft. Or the traveller might be an index kept revolving by the shaft of a water wheel, with a motion depending on variable conditions of rain-fall and stream-flow.

For purely kinematic considerations as to relative motions of points or bodies we have no essential concern with true time, nor with true velocities, understood as velocities of motions relative to a frame, and specified quantitatively as true time rates.

Now, reverting to the essential difficulty already mentioned as to our forming a distinct conception either of rest or of uniform rectilinear motion, we may go forward to some further considera-tions and scrutinies as to what men can imagine or can really know, through observation and experience, respecting motions of bodies in the universe of space.

We may have a firm persuasion, even without perfect under-standing, that in the nature of things there must be a reality corresponding to our glimmering idea of motion of a body along a straight course with changeless velocity, and that there must be

an essential distinction between such motion and motion along
a curved course or motion with varying velocity. We cannot,
however, specify such motions relatively to unmarked space and
unmeasured passage of time. We cannot specify them as to any
condition of absolute rest. We can only specify them as to part
of their characters, or conditions, or distinctions. We can do so
only in so far as qualities or distinctions of motions of one or more
bodies can be ascertained through knowable relations between
these motions and the motions of one or more other bodies.
Briefly, we can deal only with relative motions or relative rest;
not with absolute motions nor absolute rest.

Sir Isaac Newton sets forth, under the designation of the
FIRST LAW OF MOTION, the statement that—*Every body continues
in its state of resting or of moving uniformly in a straight course,
except in so much as, by applied forces, it is compelled to change
that state.*

A most important truth in the nature of things, perceived
with more or less clearness, is at the root of this enunciation, but
the words, whether taken by themselves or in connection with
Newton's prefatory and accompanying definitions and illustrations,
are inadequate to give expression to that great natural truth.
In attempting to draw from the statement a perfectly intelligible
conception, we find ourselves confronted with the preliminary
difficulty or impossibility as to forming any perfectly distinct
notion of a meaning in respect to a single body, for the phrase
" *state of resting or of moving uniformly in a straight course.*"
Newton's previous assertion *that there exists absolute space which,
in its own nature, without reference to anything else, always remains
alike and immovable,* does not clear away the difficulty. It does
not do so, because it involves in itself the whole difficulty of our
inability to form a distinct notion of identical points or places in
unmarked space at successive times, or of our inability to conceive
any means whatever of recognising afterwards in any one point of
space, rather than in any other, the point of space which, at a
particular moment of past time, was occupied by a specified point
of a known body.

To aid in the apprehension of the underlying truth referred to,
and also as an aid to the understanding of the enunciation about
to be given in the present paper as *The Law of Inertia* and
Principle of Chronometry, some purely kinematic principles will

now be adduced for consideration. Thus the question is to be opened up as to what may be the nature of relative motions of various bodies, which can in any sense truly be regarded as uniform rectilinear mutual motions. Explanations are to be given *on such motions of points in unmarked space, as can have a reference frame and reference dial-traveller relatively to which jointly those motions are rectilinear and are uniform in the sense of being changeless in quasi-velocity.* In other words, quite to the same effect—Explanations are to be given *on such motions of points in unmarked space as can have a reference frame relatively to which those motions are rectilinear and are changeless in mutual rate; or what is the same, are mutually proportional in their simultaneous progress.*

Let us imagine a reference frame, rigid in its configuration, and for simplicity let it be taken as including three rectangular reference planes firmly connected. Let several points or small bodies be kept moving by geared mechanism, such as that of toothed wheels on variously inclined axles, and toothed straight sliding racks with pinions, all carried or guided in bearings firmly attached to the reference frame, the arrangements being such that those moving points shall be made simultaneously to travel over mutually proportional lengths along straight lines fixed in relation to the reference frame. The motion may be given by a winch handle like that of a barrel organ, fixed on one of the axles; and, for help in consideration of the subject, we may imagine a uniformly graduated dial surrounding the winch handle axle, and an index attached to the axle so as to project radially outwards like a hand of a clock, and to travel round the dial keeping pace in angular motion with the winch handle. Thus the simultaneous travels of the various small bodies along their straight courses are to be mutually proportional, and they are also to be proportional each of them to the simultaneous travel of the index on the dial. Now, if any other frame of three co-ordinate planes be arranged to exist with the point of their intersection keeping at any one of the moving points, and with the three planes maintaining changeless angles with the original three reference planes, any one of the moving points will either be at rest relatively to the new set of reference planes, or will generate, in relation to them, a straight line, and the simultaneous lengths traversed by the various points relatively to these new reference planes will be mutually proportional. It is convenient to notice, in preparation for subsequent reference of motions to true time or to a truly chronometric clock, that the simultaneous lengths traversed by the various points relatively to the new reference planes, and of the winch handle index relatively to its dial, will be mutually proportional. We may thus see that for the established set of motions of the points, there can exist as many sets of reference planes or frames as we please, differently moving and differently inclined, in reference to each of which every one of the points will generate a straight line with a quasi-velocity (or rate per dial-traveller progress) proportional to the quasi-velocity of every other along its own line. We are now perfectly entitled to speak of the motions of all these points as referred to any one of the frames and the original dial-traveller, as being uniform recti-linear motions. The word uniform, it should be noticed, has, neither in its origin nor in its customary employment, any essential connection with progress of time. The notion besides of the dial-traveller as a standard to which the simultaneous travels of the various points may conveniently be referred, or rated, is not at all essential. We would be quite entitled, without knowledge of chronometry, and without having recourse to the quasi-time indicated by the dial-traveller, to speak of the motions of all the points relatively to all the reference frames, as being uniform rectilinear motions. The uniformity in rate of progress would be in respect to rate of travel of any one of the points per simultaneous travel of any other one of them. In all that has been said in this matter no assumption has been made as to any particular condition of rest or motion having belonged to the original reference frame. It may have been firmly attached to the surface of the earth, or it may have been firmly attached to the floor and side walls of the cabin of a ship sailing in devious courses over the sea and tossing on the waves. Notwithstanding any such motions, or any motions whatever, belonging to the original reference frame, the mutual motions of the points will possess the character that they admit of having reference frames, as many as we please, relative to which they will be rectilinear and mutually proportional (or, in other words, they will be uniform rectilinear motions, by mutual reference without reference to time). If the moving points alone were available to us for progressive observation or measurement it might be a difficult, perhaps an extremely difficult, geometrical

T.

or kinematical problem* to find from them a reference frame accomplishing the stated condition; but this does not hinder us from easily and distinctly understanding that such a frame is geometrically or kinematically possible. On the other hand, for a set of points moving at random like flies in the air, or for a set of points having uniform rectilinear motion as already described, together with others revolving like satellites round some of them, no reference frame to accomplish the conditions stated would be possible. For a single fly moving anyhow, reference frames would be possible, relative to any one of which the motion of the fly would be rectilinear, and would be uniform in rate of progress relatively to true time, or to any assumed standard whatever for rate of progress; but for two flies, or any greater number, no such frame would be possible. Reasons for this are so obvious as scarcely to require statement. Briefly, however, it may be mentioned that any two flies might in their mutual motions come into contact once and then separate, and then come into contact again; but no second meeting could occur with points moving straightly and mutually proportionally in relation to any frame whatever. Or the two flies might be increasing their distance apart and afterwards diminishing it; but no approach after recession is possible for points moving straightly and proportionally in relation to any frame whatever.

The explanations now given are sufficient to show that there can be mutual motions of various bodies, so related as to have a property of being uniform rectilinear mutual motions, and to explain the nature of that mutual relation. This is quite irrespective of any idea of chronometry, or any idea of absolute rest or motion in the universe, or of any idea of absolute clinural rest or absolute rotation, and of any distinction whereby one body might be said to be in absolute rotation and another devoid of absolute rotation. The mutual relation described has been purely kinematic, and will

---

* *Postscript Note, May* 1884.—On the evening of the reading of the paper (March 3, 1884), just after the close of the meeting of the Society, the author inquired of Professor Tait whether he could see how the problem referred to here in the paper as being perhaps extremely difficult, could be solved. Professor Tait replied that he could solve it very briefly by use of quaternions. The author, not being at all acquainted with quaternions, has since seen his way clearly to the solution by an easy method of mechanical adaptations. The mechanical method is merely for intellectual use, not for practical application. The ideal mechanism can serve as an instrument for use in reasoning, though friction, and elasticity of materials, &c., might render it incapable of complete practical realisation. [Cf. *infra*, p. 401.]

not be at all altered by the superposition of any new motion whether of translation or of rotation, the meaning of this statement being rendered intelligible by consideration of the attachment of the original reference frame to the floor and side walls of the cabin of a ship at sea, already mentioned.

Now, to pass from mere geometric or kinematic motions, governed mutually by connecting mechanism to the motions of bodies existing in space free from any such governance, we are to accept as an established law of nature, established through multitudinous observations and speculations, together with theories confirmed by multitudinous agreements, the following, which may be called the law of inertia.

The Law of Inertia.

For any set of bodies acted on each by any force, a REFERENCE FRAME and a REFERENCE DIAL-TRAVELLER are kinematically possible, such that relatively to them conjointly, the motion of the mass-centre of each body, undergoes change simultaneously with any infinitely short element of the dial-traveller progress, or with any element during which the force on the body does not alter in direction nor in magnitude, which change is proportional to the intensity of the force acting on that body, and to the simultaneous progress of the dial-traveller, and is made in the direction of the force.

Principle of Chronometry.

From the foregoing law it is readily deducible, as a corollary by elementary mathematical considerations, that—

Any dial-traveller which would accomplish the conditions stated would make progress proportionally with any other dial-traveller, obtained likewise from the same set of bodies, or any other set of bodies with the same or any other reference frame. Then, in view of this remarkable agreement, we define as being equal intervals of time, or we assume as being somehow in their own nature intrinsically and necessarily equal intervals of time, the intervals during which any such dial-traveller passes over equal spaces on its dial. Thus, any dial-traveller which would accomplish the conditions stated would constitute a perfect chronometer.

This gives us the ideal of a perfect chronometer. It remains for men to aim at approaching as near as they can towards that ideal in the practical realisation of good chronometry.

For good and long-enduring realisations of chronometry, astronomical methods are alone available. None of these present any simple method of procedure. They require hypothetical assumptions of supposed forces acting on the bodies considered, and, above all, there is involved in them the assumption, and after multitudinous tests, accompanied by multitudinous confirmations, the discovery of the Law of Universal Gravitational Attraction—the grandest of the discoveries of Sir Isaac Newton.

Principle of Absolute Clinural Rest and of Absolute Rotation.

Any straight line fixed relatively to any reference frame which accomplishes the conditions specified in the statement of the law of inertia has absolute clinural rest. If another straight line fixed in any other such reference frame be parallel to that former line, the two lines will continue parallel, so that by either of them the one same absolute clinure is permanently preserved. The principle here called that of absolute clinural rest is clearly enunciated in Thomson and Tait's Natural Philosophy, § 249, under the designation of "Directional Fixedness." It is there exhibited by a very simple device, and here by a somewhat different method.

Any body which has no rotation relative to a framing which accomplishes the conditions stated is devoid of absolute rotation, and if a body rotates relatively to any such frame it has the same rotation absolutely.

The Law of Inertia here enunciated sets forth all the truth which is either explicitly stated or is suggested by the First and Second Laws in Sir Isaac Newton's arrangement.

By applying the Law of Inertia to the case in which the forces acting on the bodies vanish, the law becomes a remodelled substitute for the statement set forth by Sir Isaac Newton as the First Law of Motion in his arrangement.

58. A PROBLEM ON POINT-MOTIONS FOR WHICH A REFERENCE-FRAME CAN SO EXIST AS TO HAVE THE MOTIONS OF THE POINTS, RELATIVE TO IT, RECTILINEAR AND MUTUALLY PROPORTIONAL.

[From the *Proceedings of the Royal Society of Edinburgh*, 7th July, 1884.]

IN a paper read in this Society on the 3rd of March last*, "On the Law of Inertia," &c., I had occasion to adduce for consideration a problem to the following effect:—

Relatively to the reference-frame which may itself have any motion whatever (but which is to be regarded as unknown or as disallowed for any use in observation or measurement), a set of points are known to have motions which are rectilinear and mutually proportional in simultaneous progress. From observations or measurements on successive simultaneous configurations of the set of points merely, to find a reference-frame relatively to which their motions will have that same character.

On the suggestion of this problem for solution being made to Prof. Tait, on the evening of the meeting already referred to, he promptly replied that he could solve it very briefly by quaternions. [See *ante*, page 386, footnote.] I myself soon after succeeded in devising a solution, and my object in the present paper is to offer that solution to the Society. Prof. Tait too, I trust will submit his quaternion solution this evening to the Society.

Let us take the case of three points moving rectilinearly and mutuo-proportionally relatively to a frame. Three points are enough, but we might use more, and might so bring out varied solutions; and besides it may be mentioned here that a distinction of importance will be found to exist between the results attainable for three points only, and for a greater number than three. This will be referred to at a later stage [near the end of the paper, pages 399 and 400].

Let these three points to be used be called in general, *A*, *B*, and *C*, irrespectively of changes in their mutual configuration, or in their situations relative to any frame. Let us proceed to find a frame relatively to which any one of these three points, say the

* [See *supra*, p. 379.]

point C, shall be at rest, and the other two shall move rectilinearly and mutuo-proportionally. Let successive simultaneous situations of A and B at instants of measurements be designated as A_1 and B_1, A_2 and B_2, A_3 and B_3, &c. We may thus bring into consideration and into use a sort of portable triangles A_1CB_1, A_2CB_2, A_3CB_3, and more if wanted, representing severally in forms and dimensions the likewise designated original triangles; or, it may be, representing the originals in form, while constructed on any convenient scale of dimensions. The use of such altered scale is to be understood as available if the full original size would be inconvenient for use in a kinematical diagram, or mechanism, soon to be explained for construction or ideal contemplation. After this mere mention of allowable change of scale, the explanations will generally be given, for brevity and simplicity, as if the lengths of lines in the diagram, model, or mechanism, were identical with those of the corresponding original lines, rather than on an altered scale. We may denominate these several portable triangles in succession as templet 1, templet 2, &c.

Let us take the corners C of the three templets together, and take any plane passing through C, and bring the three lines CA_1, CA_2, CA_3, into that plane. Then take a straight line to be called AA, movable in that plane; and, by motions of the lines CA_1, CA_2, CA_3, together with the motion of the line AA itself if necessary, bring the templet points A_1, A_2, A_3, into the line AA. Keep, until further notice, the line AA at a constant distance from C. This last bondage temporarily applied (that of keeping AA at a constant distance from C) is introduced merely as an aid in the reckoning of freedoms, and of their successive abolitions. It is not essential, and for some possible varied modes of thought it may well be omitted.

The figure here illustrates the arrangement so effected. It is to be understood that we are to suppose ourselves free from any doubt as to whether, at the instant of any measurement, the moving point A or B, as the case may be, is diminishing or increasing its distance from C; for instance, we are to suppose that we are fully aware at which side of M in the line AA we are to place the point A_1; the line CM being the line of shortest distance from C to AA.

Next take another straight line to be denominated as BB, and place it passing through B_1 and B_2 wherever these templet points

may happen to be. Then rotate templet A_3CB_3 round its side
CA_3 till its side CB_3, regarded as an interminate straight line,
comes to meet the line BB which itself is movable, and may, if
necessary or desirable, be shifted while retaining the points B_1
and B_2 in it. This operation may be stated in other words as
being the operation of bringing the templet line CB_3 into the
plane of B_1 and B_2 and C, or, what is the same, into the plane of
the point C and the line BB holding in it the points B_1 and B_2.
Now, in general, on the accomplishment of this, we shall have the
three points B_1, B_2, and B_3, not in one straight line, though all in
the plane of B_1, B_2, and C. That is to say, the line B_1B_2 will cut
the interminate line CB_3, but generally not in its point B_3. But
now, rotate templet 1 round CA_1 or templet 2 round CA_2 till B_2

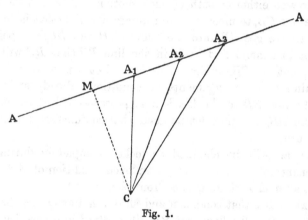

Fig. 1.

kept in the plane B_1CB_2 comes into the interminate straight line
B_1B_2, or, what is the same, into the line BB. So now we have
attained to the following state of things, *videlicet*:—

Firstly. Line AA is kept at an unchangeable distance from
the point C.

Secondly. Three templet points A_1, A_2, A_3, are kept in the
straight line AA; and three other templet points B_1, B_2, B_3, are
all situated in one straight line BB.

Under these conditions, without departing from them, and
while considering the point C and the line AA, and consequently
CA_1, CA_2, CA_3, as being all at rest, we can move any one of the
three templets by rotation round its line CA; and the other two
will be bound to move with that one, and to assume fixed places

when that one is fixed;—or, in other words, motion or rest of all the three is exactly decided by motion or rest of any one templet.

TWO VARIED METHODS FOR CONTINUATION.

Having arrived at this stage, we may go forward to accomplish a solution by either of two branch methods which will be stated now successively.

METHOD I.—The state of things already arrived at being maintained, if now further we introduce one more templet A_4CB_4, placing its point C to coincide with the C of the previous templets, and bringing its point A_4 into the line AA; and if we rotate this new templet round the fixed side CA_4 as an axis, and move also, if we please, or if necessary, the line BB in the freedom it has, and carry on either or both of such motions till we get the interminate line CB_4 to meet the interminate line BB (that is, in other words, till we get CB_4 into the plane of C and BB) the point B_4 will not in general find itself in the line BB (but BB will meet some point of CB_4 other than B_4). Let us, however, while maintaining the arrangements or conditions already arrived at, shift the line BB in the freedom it possesses till B_4 comes into that line BB, and then let us make that conjuncture binding for the future.

Now all will be clamped or locked completely during our maintenance of the temporarily employed condition of AA being kept at an unchanging distance from C.

Next release that condition and shift AA to any new distance from C, and it will follow that BB will be fixed in a new situation; and, for AA moving, BB will be bound to an exact movement accordingly.

But now introduce one more of the templets, templet A_5CB_5, with its point A_5 brought into the line AA and kept bound to remain in that line, and with its line CB_5 brought into the plane of C and BB. This being done, the point B_5 will generally not find itself in the line BB; but we can shift AA towards or from C, moving consequently the plane CBB and the line BB in that plane till the line BB gets B_5 entered into it.

Thus all is completely clamped; and the two straight lines of motion of the points A and B relative to the desired frame in which C is at rest, are found relative to each one of the configurations ABC at the successive instants of the measurements. But those *past* configurations of the points A, B, and C, are already lost to us; and so we must proceed to find the lines AA and BB relative to one or more future configurations of those three points which are the only things that are to be available to us to make measurements from. It is here to be distinctly noticed that the lines of motions AA and BB, being fixtures in the desired frame, may perfectly well be accepted as constituting that frame; or even *one* of them along with the point C (which is also a fixture in the desired frame) might equally well be accepted as constituting the desired frame.

Now we have to observe that, from the nature of the data, and of the operations performed in the kinematical model or mechanism, and result arrived at in it, it must be the case that the lengths A_1A_2, A_2A_3, A_3A_4 &c., and B_1B_2, B_2B_3, B_3B_4, &c., found in the model as representatives of simultaneous travels of the real points A and B relative to the desired frame, must be mutually proportional. Hence, by continuing forward for the future a like proportional division along BB for any arbitrarily marked graduation along AA, we can by measurements from the model foretell future instantaneous configurations of the points A, B, and C, and the situations of the fixed lines of motion AA and BB relative to those future configurations. So at any future instant when one of those prophesied configurations arrives, we shall have the means of specifying relative thereto the lines AA and BB; and these, as already explained, are to be accepted as the desired frame; that is, as being a frame accomplishing all the requisites of the problem.

METHOD II.—Reverting now to the stage arrived at just before our commencement of branching out into the two varied methods, and maintaining the state of things there attained to, we may proceed by a second alternative method as follows:—It is to be recollected that, in the state of things attained to, we have three templet points A_1, A_2, A_3, all in one straight line AA; and three other templet points B_1, B_2, B_3, all situated in another straight line BB; and that, while keeping the line AA at a changeless distance from the point C, and regarding the line AA and the point C as being at rest, and consequently regarding the points A_1, A_2, A_3, as being also all at rest, we can keep moving the line BB, in the one freedom it possesses, and so we can alter continuously the distances B_1B_2 and B_1B_3, and we may readily further see that we can alter continuously the ratio of either of these distances to the other. So let us ordain the requirement that we are thus to keep moving the line BB in the freedom it possesses till we attain the condition that

$$\text{As } A_1A_2 : B_1B_2 :: A_1A_3 : B_1B_3.$$

For the purpose of accomplishing this requirement we may use (ideally at least) a mechanism of parallel rulers or parallel projectors, &c., which will now be briefly described. For guidance in geometrical principles towards formation of the conceptions intended, imagine a straight line placed intersecting the lines AA and BB. Let us name this line as the transversal, or refer to it as the line TT. Imagine the points A_1, A_2, A_3, to be projected to the transversal by parallel projectors; and imagine the points so found on the transversal to be further projected to the line BB. Call the points so found on BB, the points b_1, b_2, b_3, in correspondence with A_1, A_2, A_3, from which they have been respectively derived. Now obviously we have by geometry,

$$A_1A_2 : b_1b_2 :: A_1A_3 : b_1b_3.$$

Further it may easily be seen that we can gradually change the directions (or clinures) of the two sets of parallel projectors until we get the point b_1 to coincide with B_1, and maintaining that coincidence we can go on changing the directions of both sets of parallel projectors till we get also b_2 to coincide with B_2. This being done we can, by observing whether or not b_3 coincides with B_3 ascertain whether or not it be the case that

$$A_1A_2 : B_1B_2 :: A_1A_3 : B_1B_3.$$

The general principle thus indicated can be applied to the case immediately before us in a simplified combination by choosing to make the transversal pass through the points A_3 and B_1 as in fig. 2, where TT represents the transversal. The figure is to be understood as being a pictorial representation on the paper, of lines and points not themselves situated in the plane of the paper, and not all existing in any one plane. Then take a pair of straight lines kept parallel by mechanism (as for instance is the case in some commonly used kinds of parallel rulers) and place one of these lines so as to pass through A_1 and B_1 and make the other pass through A_2. This second line being parallel to A_1B_1 (see fig. 2) must be in a plane with it and with the line AA and consequently with the transversal. So it meets the transversal. These two parallel lines are represented in the figure as A_1B_1 and A_2t_2. So now we have, in the figure,

$$A_1A_2 : B_1t_2 :: A_1A_3 : B_1A_3.$$

Take two other straight lines kept parallel by mechanism, and place one of them across from t_2 to B_2 and make the other pass through A_3, and then it will necessarily meet the line BB. So these two parallel lines will be represented in the figure as t_2B_2 and A_3b_3; and we shall have

$$A_1A_2 : B_1B_2 :: A_1A_3 : B_1b_3.$$

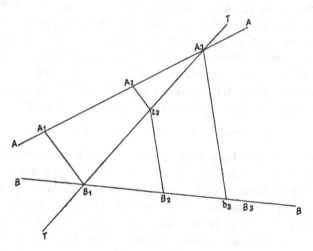

Fig. 2.

Now in general it will result that b_3 so found will not coincide with B_3; but let us now keep the line BB moving in the freedom it possesses, and keep the two pairs of parallel lines maintaining the conditions to which they have been set, and continue the motion until the points b_3 and B_3 come to coincide, and bind these two points to remain together; or, in other words, bind the line A_3b_3, when brought to pass through the point B_3 hitherto movable along the line BB, to continue holding the templet point B_3. Thus all will be clamped as long as the temporarily imposed condition of changeless distance from the templet point C to AA is maintained.

But if now we change that distance, and fix on a new changeless distance instead, while maintaining all the conditions already attained to, the whole system will take up a new configuration and will become clamped therein.

Or we might express this by saying:—Fix AA at an altered distance from C, and by going through the same process as before we get one fixed configuration for the whole system. So if we relax the condition of changeless distance of AA from C, the whole system has one and only one freedom.

For the next step we do not require another complete templet. We may use merely the lengths CA_4 and CB_4 got by measurement. That is let us have a portable triangle with two sides CA_4 and CB_4 given, but the angle between them left unknown, and that, for our present operations, is the same as to say:—left variable. Put the vertex C, or joint C, of that portable triangle at the point C of our model. Bring CA_4 into the plane CAA, and swing it round till A_4 comes into the line AA. Also put the rod or bar CB_4 into the plane CBB; and swing it round till B_4 comes into the line BB. This being done, the proportionality wanted,

$$\text{that } A_1A_2 : B_1B_2 :: A_1A_4 : B_1B_4$$

will generally *not* have been instituted. But now, while maintaining all the conditions already attained to, move AA varying its distance from C, till that proportionality takes place as indicated by the mechanism of parallel rulers, &c., already used, but with some obviously necessary additions sufficiently suggested by the explanations already given. Now the whole system becomes clamped, and the problem is solved, or the rest is easy as in Method I, the two straight lines AA and BB, being lines fixed in the sought-for frame, and it being possible to find them for any prophesied future configuration of the set of three points A, B, and C, as has been explained at the close of the explanations for Method I.

It becomes now convenient and desirable to examine into some questions as to how many distinct elements of data from measurement, or how many ascertained conditions from measurement, are required for the solution of the problem; and as to whether there are more essentially distinct solutions than one in various cases of the number of points used and the number of conditions from measurement ascertained.

It is to be recollected, as was pointed out near the end of the explanation of Method I that the lengths A_1A_2, A_2A_3, A_3A_4, &c., and B_1B_2, B_2B_3, B_3B_4, &c., found in the model as representatives of simultaneous travels of the real points A and B relative to the desired frame, must be mutually proportional. But the conditions of this mutual proportionality have been left unused in the solution, and so we may see that in Method I we have used redundant data. It is to be understood that the introduction of any one complete templet brings in just *two* not *three* new conditions. Though for it there are three sides measured, yet the only conditions thereby ascertained are that when some one side has a certain stated length, a second side has another ascertained length, which is one condition, and that also at the same instant the third side has another ascertained length which is another condition; and this makes with the previous one only *two* in all.

In Method I after the stage up to which the procedure is common to both Methods I and II; we have—(*a*) templet 4 introduced involving two additional conditions from measurement, and (*b*) templet 5 introduced, involving two additional conditions from measurement. So in respect to these two templets 4 and 5 we have four conditions introduced.

And in the whole of Method I there are the following known conditions unused:—

$$\text{That} \quad \frac{B_1B_2}{A_1A_2} = \frac{B_1B_3}{A_1A_3}$$
$$\text{that} \quad \frac{B_1B_2}{A_1A_2} = \frac{B_1B_4}{A_1A_4}$$
$$\text{and that} \quad \frac{B_1B_2}{A_1A_2} = \frac{B_1B_5}{A_1A_5}$$

That is, in all, three known conditions unused. So instead of ascertaining 4 conditions by measurement and neglecting 3, we might get the result by $4-3$, that is one new condition from measurement. Now in Method II we do demand from measurement just one new condition, and we have no redundant information. The one new condition so introduced is the condition brought into use by the incomplete templet 4; *videlicet*, that when CA has a certain stated length CB has another certain ascertained length.

So, on the whole, in Method I we have from measurement 5 templets supplying two conditions each, that is, we have 10 conditions from measurement; and, as shown already, we are thus supplied with 3 redundant conditions; and so $10 - 3$ or 7 conditions from measurement, or independent data, must somehow be enough.

Passing to Method II we see that in it we have 3 complete templets supplying 6 conditions, and 1 incomplete templet supplying 1 condition; and we have got no redundant conditions; but we have just 7 conditions found necessary and brought into use.

So the two methods agree in showing that the number of independent conditions from measurement necessary to be supplied is *seven*.

There is yet a curious matter which I have to adduce for consideration. Many matters of interest may indeed remain for further consideration or for mathematical investigation in connection with this subject; but I do not at all profess to exhaust the subject, and I shall only be glad if it be found that I escape from offering inadvertently any importantly erroneous views.

What I do propose now further to put forward is some scrutinies as to whether the problem entered on, in some of its varieties, admits of duplicate or multiple results for accomplishing its conditions; and to open out some views in relation to this matter which appear to be true and to be of interest.

It is to be noticed that when we begin putting together our templets for the kinematical model or mechanism for three original moving points A, B, and C, we have no means available for knowing on which side of templet 1 we are to place the point A_2 of our templet 2, and on which side of the same we are to place the point B_2 of our templet 2. Further, it is to be noticed that, under the restriction of our measurements being confined to three of the original points only, we have no means whatever for making the extra measurements, or taking the extra observations that would supply us with means for choosing one side rather than the other of templet 1, as that on which we ought to place either A_2 or B_2. To help our conceptions let us imagine among the original moving points a reference frame relative to which the original point C shall be at rest and which shall have no rotation relative to the original secret frame; and let us name this as the *vice-original frame* and designate it briefly by the letter Φ. Let us imagine the original triangular plane ACB as having one of its two faces red and the other blue; and imagine its face which at the point A is anterior in its motion relative to the vice-original frame to be red, and the one which at that point is posterior to be blue. In respect to this it is to be noticed that the red face thus specified though anterior at A may happen to be the posterior one at B: but this need not give us trouble, and for brevity we may speak of the red face which at A is anterior as being *the anterior face*. Let the faces of all the templet triangles be coloured red and blue correspondingly.

By going forward with considerations readily suggested by what has just been set forth, we may obviously find that the process of solving the stated problem for the case of only three original points by the kinematical mechanism can bring out two straight lines AA and BB really at rest relatively to the vice-original frame, and consequently having all their points either at rest or moving rectilinearly and mutuo-proportionally in relation to the original secret frame: and further that the same process of solving the stated problem can bring out another real true solution in finding two straight lines which we may call $A'A'$ and $B'B'$ which will be the images of AA and BB in a plane mirror whose plane always passes through the three points A, B, and C. The two straight lines $A'A'$, $B'B'$ so found may be taken as lines fixed in a frame which we may designate as Φ', and which will rotate relatively to the vice-original frame Φ, as also relatively to the original secret frame. Now, as the motion of the original points goes on making their distances apart increase unlimitedly, this relative rotation between the frames Φ' and Φ will be becoming evanescent, and the two frames will be approaching unlimitedly towards relative rest. So the solution which brings out the frame Φ' approaches ultimately to identification with that which brings out the frame Φ which is in agreement with the original secret frame.

If now instead of using the three points C, A, and B, we use a different group of three points C, A, and D, these will bring out for us as solutions two frames Φ and Φ'', of which the one Φ will be identical with the frame Φ already found by the three points C, A, and B. It follows from this (and it seems very obvious that it could be brought out in various other ways), that for four original points no frame in general could be brought out as a solution except one in agreement with the original secret frame;

that is to say, a frame either at rest relatively to that original
frame, or having all its points moving rectilinearly and mutuo-
proportionally relatively to that original frame. One reason which
seems very decisive in favour of this conclusion is:—that if any
three of all the original points be retaining their distances apart
unchanging, then these three will themselves constitute a frame
Φ which will be in agreement with the original secret frame : and
then for any other one or more of the original points taken along
with these three, no frame will be possible to serve as a solution
except such as shall be in agreement with that one.

POSTSCRIPT.

The ideas noted in what follows had not completely occurred
to me till after the evening of the meeting of the Society when
the paper was read, and a suggestion towards their development,
came from Professor Tait's paper of the same evening "On
Reference Frames." Thus it seems suitable to annex them here
as a postscript.

It may be noticed that in the case referred to in the last
sentence of the paper just before this postscript—the case of three
original moving points retaining their distances apart unchanging
—the method of procedure employed throughout the paper would
collapse because the points A and B would be at rest relatively to
the vice-original frame, and so the straight lines of motion AA
and BB previously used would become only two points. Yet a
solution is in this case even more readily available than in the
previously considered cases. It thus becomes desirable to find
some way of harmonising the two modes of thought or of procedure
so as to bring them into connection; rather than to be content to
suppose that, in passing from one to the other, we should have
quite to abandon the one mode of thought, and take up another
and quite unallied mode instead.

A satisfactory connection between the two presents itself, if,
instead of taking from the kinematic model only the lines of
motion AA and BB, as the basis for our desired frame, we take
from that model also the points A_1 and B_1 on those lines, getting
them known by measurement of their distances from any future
points A_n and B_n, so that when, among the three original points
A, B, and C, the particular configuration A_nCB_n found in antici-

pation from the model shall come to exist, the old triangular frame A_1CB_1 shall become known to us relatively to the then existing frame A_nCB_n. In this way of procedure, the solution, for the case of A and B being, as well as C, at rest in the vice-original frame, will come out simply by the distances A_nA_1 and B_nB_1 being each zero in length, and by the concomitant of this, that the old frame A_1CB_1 is to be found as being coincident with the new momentarily existing and known frame A_nCB_n.

NOTE ON REFERENCE FRAMES. By Professor TAIT.

As I understand Prof. J. Thomson's problem [cf. *ante*, p. 386] it is equivalent to the following:—

A set of points move, Galilei-wise, with reference to a system of co-ordinate axes; which may, itself, have any motion whatever. From observations of the *relative* positions of the points, merely, to find such co-ordinate axes.

It is obvious that there is an infinitely infinite number of possible solutions; because, if one origin moves Galilei-wise with respect to another, and the axes drawn from the two origins have no *relative* rotation, any point moving Galilei-wise with respect to either set of axes will necessarily move Galilei-wise with respect to the other. Hence any one solution suffices, for all the others can be deduced from it by the above consideration.

Referred to any one set of axes which satisfy the conditions, the positions of the points are, at time t, given by the vectors

$$\alpha_1 + \beta_1 t \text{ for } A, \quad \alpha_2 + \beta_2 t \text{ for } B, \text{ \&c., \&c.}$$

But it is clear, from what is stated above, that we may look on the pair of vectors for any *one* of the points, say α_1 and β_1 for A, as being absolutely arbitrary:—though, of course, *constant*. We will, therefore, make each of them vanish. This amounts to taking A as the origin of the co-ordinate system. The other expressions, above, will then represent the relative positions of B, C, \&c., with regard to A.

The observer on A is supposed to be able to measure, at any moment, the lengths AB, AC, AD, \&c.; the angles BAC, BAD, CAD, \&c.; and also to be able to recognise whether a triangle, such as BCD, is gone round positively or negatively when its corners are passed through in the order named. What this leaves undetermined, at any particular instant, is merely the absolute direction of *any one line* (as AB), and the aspect of *any one plane* (as ABC) passing through that line. These being assumed at random, the simultaneous positions of all the points can be constructed from the permissible observations. But it is interesting to inquire how many observations are necessary; and how the βs depend on the αs.

Thus, at time t, whatever be the mode of measurement of time, we have equations such as follows:—

$$- a = \alpha_2^2 + 2S\alpha_2\beta_2 \cdot t + \beta_2^2 t^2,$$

$$- b = S\alpha_2\alpha_3 + S(\alpha_2\beta_3 + \beta_2\alpha_3) \cdot t + S\beta_2\beta_3 \cdot t^2,$$

$$- c = \alpha_3^2 + 2S\alpha_3\beta_3 \cdot t + \beta_3^2 t^2,$$

$$\ldots = \ldots\ldots\ldots\ldots\ldots\ldots\ldots\ldots$$

For any one value of t we have n equations of each of the 1st and 3rd of these types, and $n(n-1)/2$ of the 2nd, $n+1$ being the whole number of points. In all, $n(n+1)/2$ equations.

The scalar unknowns involved in these equations are (1) the values of t; (2) α_2^2, α_3^2, &c.; (3) β_2^2, β_3^2, &c.; (4) $S\alpha_2\alpha_3$, &c.; (5) $S\beta_2\beta_3$, &c.; (6) $S\alpha_2\beta_2$, $S\alpha_3\beta_3$, &c.; and (7) $S(\alpha_2\beta_3 + \beta_2\alpha_3)$, &c. Their numbers are, for (2), (3), (6), n each; for (4), (5), (7), $n(n-1)/2$ each; in all $3n(n+1)/2$. Suppose that observations are made on m successive occasions. Since our origin, and our unit, of time are alike arbitrary, we may put $t=0$ for the first observation, and merge the value of t at the second observation in the tensors of β_2, β_3, &c. This amounts to taking the interval between the first two sets of observations as unit of time. Thus the unknowns of the form (1) are $m-2$ in number. There are therefore

$mn(n+1)/2$ equations and $3n(n+1)/2 + m - 2$ unknowns.

Thus $m=3$ gives an insufficient amount of information, but $m=4$ gives a superfluity.

In particular, if there be three points only, which is in general sufficient, 3 complete observations give

9 equations with 10 unknowns;

while 4 complete observations give

12 equations with 11 unknowns.

Thus we need take only two of the three possible measurements, at the fourth instant of observation.

The solution of the equations, supposed to be effected, gives us among other things, α_2^2, α_3^2, and $S\alpha_2\alpha_3$. *Any* direction may be assumed for α_2, and *any* plane as that of α_2 and α_3. From these assumptions, and the three numerical quantities just named, the co-ordinate system required can be at once deduced.

This solution fails if $(S\alpha_2\alpha_3)^2 = \alpha_2^2\alpha_3^2$, or $TV\alpha_2\alpha_3 = 0$; for then the three points A, B, C, are in one line at starting. But this, and similar cases of failure (when they *are* really cases of failure) are due to an improper selection of three of the points. We need not further discuss them.

But it is interesting to consider how the vectors β can be found when one position of the reference frame has been obtained. Keeping, for simplicity, to the system of three points, we have by the solution of the equations above the following data:—

$$S\alpha_2\beta_2 = e,\ S\alpha_3\beta_3 = e',\ S(\alpha_2\beta_3 + \beta_2\alpha_3) = f,\ T\beta_2 = g,\ T\beta_3 = g',\ S\beta_2\beta_3 = h;$$

where e, e', f, g, g', h are known numbers; which, as the equations from which they were derived were not linear, have in general more than one system of values. The second, third, and sixth of these equations give

$$\beta_3 S . \alpha_2\alpha_3\beta_2 = hV\alpha_2\alpha_3 + (f - S\beta_2\alpha_3)\ V\alpha_3\beta_2 + e'\ V\beta_2\alpha_2.$$

Provided β_2 is not coplanar with α_2, α_3, this equation gives, by the help of the fifth above, a surface of the 4th order of which β_2 is a vector. But β_2 is also a vector of the plane $S\alpha_2\beta_2 = e$, and of the sphere $T\beta_2 = g$. Hence it is determined by the intersections of those three surfaces.

But if $S . \alpha_2\alpha_3\beta_2$ vanishes, the equation above gives (by operating with $S . V\alpha_2\alpha_3$)

$$0 = h(V\alpha_2\alpha_3)^2 - (f - S\beta_2\alpha_3)\ S . \beta_2 V . \alpha_3 V\alpha_2\alpha_3 + e'S . \beta_2 V . \alpha_2 V\alpha_2\alpha_3,$$

which gives a surface of the second order (a hyperbolic cylinder) in place of the surface of the fourth order above mentioned. This may, however, be dispensed with:—for β_2 is in this case determined by the planes $S\alpha_2\beta_2 = e$ and $S . \alpha_2\alpha_3\beta_2 = 0$, together with the sphere $T\beta_2 = g$.

59. Safety under Repeated and Varying Stresses.

[The British Association appointed a Committee consisting of Mr W. H. Barlow, Sir F. J. Bramwell, Prof. James Thomson, Sir D. Galton, Mr B. Baker, Prof. W. C. Unwin, Prof. A. B. W. Kennedy, Mr C. Barlow, Prof. H. S. Hele Shaw, Prof. W. C. Roberts-Austen, and Mr A. T. Atchison for the purpose of obtaining information with reference to the Endurance of Metals under repeated and varying stresses, and the proper working stresses on Railway Bridges and other structures subject to varying loads.

They reported in 1887 (see *Brit. Assoc. Manchester*, 1887, report, page 424). It is understood the two passages here printed from the report were drafted by Professor Thomson and sent to Mr Barlow with the following letter.]

Craigpeaton, Cove,
Dumbartonshire,
23 *Aug.* 1887.

My dear Mr Barlow,

I enclose two paragraphs such as you proposed that I should send to you:—one relating to my views put forward in my paper of Nov. 1848, and the other relating to my brother's discoveries of May, 1865 as to Viscosity in Elasticity and recovery by rest from some effects of fatigue. I may mention that my own view of 1848, following on Hodgkinson's experimental results, was amply verified by various easily made experiments which I made soon after the paper was published. These were never actually published in any printed paper by me but they were made publicly known often in lectures and otherwise by myself: and any person interested could easily test the matter by experiments for himself.

I may also mention that I would despair of finding how to give any brief recommendation for good new amended regulations to be suggested to the Board of Trade.

The subject of the regulations if they are to be made much better than at present would I think become very complicated. You know of my advocacy of a change of practice by use of more of the method of Force Tests.

Excuse my writing now briefly, being in great haste for post.

Yours truly,
JAMES THOMSON.

[From the *Report of the British Association*, 1887, p. 424.]

In a report to the British Association in 1837, on strength and other properties of cast iron, Mr Eaton Hodgkinson (*Brit. Assoc. Report* for 1837, Part I. pages 362 and 363) made announcements, from his experimental researches, to the following effect:—That in various experiments on transverse loading of bars he had found visible permanent sets produced by such small loadings as 1/30, 1/52, and 1/80 of the breaking weight; showing, he said, '*that there is no weight, however small, that will not injure the elasticity*'; and as a conclusion that '*the maxim of loading bodies within the elastic limit has no foundation in nature.*'

Again, in the *Brit. Assoc. Report* for 1843, Part II. page 24, Mr Hodgkinson, after detailing further experiments on the same subjects, says:—" It appears from the experiments that the sets produced in bodies are as the squares of the weights applied, and that there is no weight, however small, that will not produce a set and permanent change in a body, and that bodies when bent have the arrangement of their particles altered to the centre; and when bodies, as the axles of railway carriages, are alternately bent, first one way and then the opposite, at every revolution, we may expect that a total change in the arrangement of their particles will ensue."

Such assertions as those in Mr Hodgkinson's two communications here referred to, if accepted in full, must necessarily induce very uncomfortable feelings as to endurance of engineering structures. Mr (now Professor) James Thomson, however, in a paper published in the *Cambridge and Dublin Mathematical Journal*, vol. III. p. 252, Nov. 1848*, without abandoning the idea of there being some real foundation in nature for prevalent opinions as to limits of elasticity, showed how the elastic range of change of form might, in many of the ordinary cases of materials newly prepared by manufacturing processes, be found to be very narrow on account of the existence of mutual strains or stresses among the particles composing them—that thus permanent sets might be met with on the application of very small loadings—that in this way, through the ductile yielding of the more severely stressed parts, the range of elastic action, or range of action within elastic limits, would be greatly widened, and that after the application of a heavy load,

* [See *supra*, p. 334.]

which the material could properly bear, subsequent applications of any smaller loads would produce no new permanent set or alteration —none, at any rate, in any way corresponding to those great and alarming alterations indicated in Mr Hodgkinson's announcements. (That paper of Professor Thomson's came, besides, under the notice of practical men through its having been republished in one or more of the engineering journals of the time; and it has recently been republished in the article on "Elasticity" by Sir William Thomson in the *Encyclopaedia Britannica*.)

*   *   *   *   *

In a paper by Sir William Thomson, entitled "On the Elasticity and Viscosity of Metals," published in the *Proceedings of the Royal Society* for May 18, 1865*, an account is given of experimental researches instituted by him and conducted in his laboratory in the University of Glasgow, through which some new and previously unsuspected properties in the elasticity of metals were discovered. These cannot be fully described here in detail, but it may be mentioned that the new results, of greatest interest and probably of greatest practical importance, related to temporary and gradually subsiding effects left in wires by previous elastic oscillations. Energy was expended (dissipated) much more in any one torsional oscillation of a wire which had for some time previously been kept actively oscillating, than in a like oscillation either of the same or of a different but similar wire after having been for some time previously in a state of rest or of less active oscillation.

In the continuation of these experimental researches (after the publication of the paper, it would seem) the effects of the kind of fatigue and rest here referred to manifested themselves very remarkably in the oscillation of wires kept almost constantly in activity, during most days of the week, but getting rest usually from Friday evening till Monday morning. The successive oscillations diminished in their amplitude, by internal resistance or some condition like viscosity in their elasticity, much less on the Monday mornings, after their Sunday rest, than at other times, succeeding closely to previous activity.

* [See Sir Wm. Thomson, *Math. and Phys. Papers*, Vol. iii. pp. 22—26, §§ 30—34.]

GEOLOGICAL

60. On the Parallel Roads of Lochaber.

[Read before the Royal Society of Edinburgh, 6th March, 1848. Reprinted from the *Edinburgh New Philosophical Journal* for July, 1848, Vol. XLV. p. 49. On reading this paper, consult the Map of the Shelves or Parallel Roads of Lochaber, in Vol. XLIV. of this Journal.—Reproduced on next page.]

THE Parallel Roads, Shelves, or Terraces, of Lochaber, constitute a wonderful inscription, traced by the hand of Nature, over the surface of a wide range of mountains and glens. To interpret this writing, and to disclose the story which these mysterious but clearly-marked characters transmit, has long been an object of much interest, as well as of great perplexity, to geologists. As yet, however, no one has succeeded in arriving at an explanation of the subject, which, after having undergone the scrutiny of others, has given general satisfaction; and scientific men are still, perhaps, as much divided in opinion as ever in regard to the nature of the operations by which they suppose these terraces to have been produced. Two papers, taking different sides on this question, have appeared in the last two numbers of Jameson's *Philosophical Journal,*—the first by Mr David Milne, and the second by Sir George S. Mackenzie. The new and very interesting discoveries which have lately been made by Mr Milne in his researches among the hills, are brought forward by both writers in confirmation of their respective theories. These discoveries, how-ever, when taken in connection with the highly important principles of the motion of glaciers recently developed by Professor Forbes, appear to me to be far more strongly confirmatory of the leading features of the explanation given by Agassiz; at the same time that they enable me to develop this explanation more fully, modifying and correcting it in some degree, so as to make it

MAP OF
PART OF LOCHABER
SHOWING THE SHELVES IN THE GLENS.

accord with the new facts and principles, and thus putting it in a form in which, to me at least, it appears so satisfactory as to leave scarcely the slightest doubt of the agency of ice in the formation of the Parallel Roads.

Mr Milne's paper may be regarded as consisting primarily of two parts,—the object of the one being to prove that the terraces are the beaches of lakes which have been maintained among the hills by barriers occupying the lower parts of the glens; and the object of the other, to shew that these barriers consisted of earthy detritus, and to explain the way in which he thinks they may have been formed, and subsequently removed.

His explanation differs from those given by Dr MacCulloch and Sir Thomas Dick Lauder in 1817 and 1818, principally in his attributing the removal of the barriers not to any violent convulsions of nature, but to the gradual operation of existing causes. These, if I fully understand his statements, he supposes to be the erosive action of the waters of the lakes themselves, combined with that of rivers and streams. On this subject he says—"My explanation of the Lochaber shelves depends entirely on the supposition that the valleys were, in the lower parts of them, filled up with detrital matter capable of being gradually worn down and washed away." Sir George S. Mackenzie, although there is much of his reasoning which I do not consider satisfactory, appears to succeed completely in confuting the explanation given by Mr Milne, so far as it depends on the supposed existence of earthy barriers. On the other hand, Mr Milne proves, I think beyond the possibility of doubt, that *the Parallel Roads are the beaches of ancient lakes, which have been maintained among the mountains by barriers across the lower parts of the glens.*

With reference to objections to the supposed existence of barriers which had previously been brought forward, Mr Milne remarks—"These objections resolve entirely into the difficulty of explaining the disappearance of the barriers, which must have dammed back the water in the valleys; but it would be no good reason for rejecting an explanation founded on the existence of barriers, even though we could not very clearly account for the disappearance of them, provided that there is direct and conclusive evidence that such barriers existed. Now, I conceive that there is such evidence furnished by the considerations before referred to." Ideas similar to these of Mr Milne had also occurred to

Sir Thomas Dick Lauder nearly thirty years ago, and, in a paper which he laid before the Royal Society of Edinburgh, they are expressed in the following terms:—" I believe it will be readily admitted, that it is much easier to suppose the existence of former barriers, than to discover the means which operated in their removal; but it must be also granted, that the difficulty of accounting for the destruction of such large masses, does not by any means imply that they never had any being at all, particularly where a number of facts remain to lead us to an opposite conclusion. From all the present appearances it is extremely probable that the barrier of Loch Roy was not only very thin, but of soft materials, at the two parts which have been removed."

Thus, both Sir Thomas Dick Lauder and Mr Milne have come decidedly, and, I think, with good reason, to the conclusion that barriers did exist; but then we are by no means obliged to assume, that these were composed of earthy materials. It is in this assumption, in fact, that all the difficulties connected with the explanations given by these two writers are involved; and to me it seems perfectly clear that the barriers in reality were formed of glaciers.

The glacial explanation of the Parallel Roads given by Agassiz in his paper in Jameson's *Journal* for 1842, was necessarily imperfect in its details. Sufficient facts in regard to the phenomena of the terraces themselves, and true principles of the motion of glaciers, were then wanting. Had these been within the reach of Agassiz, he could easily have modified his explanation so as to remove all valid objections which have been brought forward against it, and could have shewn the invalidity of others which are still adduced, but which, I think, will not be admitted by those who have duly appreciated the principles of the viscidity of glaciers, as developed in the theory of Professor Forbes. The object of Agassiz, however, at that time, was probably rather to adduce the Parallel Roads as confirming his grand idea of the former extensive prevalence of ice in these latitudes, than to enter fully into the details of the mode in which the roads had been produced; and in representing his supposed glaciers on the map which accompanies his paper, his intention was, perhaps, not so much to assert that the glaciers had acted exactly in the way he indicated, as to illustrate the supposition that glaciers, acting in some such way, would be found, in the end, fully to explain all

the phenomena. Be this as it may, Mr Milne succeeds in shewing
that the explanation by means of the supposed glaciers is inconsistent with observed facts. He then goes on to assert, that
glaciers *could not possibly* have penetrated to the places where
their presence would actually have been required. This statement, of course, constitutes the turning point of the whole
argument, since, if it were correct, it would overthrow the glacial
explanation. I hope, however, to be able, in what follows, to give
good reasons against its soundness; but, in the meantime, it will
be necessary to advert to the facts which invalidate, in its details,
the explanation given by Agassiz.

Previously to the researches of Mr Milne, it had been known
that there exist three "*summit-levels*," or "*water-sheds*," in connection with three of the Parallel Shelves; but the existence of
a fourth had not been noticed, and it had even been asserted by
Mr Darwin*, that "the middle shelf of Glen Roy is not on a

* [From *Life and Letters of Charles Darwin*, Autobiography, p. 68. "During
these two years I took several short excursions as a relaxation, and one longer one
to the Parallel Roads of Glen Roy, an account of which was published in the
Philosophical Transactions [1839, pp. 39—82]. This paper was a great failure,
and I am ashamed of it.

"Having been deeply impressed with what I had seen of the elevation of the land
in South America, I attributed the parallel lines to the action of the sea; but I had
to give up this view when Agassiz propounded his glacier-lake theory. Because no
other explanation was possible under our then state of knowledge, I argued in
favour of sea-action; and my error has been a good lesson to me never to trust in
science to the principle of exclusion."

In the paper by L. Agassiz contributed to the Geological Society of London on
Nov. 4, 1840 (*Proceedings*, Vol. III. pp. 327—332), which was fundamental in the
identification of glaciation as a prime agent in geology, the following passage
occurs (p. 332) at the end of a general argument, which is the earliest statement
of a glacial theory of the Parallel Roads, and was derived from investigation on the
spot :—

"Another class of phenomena connected with glaciers is the forming of lakes
by the extension of glaciers from lateral valleys into a main valley : and M. Agassiz
is of opinion that the parallel roads of Glen Roy were formed by a lake which was
produced in consequence of a lateral glacier projecting across the glen near Bridge
Roy, and another across the valley of Glen Speane. Lakes thus formed naturally
give rise to stratified deposits and parallel roads or beds of detritus at different
levels."

For the literature of this subject see *More Letters of C. Darwin*, Vol. II. pp. 171
—2, and Lyell's *Antiquity of Man*, 1863, pp. 252—64. For a modern account of
the working out of the details of the glacial theory, which may be compared with
the views given in the text, see Sir Archibald Geikie's book on the scenery of
Scotland in relation to its geology.

Sir J. D. Hooker, writing to Charles Darwin from the Himalayas in 1849,

level with any water-shed." Mr Milne has, however, found the
wanting water-shed in Glen Glaster, a small glen which, though
branching up from Glen Roy near the bottom of it, does not
appear to have been visited, and certainly has not been correctly
described by any former observer. But this is not all. Mr Milne
has also traced the channel of an ancient river, proceeding from
the water-shed in question, down into Glen Spean, and there
terminating in a huge delta, or alluvial deposit, at the only shelf
which winds round the sides of the latter glen, thus marking the
point where the turbid waters of the river were swallowed up
under the stagnant surface of the lake which, by these same
indications, is palpably shewn to have stood in Glen Spean on a
level with the lowest shelf, at the time when Glen Roy was
occupied with water to the height of the shelf next above.

In connection with these circumstances, Mr Milne finds that
the uppermost shelf of Glen Roy does not, as was erroneously
indicated on Sir Thomas Dick Lauder's map, run round
the sides of Glen Glaster, but that it suddenly stops short in
Glen Roy, just above the entrance to that smaller tributary
glen.

From this we conclude, that the barrier which blocked up
Glen Roy, so as to occasion the formation of its highest shelf, must
have disconnected it from Glen Glaster, and thus forced it to
discharge its surplus water into the valley of the Spey by the
summit-level at its head, instead of permitting it to discharge by
the lower summit-level at the head of Glen Glaster, and down by
the ancient river-channel into Glen Spean,—a course which must
have been followed by any water occupying Glen Glaster, or
communicating with it uninterruptedly.

Now, to explain the formation of the highest shelf of Glen Roy,
Agassiz supposed one glacier, in the lower part of Glen Spean, to
have extended across from Ben Nevis to Moel Dhu ; and another,
farther up that glen, to have issued from the valley of Loch Treig;

independently identified the parallel roads as the beaches of glacial lakes, on the
basis of his experience in that region (*Life and Letters of C. Darwin*, Vol. I.
p. 376). At this time Lyell, as well as Darwin, seems to have attributed them
to marine action. In 1861 Darwin, who had by that time abandoned the early
views advanced in his *Phil. Trans.* memoir of 1839, writes to Hooker (*More Letters*,
Vol. II. p. 190), "It is I believe true that Glen Roy shelves (I remember your
Indian letter) were formed by glacial lakes." See Sir W. Thiselton Dyer's obituary
notice of Sir J. D. Hooker, *Proc. Roy. Soc.*. B, 1912.]

the two being sufficiently high and extensive to maintain the
water between them, and, of course, also the water in Glen Roy, at
the level of the shelf under consideration. In confirmation of this
supposition, he stated that the shelf is marked on the south side
of Glen Spean, between the sites of the two supposed glaciers.
Were the supposition true, the shelf should certainly be marked
in that situation, and also round Glen Glaster; but, according to
Mr Milne, it is to be found in neither of these places. The middle
shelf of Glen Roy, according to Agassiz, should also occur in Glen
Spean, between the two supposed glaciers; but Mr Milne asserts
that, in fact, it does not. Thus, then, the glaciers supposed by
Agassiz will not satisfy the conditions of the question; nor will
any other system of blockage do so, except one, according to which
Glen Glaster would, for a certain period, have been separated from
Glen Roy.

We are, therefore, if we proceed on the supposition of the
agency of glaciers, led to the conclusion, that the one which
stopped up the mouth of Glen Roy to form its highest shelf, must
have extended up that glen beyond the mouth of Glen Glaster.
It must also have blocked up Glen Collarig nearly to the place
named the Gap. Then, to explain the formation of the middle
shelf, it is only necessary to suppose that the glacier retired
a little, so as to connect Glen Glaster with Glen Roy. The water
in the latter would immediately begin to discharge itself by the
ancient river-course before mentioned, and its surface would thus
be lowered to the level of the middle shelf. Lastly, the lowermost
shelf of all would be formed when the glacier retired to near the
mouth of Glen Spean.

Mr Milne, however, asserts that, on account of the character of
the mouth of Glen Roy, in regard to levels, direction, and distance
from Ben Nevis, such a glacier as I have described could not have
existed; but there does not appear to me to be any real difficulty
in the supposition.

The following considerations will, I think, tend to render this
clear. Of all climates capable of generating glaciers, there are
two extremes which must produce two corresponding extremes in
the mode of distribution of the ice on the surface of the earth.
The one of these extremes of climate may be instanced as occurring
in Switzerland, and the other in the Antarctic Continent recently
discovered by Sir James Clarke Ross. In Switzerland the mean

temperature of the comparatively low and flat land is so much
above the freezing point, that the ice no sooner descends from the
mountains than it melts away; and it is thus usually prevented
from spreading to any considerable extent over the plains. In the
Antarctic Continent, on the contrary, the mean temperature is
nowhere so high as the freezing point. The ice, therefore, which
descends from the hills, unites itself with that which is deposited
from the atmosphere on the plains; and the whole becomes con-
solidated into one continuous mass, of immense depth, which glides
gradually onward towards the ocean. The portions which are
protruded out to sea break off, and are floated away as icebergs;
the remainder being left, presenting to the sea a perpendicular
face which rises, in insurmountable cliffs, to the height of from
150 to 200 feet above the water, and extends below the water to
the depth of perhaps 1000 feet.

Now, a climate somewhere intermediate between these extremes appears to be that which would be requisite to form the
shelves in the glens of Lochaber. The climate of Switzerland
would be too warm to admit of a sufficient horizontal extension of
the glaciers; that of the Antarctic Continent too cold to allow the
lakes to remain unfrozen. If the climate of Scotland were again
to become such that the mean temperature of Glen Spean would
be not much above the freezing point, there seems to be every
reason to believe that that glen would again be nearly filled with
an enormous mass of ice; while its upper parts, and also Glen
Roy, would be occupied by lakes, which would once more beat
upon the ancient and long-deserted beaches,—that the rivers
would resume their former channels, flowing out of the lakes by
the summit-levels between the glens,—and that the ancient aspect
of the country would, in all respects, be again restored.

It will perhaps be objected, that in imagining the ice to make
its way into Glen Roy, we are supposing it to flow up hill. A
semifluid mass, however, so long as its upper surface slopes downwards, cannot be regarded as flowing up hill, no matter what may
be the form of the bottom on which it rests. If a slightly-inclined
trough or channel have an opening made in one side, at the middle
of its length, and if a stream of thick mud be kept flowing into it
by this opening, the mud will not all turn suddenly round towards
the lower end of the channel, but a portion of it will flow in the
opposite direction, apparently up hill, till its surface comes to

meet the bottom of the channel at a level little, if at all, below
the surface of the mud at the side entrance.

In confirmation of the views just brought forward, regarding
the possible horizontal extension of the glaciers, I may refer to
the evidence given by Professor Forbes, in his *Travels in the Alps*
(page 50), of immense erratic blocks having been conveyed by
glaciers from the main chain of the Alps across all the inequalities
of the great plain of Switzerland, and deposited high on the hills
round the Lake of Neufchatel; the total distance travelled over
being 60 or 70 miles, and the total declivity due to their descent
being certainly not more than 1° 8′, and probably not half so
much.

Glen Gluoy, in regard to its blockage, seems to have been
quite independent of all the other glens to which I have as yet
alluded. A glacier occupying the present site of Loch Lochy, and
receiving supplies from the various neighbouring mountains, would
appear to afford a sufficient explanation of the phenomena observed
in this glen. Mr Milne has, however, discovered in it a shelf
which is lower than the one previously known, and which does
not appear to be in connection with any summit-level. If this be
the case, we may suppose that, while the lake was at the level of
this second shelf, its discharge took place by the present mouth
of the glen, through an elevated channel in the moraine of the
glacier. The lake would therefore have resembled almost exactly
the Lac de Combal and the Matmark See, described and figured
in the work by Professor Forbes to which I have already referred
(pp. 193 and 345).

There is, however, a circumstance connected with this shelf
which seems to me to involve some difficulty. As represented by
Mr Milne, its terminations, on both sides of the glen, are farther
from the mouth of the glen than those of the shelf above. In
fact, the upper shelf is shewn round the sides of Glen Fintec,
while the lower shelf is made to stop short without reaching the
entrance to that glen. Should this representation be really
correct, it would appear to involve the supposition, that the
glacier, when at the lower level, penetrated farther into Glen
Gluoy than it did when at the higher level. Now the question
arises,—Is it likely that this could have been the case? Perhaps
light may be thrown on the subject by some curious circumstances
connected with the Lac de Combal. The glacier which occasions

the damming up of this lake has actually retired a considerable
way down the glen in which the lake is situated, since the
deposition of that part of its moraine which now retains the
water; and yet the surface of the glacier is some hundreds of feet
higher than that of the lake. Besides this, the glacier, at a point
farther from the head of the glen, threatens to overwhelm with its
moraine the channel of the river by which the superfluous water
of the lake is at present discharged. How imminent the prospect
of this occurrence really is, may be judged from the fact, that it is
necessary annually to remove the debris thrown down by the
glacier on the road which, together with the river, winds through
the bottom of a deep ravine, enclosed on the one side by the
moraine of the glacier, and on the other by the continuation of
the hill which forms a side of the glen containing the lake.
Should the glacier force itself even a very little farther in this
direction, the surface of the lake would not only be raised above
its present level, but its horizontal extension towards the lower
part of the glen would be increased. The beach of the lake at
present existing, together with that of the new one thus formed,
would therefore exhibit exactly the peculiarities which, according
to the representation of Mr Milne, appear to exist in the two
shelves of Glen Gluoy. This fact is enough to make the difficulty
appear to be not insuperable. The simplest view, however, to
take of the subject may, perhaps, be to suppose that the glacier
which occasioned the formation of the higher of the Glen Gluoy
shelves, had at some period protruded a terminal moraine as far
up the glen as the terminations of the lower shelf; that on the
final retiring of the glacier this old moraine served as a barrier to
dam up the water to the level of the lower shelf, and that it has
been subsequently washed away by the river flowing over it.

I have thought it right to point out the foregoing difficulty for
the consideration of those who may have it in their power to gain
further information on the subject. Should any mutual action of
a glacier and its moraine have occasioned the peculiarity in
question, we might expect to find some remains of the moraine
between the terminations of the upper and those of the lower
shelf. It may here be remarked, that there is not the same
difficulty in accounting for the removal of this moraine, as for that
of the barriers supposed by Mr Milne to have existed at the
mouths of the other glens. For, in this instance, the water from

the lake of Glen Gluoy must have discharged itself over the top of
the moraine; while, in the case of the other glens, it certainly
flowed out by the summit-levels between the glens; and would,
therefore, have no power of cutting away the barriers.

There is, in the Lochaber district, still another glen, containing
a shelf, which was discovered by Mr Darwin, and described by him
in the *Philosophical Transactions of the Royal Society of London
for* 1839. This glen is situated near Kilfinnan, at the north-
eastern extremity of Loch Lochy. The shelf in it is stated by
Mr Darwin to be in every respect as characteristic as any shelf in
Glen Roy. He believes it to be perfectly horizontal; and, in
connection with it, he discovered a water-shed, similar in its
nature to those which have been already mentioned. Now, as
this author remarks, in regard to any explanation by means of
earthy barriers, "the discovery of the shelf at Kilfinnan increases
every difficulty manifold." Every additional glen containing a
shelf, in fact, requires us to assume the deposition of an additional
barrier, and the subsequent removal of this by causes which have
left the shelves undisturbed. To admit, at the mouth of even a
single glen, of a barrier of such a peculiar nature as would enable
it to stand for a long time, but at last to be swept away, although
no river flowed over it, seems difficult enough; but to imagine
that numerous glens should chance to be placed in such peculiar
circumstances, appears to be quite unnatural; no sufficient and
generally-acting cause being assigned for the repetition of the
supposed phenomenon. On the other hand, the existence of the
shelf in question is exactly what should have been expected,
according to the glacial theory I have maintained. The same
mass of ice occupying Loch Lochy, which I have supposed to have
been instrumental in forming the shelf in Glen Gluoy, would, to
all appearance necessarily, have blocked up also the glen at
Kilfinnan, and thus have produced the shelf which is really found
to exist round its sides, on a level with the water-shed at its top.
Mr Milne himself mentions the occurrence, in various parts of the
Highlands, of other glens containing shelves, none of which have,
however, been so carefully investigated as those we have been
considering. According to what I have already said, this would
appear to add to the difficulties of the explanation by means of
earthy barriers, and to confirm the one I have given, depending
on the agency of a climate such as would cover with a thick

T.   27

bed of ice almost the whole surface of the land in the neighbourhood of high mountains.

It will be unnecessary for me to enter at length into a discussion of the diluvial theory of the parallel roads, given by Sir George
Mackenzie, as, after a full consideration of it, it does not seem to
me to be capable of explaining the observed facts. I may,
however, mention some of the leading objections which I would
bring against it. During the sinking of the supposed wave, on
the arrival of its surface at each successive summit-level, there
would be no *sudden* check to the flow of the water through the
glens, nor even to the rate of depression of the general surface of
the wave; but even if some material alteration in the flow of the
water were to occur at those particular occasions, there seems to
be no reason to suppose that these vast shelves would be the
result. No attempt, besides, is made according to this theory, to
shew why the various shelves should be expected to stop short at
the particular places where, by observation, they are found to do so.

An objection which has been urged by Mr Lyell against the
glacial theory of the parallel roads must not be left unnoticed.
He thinks there are proofs to be met with in various parts of
Scotland of great changes having occurred in the relative levels of
the sea and land; and he supposes that such changes would have
destroyed the horizontality which is found to characterise the
terraces. Now, there is probably no doubt that important changes
in the elevation of the land have occurred since the *commencement*
of the glacial period, but I do not think that any proof can be
given of their occurrence since its *termination*. In other words,
I think no proof can be adduced, that, ever since the last great
disturbance of the land, the climate has been so warm as to
preclude the supposition of the existence of glaciers round Ben
Nevis. Could this, however, be proved, still it does not appear to
me that it would invalidate the glacial theory of the terraces. It
is easy to conceive that the whole of Scotland might participate
in a general elevation or depression; each part remaining unaltered in regard to inclination to the horizon; and even were we
to suppose the south of Scotland to have risen 30 feet, while the
north remained stationary, and the intervening parts moved in
proportion to their distances from the north, the utmost deviation
from horizontality which would thus be produced in the terraces
would not exceed a foot of difference between the levels of the

northern and southern extremities of any one of them; an amount which would be quite imperceptible by any mode of measurement which could be applied on surfaces so uneven.

In conclusion, I may remark, that, in calling in the aid of glaciers towards the explanation of the Parallel Roads, no gratuitous or unsupported assumption is made. So many various and independent proofs of the existence of a glacial climate in these countries, during some of the most recent geological periods, have been accumulated, especially within the last few years, that we may now regard it as an established fact, and use it like a stepping-stone to assist us in farther investigations. In addition to other proofs of a cold climate derived from organic remains, and from effects which appear to have been produced by icebergs floating at sea, indications of glaciers, in some instances of the most unequivocal character, are to be met with in various mountainous parts of Great Britain and Ireland. Such appearances, more or less satisfactory, have been pointed out by various authors, of whom it may be sufficient to mention Buckland, Lyell, Bowman, Agassiz, Maclaren, and Forbes. In the island of Skye, in particular, among the Cuchullin Hills, which have been lately explored by the last-mentioned author, Professor Forbes, there are to be seen more striking and indisputable traces of glaciers than in any other locality which has, as yet, been examined. This is in a great degree to be attributed to the durable nature of the hypersthene rocks of which those hills are composed; a property which has caused their surfaces to retain not only the general forms, but also the most minute markings produced by the glaciers; and which, at the same time, has prevented these from being concealed under a coating of decayed materials. The face of the country seems, in fact, to have retained, almost absolutely unaltered, all the appearances which it presented on the retiring of the ice.

In the Lochaber district, among other indications of the action of glaciers, Agassiz has pointed out one which is interesting in itself, and more so when taken in connection with the foregoing. At the mouth of Loch Treig, the rock consists of gneiss, intersected by veins of quartz. The quartz everywhere projects two or three inches above the gneiss, its upper surface being polished and striated, exactly as is the case with quartz veins exposed to the action of glaciers at the present day. It is clear that the gneiss and the quartz had originally been planed down to one even surface; and that the gneiss, not being perfectly durable, has since decayed away, and thus left the quartz veins standing in relief.

It would be out of place for me here to enter at greater length into the question as to the former prevalence of glaciers, or of a glacial climate. For farther details, I must refer to the authors who have discussed the subject, particularly to those I have already mentioned.

61. On Features in Glacial Markings Noticed on Sandstone Conglomerates at Skelmorlie and Aberfoil.

[From the *Report of the British Association, Southampton*, 1882, p. 537.]

When glacial striation is met with on rock faces, it may often be a matter of interest to find in which way along the direction of parallelism of the striae the abrading or polishing ice must have advanced. Casual observers, unskilled in the scrutiny of glacial phenomena, may sometimes too hastily assume, when they find the lines of the striae inclined to the horizon, that the ice has been moving over the striated rock face or hill-side in the down-hill direction of the striae. The direction of the motion of the ice along the striae, however, cannot generally be safely inferred from mere local configurations of the land; and, in some cases, the configuration of the land, whether locally regarded or considered in wide scope, may afford no decisive proof at all as to which way the ice has advanced along the striae.

In the rather limited researches which the author has had opportunity to make in respect to glacial phenomena in geology, he has felt interest in seeking for indications presented by the striated rock-faces themselves, which might conclusively shew in which way the ice must have advanced. He wishes, in the present paper, to make mention of one or two rather remarkable indications of this kind which he has met with in the last two years.

At Skelmorlie, on the Firth of Clyde, at a height of about 150 or 200 feet above the sea, he has found glacially striated surfaces, on red sandstone containing pebbles of quartz and of other kinds of stone. Many of these pebbles projected considerably out from the general smoothed and striated surface of the sandstone, and from each of such pebbles there extended to one side a ridge of the sandstone, like a tail;—the sandstone being there worn away less than at other places devoid of the protecting influence of any hard protuberant pebble. The manner in which the protection had been given must have been this:—The ice, in passing over the protuberant hard pebble, must, in virtue of its plasticity, have had a groove moulded into it by the pebble, and this groove passing forward from the pebble in the motion of the ice, must have worn away the sandstone facing to it less than would the other parts of the ice-face wear away the sandstone facing to them. The length of the noticeable tail would depend mainly on the distance that the ice could advance before the groove in it would be gradually obliterated. A pebble of the size of a bean, for instance, might often be found to have a tail visible for a length of five or six inches, or perhaps from that to a foot; and larger pebbles might be seen to have tails two or three feet in length.

Again, on the recent visit to the railway cuttings near Aberfoil, referred to by the author in his previous communication*, a remarkable example was found of a glacially worn and finely striated conglomerate rock face. The situation of this is at Ballanton, about a quarter of a mile from Aberfoil. The hard pebbles were of various sizes, and many might be of sizes such as those of beans, and eggs, and large potatoes. Some of them were worn away by the ice continuously with the surrounding sandstone matrix; but those of them which projected shewed very conspicuous tails, many of which might be four or five feet long or more.

Various other indications, presented by the striated surfaces themselves, of the direction along the striae in which the motion of the ice has been made, may occasionally be found. Doubtless none can be more strikingly remarkable than the tails extending from hard pebbles or other hard nodules, or hard veins, included in the ice-worn rock. The author wishes, however, to mention that on making minute examination, by a magnifying lens, of the scratches on worn quartz pebbles in the striated sandstone at Skelmorlie, he was able to find also in these very small markings, indications which appeared clearly to shew the direction of the motion of the ice, and that these indications were perfectly in agreement with those given by the tails extending from the pebbles along the striated sandstone surface.

---

* ["Mention of an example of an Early Stage of Metamorphic Change in Old Red Sandstone Conglomorate, near Aberfoil," *Brit. Assoc.* 1882, p. 536. Not here reprinted. It refers to distorted pebbles similar to those described in H. C. Sorby's letter, p. 252, *supra.*]

62. On the Jointed Prismatic Structure in Basaltic Rocks.

[Abstract. From the *Transactions of the Geological Society of Glasgow*, 8th March, 1877.]

The views presented in the following paper have in great part been previously brought forward and published on some other occasions, but not hitherto in any such way as to become satisfactorily accessible among geologists. The fundamental ideas and primary observations on which they were based were submitted by the author to the Belfast Natural History and Philosophical Society in a paper read on November 26, 1862, and the subject was shortly afterwards brought under the notice of the British Association at the Newcastle Meeting in 1863, and a brief but complete account of the chief points is to be found printed in the Association's Report for that year. A fuller account of the author's views, with his discussion of the prevailing views of others, is printed in the Annual Report of the Belfast Naturalists' Field Club for 1869–70, as an abstract of a paper read by him to that Society in November, 1869. He was afterwards requested to bring the subject before the Geological Society of Glasgow; and the present paper contains a fuller exposition than any previous one, and includes the discussion of a few facts which have later come under notice.

Basaltic Rocks, and rocks generally of the trappean class (which comprises most of those that have originated as volcanic lavas), are very frequently to be found divided into prismatic columns, more or less regular in form, and the columns are sometimes found to be divided into short lengths by cross fractures or surfaces of easy partition, which show a remarkable regularity of form. The columns, however, it is to be particularly noticed, are quite devoid of the kind of exact regularity that belongs to crystals. They vary in the number of their sides from three to eight or nine, and it seems that occasionally a column may be found with so many as twelve sides. The angles between the sides have no regularity whatever; those seen in the end or cross-joint or cross-parting of a column, being generally unequal among one another, and the magnitude or obtuseness of any one angle being subject often to considerable variation along the length of the column. Also, in passing along the length of a column, a face may often be found becoming narrower and narrower till at last it dies out, and the adjacent faces may be noticed to twist round, altering their angle of mutual inclination in passing along the column. For such reasons as these, a notion which has often been vaguely entertained, that the columns may be of the nature of gigantic crystals, is to be set aside at once as untenable.

The columns are often remarkably straight and parallel—sometimes they radiate, or, we may rather say, converge, like the spokes of a wheel; and sometimes, though more rarely, they are curved. They are to be met with in various sizes. Sometimes they may be found only an inch across, and sometimes so much as 9 feet; and when small in diameter they may often be short in length—perhaps a few inches or a foot or two long; while columns of larger sizes are often found extending to lengths of 50 or 100 feet, and sometimes even to 200 or 300 feet. In a horizontal layer of basalt, or trap rock, the columns are vertical; in a dyke they are perpendicular to the rock faces which have confined the basalt, trap, or lava in its molten state; and, when occurring in a trough or valley of rock, with irregular configuration of its sides and bottom, some of them may occasionally be found curved along their length. As a general rule the columns are found to abut perpendicularly against the rock faces which have confined their substance in its molten state. The descriptive particulars just given are collected from many observations and sources of information, and seem to be fairly trustworthy so far as they go. But, in the absence of far more trustworthy knowledge of the physical conditions under which the structure has originated than has as yet been attained to, it must be supposed that there may be yet other important features remaining which have hitherto escaped notice, or have been imperfectly described. The well-known and generally-admitted features referred to in the description just given seem to indicate very clearly that the direction of the columns has been somehow imparted to them by the cooling surfaces into contact with which the molten substance has been poured. It seems not unreasonable to suppose that the columns have commenced from the colder surface, proceeding thence into the cooling mass with a tendency to perpendicularity to the colder surface, and that in advancing farther forward longitudinally, they have had a tendency to proceed perpendicularly to successive isothermal interfaces in the cooling mass; or, at least, that their longitudinal configurations, whether straight or curved, and whether parallel or radiating, may have been decided in an important degree by influence of the configurations of the isothermal interfaces existing in the cooling mass during their growth as columns.

Of the opinions which have commonly been entertained among geologists as to the manner in which the jointed prismatic structure has originated, none have appeared to the author satisfactory. They have very generally involved one or other, or a combination of both of the two following principles :—

1st—Prismatic fracture by shrinkage in cooling, like the cracking which may be observed in starch or mud in drying.

2nd—An assumed spheroidal concretionary action of the lava or basalt in solidifying from the molten state.

He gave sketches in fuller detail of the views put forward by several of the leading geologists in various parts of the world, as expressed by those writers themselves, who have either proposed them or have at least promulgated them with favour, though in some cases with vagueness—the natural result, he said, of attempting to explain intrinsically untenable views. Sir Charles Lyell, writing, for instance, in his *Elements of Geology*, under the heading "Columnar and Globular Structure," while appearing to put forward with favour the spheroidal concretionary theory, seems also to be affected with this character of vagueness. Dr Daubeny, Beete Jukes, and several others, with much more boldness maintain the spheroidal concretionary theory under various modifications of its details. Dr Daubeny*, for instance, believes that the cause for the columnar arrangement of trap is to be found in the spheroidal or globular structure often shown in these rocks during their disintegration by exposure to the weather. He then says, "It is evident that a series of globular concretions of trap placed in close contact whilst in a pasty condition, or in a state of transition from fusion to solidity, would be by mutual pressure converted into a succession of jointed columns, which, owing to slight differences

* Daubeny, *Volcanos*, 2nd edition, 1848, pp. 660 and 661.

in the compactness and consequent softness of the several parts of the mass, would rarely be exact in their sizes and in the number of their sides, but would exhibit all those variations which in that respect columnar basalt commonly displays." Dr Daubeny, in connection with these views, describes a "natural grotto," called the Käse Keller, at Bertrich, near the Mosel, between Coblenz and Trèves [see Fig. 1, p. 426]*. It is called the Käse Keller (Cheese Cellar) from the resemblance to Dutch cheeses piled up in columns which is presented by the roughly spheroidal joints of which its columns are made up. He says "It beautifully illustrates the origin of the jointed columnar structure which this rock so often assumes, since a little more compression would have reduced these globular concretions into a prismatic form, each ball constituting a separate joint in the basaltic mass†."

Dr Daubeny supposes that in this case the molten lava had been free to flow partly away from among the supposed globular concretions during the time of their solidification and growth, and that thus the supposed globular structure had room freely to develop itself—the surrounding vacancies affording to the balls freedom from much compression, and leaving them free to retain their original globular figure‡. Dr Daubeny contrasts the compact prismatic columns at Fingal's Cave at Staffa with the columns, resembling piles of nearly globular cheeses, in this German cave; and accounts for the difference by supposing that the basalt of Staffa was solidified as a continuous bed without there having been freedom for the molten lava to escape from among the supposed balls during their supposed solidification and growth; and he attributes the hollowing out of the Staffa cave in the solid columnar rock to the action of the sea, while the Käse Keller was, he believes, an original cavity in the rock whilst still hot and only partially solidified—a cavity formed by the flowing away of liquid basalt from among the supposed piles of globular concretions. In closing his discussion of the subject, Dr Daubeny distinctly announces, as the conclusion he arrives at, "that the spheroidal structure will be found to be one of the most prevalent in rocks of

---

* This drawing of the Käse Keller is copied from one in Lyell's *Manual of Elementary Geology*, 3rd edition, 1851, p. 386, as that seems to be a better drawing than the sketch at p. 79 in Daubeny's book.

† Daubeny, p. 78.   ‡ Daubeny, p. 78.

Fig. 1. Käse Keller.

Fig. 2.

Fig. 3.

Fig. 4.

Fig. 5.

Fig. 6.

Fig. 7.

Mary H. Thomson, del.

the trap family, and that the prismatic structure is in general only a consequence of the spheroidal*."

The views put forward in detail by Dr Daubeny have here been briefly cited, as they constitute a fair enough specimen of the general character of views offered, though with many variations, by different geologists of note. Mr Scrope, however, repudiates the spheroidal concretionary theory, denying that the columnar structure could have had its origin in spheroids pressing against each other, pointing out, that imbedded knots of olivine are often to be found severed apart by the seam which separates two contiguous columns; and he thence argues, "that it is impossible to talk any longer of the columnar structure being occasioned by the mutual pressure of spherical concretions †." He then himself maintains that the columns have originated through "fissuring" of the hot rock by "contraction" during its process of "consolidation" or "refrigeration," and he supposes the cross-joints to have been formed contemporaneously at each part of the length of the column with the advance of the prismatic fissures‡ at that same place, the cross-joints being supposed to be successive bounding faces between the solidified end of the column and the as yet molten lava into which the solidification is advancing, and the prismatic fissures being supposed at each period to extend quite forward to the molten lava§. Mr Scrope's view, as he himself states in the passage referred to, comprises the supposition that the concavity of the ball-and-socket-like cross-joints ought to be always directed upwards—that is, nearly in his own words, in the direction facing away from the

* Daubeny, p. 661.

† See citation from Scrope in Daubeny's second edition, page 65, note. (A passage nearly to the same effect as this citation is to be found in Scrope, *Volcanos*, revised and enlarged second edition of date 1872, pages 103 and 104, foot-note.) Daubeny, in his passage just referred to, follows up his citation from Scrope about the knots of olivine by dissenting from Scrope's conclusion, and still maintaining that the columns have been formed by spherical concretions which, while forming themselves by mutual pressure into columns, would be still in a soft and pasty state; but asserting that the seams or severances between the columns, and cutting through the knots of olivine, might take place by shrinkage after consolidation. The general tenor of the arguments in the controversy between Daubeny and Scrope exhibits strikingly an utter bewilderment pervading the minds of both on this subject.

‡ Scrope, *Volcanos*, 2nd ed., 1862, page 104.

§ *Ibid.* Much, if not all, of the same remains in Scrope, revised and enlarged second edition, 1872.

surface of refrigeration, which he believes to be uniformly the bottom or base of the lava bed; or, in other words, that each separate piece of the jointed column ought, according to his supposition, to have its bottom convex and its top concave. This supposition, the author (Professor Thomson) remarks, is not veri- fied, but is decidedly controverted by the basaltic columns of the Giant's Causeway, the cross-joints being often concave upwards, and often concave downwards, and often nearly flat: and this one circumstance, even by itself alone, Prof. Thomson considers must set aside the suppositions maintained by Mr Scrope. That the cross-joints as exposed to view in the Giant's Causeway are some of them convex upwards, and some concave upwards, is very clearly shown in Figs. 8 and 9, which are accurately copied from excellent photographs selected by the author on a visit to the Causeway when he was scrutinizing the stones themselves, and found the photographs to give good representations of various features important for the geological considerations under dis- cussion in the present paper. It may be noticed, that in each of these two drawings some of the exposed joints forming the tops of the columns left remaining are convex upwards, and that others are concave upwards; and that some of the concave ones in Fig. 9 are distinctly indicated by their having rain water lying in their hollow tops.

Professor Thomson then gave a discussion of the spheroidal concretionary theory. He showed many objections to it, any one of which separately, he said, might be sufficient to set it aside. He believes it to be founded from the outset on a total mistake. He regards the spheroids so often met with in decaying basalts or lavas as being not concretions at all, but as being the results of decay or decomposition penetrating from without inwards in blocks, into which the rock has been divided by fissures, which may have arisen from various causes. Examples of such spheroidal exfoliations are shown in Figs. 2 and 3, p. 426. Fig. 2 is a drawing from an actual stone picked up on a hillside, and which may be assumed to have been long exposed to weathering. The stone itself when scrutinized shows at least six distinct shells or coats surrounding a nucleus. Fig. 4 is given as a drawing from Prof. Thomson's own observations of exfoliation proceeding inwards in blocks of shattered trap noticed *in situ* in a quarry near Largs, or of exfoliation such as may often be seen elsewhere in shattered

Fig. 8.

Fig. 9.

Mary H. Thomson, del.

basaltic or trap rocks *in situ*; and as indicating clearly that the positions of the several foliations and nuclei are determined, not by any original concretionary action, but by the casual cracks produced, it may be supposed, by earthquakes at various times. In reference further to the concretionary theories of basaltic columns, he pointed out that even if concretions were forming in a cooling mass of molten basalt or lava, we have no right to suppose that they would arrange themselves in straight, or nearly straight, rows, like beads on a vast number of parallel strings; nor that, if they did exist, and were so arranged, those in any two adjoining rows would so meet as, in pressing against each other, to form continuous columnar faces; and he gave various other reasons, though those here cited may amply suffice. He believes that no credence whatever ought to be given to any form or modification of the spheroidal concretionary theory of the jointed prismatic structure, but that it ought to be utterly and promptly rejected from the science of Geology, without any vestige of it being retained to confuse the investigation of the subject in the future.

Professor Thomson proceeded to give his own theory, which may be briefly sketched out as follows:—He supposes that the division into prisms has arisen by splitting, through shrinkage, of a very homogeneous mass in cooling, and thus far his explanation is in agreement with parts of the views of some of those who have previously speculated on the subject. Beyond the mere attribution of the mode of origin of the prismatic columnar structure to splitting through shrinkage in cooling, like the cracking which may be observed in starch or mud in drying, he offers some explanations as to what conditions, affecting the stresses in the interior of the mass, and involving the influences of contemporaneity or of sequence in time of different parts of the fissure system, must accompany the progress of the system through the interior of the mass. He shows how the splitting action in the interior is distinct in essential features from any such cracking as is exhibited *on the surface* of mud in drying, and from the cracking into a tesselated pattern, of the glaze on porcelain, or of the varnish on terrestrial or celestial globes. On this subject, briefly mentioned at this stage, his remarks will be cited more fully further on in the present abstract. He thinks there yet remains ample scope for much additional scrutiny, both observational and theoretical, towards fuller elucidation of the conditions accompanying the remarkable results presented by the prismatic structure often met with in basaltic and various other rocks, and other substances.

The "cross-joints," he specially remarks, are in no way essential to the formation of the prismatic structure; for the columns are not always jointed, but are sometimes found quite continuous for lengths great relatively to their diameters, so as to constitute long bars of stone. In excavations, for instance, which were made for the formation of the new Cemetery at Belfast, columnar structure in basalt or other trap rock was met with, the columns being often many feet in length, and only a few inches in diameter. Fig. 5, p. 426, shows a portion of one of those long columnar bars which were devoid of the so-called cross-joints.

The chief new view, which had originated with himself in connection with this subject, and which he offers as a contribution towards the development of any true theory, depends on a supposition which had occurred to him in August, 1861, through careful scrutiny and consideration of appearances presented by stones in the Giant's Causeway. This supposition is that the so-called "cross-joints" are fractures which have commenced in the centre of the column, and have advanced to the outside as a circle increasing in diameter. This mode of fracture, he thinks, is evidenced by various markings which he has specially observed in many stones of the Giant's Causeway. These stones usually show a remarkably symmetrical conformation round the outer parts of their cross-joint faces, a conformation which has commonly been associated in idea with ball-and-socket jointing, and has often been attributed to spheroidal concretionary growth; but which to him had presented appearances [see Figs. 6, 8, 9, 10, and 11, pp. 426, 429, 435] suggestive of a connection, in nature and mode of origin, with those appearances which are remarkable in an ordinary conchoidal fracture—subject to the important distinction, however, that while the conchoidal fracture, in ordinary cases, has emanated from an external spot of the stone where the blow has been struck, the supposed conchoidal fracture across the column appears to have started into existence at the centre, or at some internal spot in the column, and to have flashed thence with a complete circular advancing front towards the exterior of the column. The kind of fracture thus imagined may be briefly designated as a *circular conchoidal fracture*.

On the cross-joint faces, or surfaces of transverse partition, in numerous instances he has noticed roughly figured rays spreading out from or converging towards the centre or some other internal spot in the face. These appeared to him, from the time of his first observing them, in or before 1863 *, to be of the same nature as the well known fan-like radial markings of a conchoidal fracture.

Following up the train of consideration and scrutiny thus suggested, he remarks that there are not many very distinct ways in which we can suppose a fissure to have spread across a column or prism of solid stone. First, if we for a moment suppose the fissure to have begun at one side of the column, and to have advanced across to the opposite side, we must expect to find the resulting fracture quite unsymmetrical, and presenting very different appearances at the places where it entered the previously unbroken stone prism, and where it came to its termination, leaving the column broken behind its advancing front. No such appearance is to be found in any of the ordinarily shaped cross-joints; but, on the contrary, there is commonly a very remarkable appearance of approximate symmetry of character in the cross-joint with respect to the different sides and angles of the column. Perfect symmetry is, of course, not to be expected, as the columns themselves are often far from being of any regular or symmetrical form; but, so far as Professor Thomson's observations of the stones in the Giant's Causeway have extended, he believes no appearance is to be found indicating an advance of the fissure across the column from one side to the opposite, in any of the joints which exhibit, in other respects, the usual remarkable features. There may, no doubt, be numerous cases of fracture due to shattering, by causes different from those which have produced the ordinary remarkable transverse severances. Next, any idea that the cracking of the column could have simultaneously begun all round the circumference, and advanced to terminate in the centre, requires little more than to be brought before the mind for consideration to be rejected as untenable. There seems, then, to

---

* Mention of these rays as noticed by the author is to be found published in an abstract of a paper by him, in the *British Association Report*, Newcastle Meeting, 1863, p. 89 of part 2 of the *Report*.

remain nothing to suppose but that the ordinary cross-joint fissures first came into existence in the interior of the column, and then flashed out thence towards the circumference.

In order to produce the cross fractures commencing in the centre, he supposes that a longitudinal tensile stress must have existed in the middle of each column previously to the cracking of the cross-joints. To account for such a tensile stress, he suggests, as a probable hypothesis, that after the column was formed, chemical action, caused by infiltration of water, might cause an expansion of the outside of the column, and that the outer part, thus growing longer, would pull the internal part more and more intensely, until at last the internal part would give way, and break into short lengths. The fissures thus formed, it is obvious, must stop short without extending quite to the outside of the column, as the pull causing the fracture in the interior is due purely to longitudinal push in the outer part of the column. That outer part, therefore, will not be subjected to the pull at all, and so the enlarging circular conchoidal fracture should be expected to stop short without penetrating to the outside of the column, especially at the angles. On the event of the central part cracking, and so ceasing to bear a pull, the outer part, being less resisted than before, would increase in length in the immediate neighbourhood of the new internal fissure, and so would bring parts nearer the circumference than before into the condition of being subject to a pulling stress. Also, the reverberation or tremor at the instant of the cracking might, it seems reasonable to suppose, carry the advancing circular edge of the fissure somewhat further out than the region which would be subjected to a pull if the action were slow instead of being by a start. The appearances of the cross-joints—with the central area of each like a circular or oval flattish face, or like the convex or concave form of a watch-glass, but not extending out quite to the angles, and usually not quite out to the sides of the columns—seem to be in accordance with the suppositions here made, and to give considerable corroboration to them (see Figs. 6, 8, 9, 10, and 11).

The cracks, if formed as supposed, without extending quite to the outside of the column, would constitute places of weakness, from which, under the shattering influence of earthquakes or other causes, fresh fractures would readily proceed quite to the outside, severing the columns completely across; but these fresh fractures occurring in ways quite different from those in which the original circular ones had done, could not be expected to be in continuity with the supposed original circularly terminating fissures. Thus is accounted for the approximately circular outer boundary to the flattish or lunette-shaped middle part of the cross-joint, which is very commonly to be seen.

On a visit to the Giant's Causeway in the summer of 1869, Professor Thomson had noticed some phenomena tending to confirm his previously formed views. He met with several instances in which a small mass of stone, different in texture and in hardness from the rest of the basalt, showed itself in the cross-joint of the column, and in which the joint presented to his view the appearance as if the cross fracture had originated at, and spread out from, this spot of irregular quality. When this extraneous or irregular lump happened to be near the middle of a column, there appeared to emanate from it in all directions approximately straight but roughly formed rays (see Fig. 10, and Fig. 8, with the explanation in footnote)*; and when the lump happened to be near one side of the column, the rays emanating from it spread out in curved forms like a brush, and the several rays in proceeding outwards seemed to bend gently somewhat towards the nearest external face of the column (see Fig. 11, and its explanation in footnote)†. This seemed as if they had tended to run so as at each moment to be

* Fig. 10 shows the general character of features frequently to be noticed with approximately straight rays emanating from about the centre of the column. In this figure, in one or more of the columns, a nucleus or lump of a different kind of stone is shown in the centre, in imitation of what has been noticed by the author in several cases in which a distinctly observable nucleus appeared to him to have originated the fracture. Fig. 8 is accurately drawn from a photograph of the group of stones ordinarily shown by the Causeway guides to visitors as the "Ladies' Wishing Chair." Of the columns here shown the one at the left front of a person sitting in the chair shows faintly in the figure, but shows, or at least when last seen by the author, did show very distinctly the appearance, on its top, of radiating striæ or roughly formed rays, which seemed to betoken a conchoidal fracture.

† Fig. 11 is copied accurately in outlines and all essential features from a drawing of an actual column in the Giant's Causeway. The original drawing was made by the author (Professor Thomson), not from memory, but while standing over the stone, and imitating carefully all the main features. The nucleus from which the brush emanates was of a very hard, dark-coloured stone. The original of this drawing he has carefully preserved for security against casual development of modified features in the making of copies.

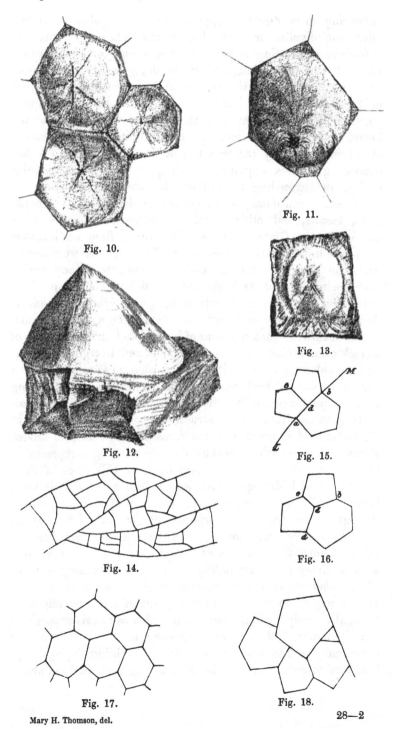

Fig. 10.

Fig. 11.

Fig. 12.

Fig. 13.

Fig. 14.

Fig. 15.

Fig. 16.

Fig. 17.

Fig. 18.

Mary H. Thomson, del.

28—2

advancing in a direction approximately perpendicular to the advancing circular or oval edge of the enlarging fissure. If a fracture originating at one side of a column were to advance across to the other side, and in so doing were to cut across any irregular lump in the mass, that lump would leave a kind of tail extending from itself forward in the direction of propagation of the fissure; but the part of the fissure formed before arrival of the fissure's edge at the lump would be scarcely at all influenced by the presence of that irregularity. A tail emanating in this way from an irregular lump or a vesicular cavity, and extending forward in the direction of advance of the crack, is continually to be noticed in the breakage of flints, glass, basalts, and other brittle substances. Now, the cases noticed at the Giant's Causeway, in which, from an included lump, the lines radiated out in various directions, and were curved when the lump was eccentric, tend to corroborate the supposition that the fissure had its beginning at the irregular lump, where some local weakness or overstraining might exist, and that it flashed out from thence towards the circumference of the column.

He mentioned that, in favour of his supposition of possibility of a crack commencing at some central point, and diverging outwards with an enlarging circle as its front or edge, so as to constitute a circular conchoidal fracture, a remarkable confirmation had, long after the publication of his views as to the nature of the cross-partings in basaltic columnar structure, presented itself to him in the features of a broken flint found casually in a heap of broken stones. This flint, which he showed at the meeting, is represented in Fig. 12. The appearances presented by this piece of flint, in its cone with divergent rays, lead to the belief that it must have received the remarkable conical part of its configuration through a blow struck, on an originally larger mass of flint, at the spot where there is now the vertex of the cone; and that, from the point of impact, a fissure must have flashed out in conical form, with a divergent or expanding circle as its advancing front or edge. That any uncrushed material should remain at the place of violent impact to form the smooth point of the cone, might be thought scarcely credible; but confirmation as to this supposition is to be obtained through noticing divergent cracks, which may occasionally be seen in solid glass balls used as children's playthings, like boys' marbles. These balls, after having been knocked about, sometimes show cracks, with circular margins, penetrating into the interior from points of impact, and which may be characterized, along with such as the one just referred to in the flint, by calling them *conical conchoidal fractures.* These fractures have resemblance in shape to limpet shells. An example of these in the glass balls he showed at the meeting, as also another broken flint, which he had subsequently found, showing a like conical feature to that of the one first noticed, and indicating that the first one is not to be regarded as an isolated result of casual circumstances unlikely to recur; but that there must exist some tendency, or readiness, to the formation of such fractures emanating from spots of violent straining in some cases of impact.

He adduced another confirmation of his view as to the cross-partings in the columns—a confirmation which had been found and carefully considered by Mr W. Chandler Roberts*. It was in a piece of indurated clay or red ochre that the confirmatory phenomena were met with. Mr Roberts, when on a geological excursion to the Giant's Causeway, at the close of the Belfast Meeting of the British Association in 1874, just after attention had been freshly brought to bear on the subject by papers read by himself and the author at that meeting, happened to find the piece of red ochre, and was struck with its features as being remarkable in their resemblance to those specially noticed under the views of Professor Thomson in regard to the cross-partings in basaltic columns. Mr Roberts carefully preserved the specimen, and, by his kindness, Professor Thomson was enabled to show it at the meeting of the Glasgow Geological Society; and it is well represented in Fig. 13, p. 435, where it is shown on a slightly enlarged scale, the lineal dimensions being increased to about $1\frac{1}{2}$ of the corresponding lengths in the little block of ochre itself, which is of a roughly cubical form, and about the size of a hazel nut. One face of this little block shows very remarkably the same kind of features as those which, met with in basaltic columns, are specially brought into notice in Figs. 10 and 11, p. 435. There is a speck or granule from which, as it appears, a divergent circular conchoidal fracture has originated; there is a lunette-shaped portion bounded by a circular or somewhat oval margin, where some delay or vibration in the propagation of the advancing

* [Deputy-Master of the Mint—afterwards Sir Wm. Roberts-Austen.]

fissure may be supposed probably to have occurred, and to have left its mark. Outside of this margin, very distinctly marked, radiating, brush-like striæ are situated, and these seem to show that, after some delay or modification which left the oval margin marked, the fissure had continued to advance outwards. This wonderful miniature occurrence, seemingly of the same kind of fracturing process as appears to have occurred in a large way and with multitudinous repetition in the rock of the Giant's Causeway itself, is not to be supposed to have any special connection, in virtue of locality, with the basaltic structure there. The little block of ochre, though found near the Causeway, is not to be supposed to be part of a tiny column, such as those of the Causeway, and due to the same set of localized causes. Its occurrence there, and its discovery there, must be attributed merely to casual coincidences.

In respect to the manner in which the fissuring of the rock longitudinally into columns may be supposed to have occurred, the author drew special attention to distinctions in the circumstances, on the one hand, under which a tesselated system of cracking can spread out over an extensive surface of mud in drying, as in cases when the water is drawn off from a pond having a smooth, muddy bottom, or under which tesselating cracks often spread over the surface of the glaze of porcelain, or over the varnish of terrestrial or celestial globes; and on the other hand, under which a fissure system can penetrate with columnar configuration through the interior of a mass of rock. In the spreading of a system of tesselating cracks over a surface of shrinking mud, or over the glaze of porcelain—which, probably by penetration of moisture, or from some other cause, may be supposed to swell more than the film of glaze on its surface can endure—or over the varnish of a globe, it is usually to be noticed that a few long and often approximately straight or moderately curved fissures first break out, and that other fissures break transversely to them, but generally or always stop on abutting against any previously formed fissure; and that the subdivision or tesselation may, by successive transverse cracks, each stopping against a previous one, be carried on more and more. This is exemplified in Fig. 14, p. 435, where the order of sequence of the successive breakages in many cases can readily be judged of by noticing that, when one crack abuts against another, which extends continuously across the point of abutment, the abutting one is the posterior of the two. This kind of tesselation is, however, very essentially distinct from that which is to be seen in walking over the Giant's Causeway, or in examining the configuration of basaltic rocks in general. In the tesselation occurring in basaltic rocks (see Fig. 17, p. 435), we never find any of the regularly formed fissures of the column side faces abutting against any flat column face. No such configuration is met with as is sketched in Fig. 18. Wherever a separating fissure between two columns meets another column, at that place there is an edge or arris line of meeting of two faces of that other column—that is to say, that regularly or ordinarily three fissures separating three columns radiate out from one common arris line for all the three (see Fig. 17, also Fig. 16). It seems that even if, from any cause —as, for instance, in the case of a mud surface in drying—there were formed a long, straight, or curved crack, such as *LM*, Fig. 15, and if another crack, *cd*, were to terminate against it, as at *d*, the part *ab* being straight or curved, a change would take place during the inward advance of the fissuring into the body of the mass, under which *ad* and *db* would cease to be continuous, and would come to form an angle, *adb*, somewhat as shown in Fig. 16.

He called attention to a class of facts which within recent years had been brought prominently into notice in connection with opinions which may or can be held regarding the mode, or possibly varied modes of formation of columnar structure in rocks. Mr W. Chandler Roberts, at the Belfast Meeting of the British Association in 1874, had pointed out that a prismatic structure, closely resembling that of basalts, can be produced artificially at will in certain brick-like masses of fire-clay and sand, by heating them under certain conditions to redness—a temperature far below that at which they would fuse. This indicated still more strikingly than had previously been done by the columnar cracking of starch, and some other substances not molten, that a process of cooling from the molten state is not at all essential to the formation of columnar structure resembling that of basalts. In the case of basalts, undoubtedly, cooling from the molten state has occurred, and we may suppose that the columnar fissuring took place at some stage in the process of cooling.

He also referred to the existence of columnar structure in sandstones, which had recently been brought into notice among

geologists in a paper by Mr David Corse Glen, F.G.S., printed in
the Society's *Transactions*, Vol. V., p. 154, 1873, relating chiefly to
columnar sandstone at Kilchattan, in the island of Bute.   In the
development and discussion of theories as to the columnar structure
in basalts, previously to this paper by Mr Glen, the existence or
columnar structure in sandstones had, so far as Professor Thomson
knew, been quite unnoticed; and, if at all mentioned, he believed
had not been introduced in any important way into consideration.
The notion of the columnar structure having instituted itself by
some concretionary or other action within the material undergoing
solidification, but barely solidified from the molten state, had
indeed commonly exercised a too dominant influence in the consideration of the subject.   It is important to notice that the
sandstone columns of Kilchattan, in Bute, are quite devoid of any
such cross-jointing, or cross-parting, as that which is remarkable in
the columns of the Giant's Causeway.   The longitudinal direction
of the columns is nearly vertical, and they cross the planes of
stratification obliquely at an angle of about 60°.   Their direction
and configuration seem to have been determined quite independently
of the stratification.   Fig. 7, p. 426, is a drawing of a portion
taken from one of these sandstone columns of Bute.   The strata
show themselves passing obliquely across the column, being distinctly indicated through varieties of colouring and grain; and they
are represented in the drawing.   This column is about five inches
in diameter; and the set of columns are stated by Mr Glen as
varying in size from six or seven inches diameter down to half an
inch.   The manner in which these columns have originated, and
the circumstances which have determined their sizes, or some
average sizes which they might tend to assume rather than other
very different sizes, must remain as important, and probably as
very difficult, matters for research and for theoretical speculation.

MISCELLANEOUS PAPERS

63. On Atmospheric Refraction of Inclined Rays,
    and on the Path of a Level Ray.

[From the *Report of the British Association*, Section A,
Brighton, 1872, page 41.]

Many years ago, in considering, from a civil-engineering point
of view, the path of a level or nearly level ray of light through
the atmosphere, with special reference to corrections in observations
with the levelling-instrument, the author found himself unable to
rest satisfied with the views put forward on the subject in books
on Practical Geodesy, or in any writings with which he was
acquainted. The only views which he then met with were to the
following effect:—

The atmosphere was regarded as consisting of an infinite
number of infinitely thin horizontal laminæ, with a gradual
increase of density in passing downwards through these laminæ,
so that the density in each lamina would differ only in an infinitely small degree from that in the one immediately above it, or
from that in the one immediately below it. It was then inferred
that a ray of light, passing obliquely downwards through the
laminæ, must, at each successive transition from one lamina into
the denser one next below, suffer refraction so that its course must
make a less angle with a normal to the laminæ in the denser
lamina than it did with the same normal in the rarer one immediately above, and that the path of the ray must therefore
be curved with the concave side downwards. From this reasoning,
without noticing that its whole foundation, in oblique transition of
the light across laminæ with gradual change of density in those
successively traversed, vanishes in the case of a horizontal ray,

authors have tacitly assumed that a ray proceeding through the
atmosphere, so as to enter a levelling-instrument horizontally,
should be expected to be curved with its underside concave. In
one sense such a conclusion, in connexion with the mode stated in
which it has been inferred, may be partly justified—that is, if the
consideration be that a ray coming from a considerable distance so
as to enter an instrument horizontally must have previously been
descending obliquely through the nearly spherical level laminæ of
the air which are rounded in correspondence with the figure of the
earth. Rays arriving level at an observer's station from the rising
or setting sun afford an instance of what is here referred to, and
one in which the light has descended obliquely through the whole
depth of the atmosphere. It may readily be admitted, from the
usual reasoning cited above, that any such ray will be curved and
concave downwards at all parts of its course where it is sensibly
descending; but as the advancing ray gradually approaches to the
level position with a gradual diminution down to cessation of
oblique descent through the laminæ, it might still, so far as that
reasoning would indicate, be held an open question whether the
curvature of the ray would approach towards zero, or whether it
would approach towards a maximum, or generally what might be
the condition as to curvature or straightness of the ray, as the ray
comes to be level.

The author proposed the question in 1863 to Professor Purser,
of Queen's College, Belfast; and Prof. Purser, on the moment,
made out an analytical investigation which depended on the
proportionality of the sine of the angle of incidence to the sine of
the angle of refraction holding good for infinitely thin laminæ
differing infinitely little in density, and holding good to the extreme
case in which the ray becomes parallel to the laminæ. This
investigation appeared to the author of the present paper to be
consistent with all physical conditions; and he regarded it as an
hypothesis likely to be fully confirmed by experimental investigations, if at any time experiments bearing on the subject should be
found practicable. From direct experiments, however, on the
curvature of a ray of light in the atmosphere, no accurate results
are to be hoped for, on account of the great and constantly varying
disturbances to which the ray is subject, through changes in the
distribution of heat and moisture in the air, and movements of its
parts among one another, and other varying influences.

Prof. Purser's investigation, which from the first has been deemed by the author of the present paper to be of much interest and value, was to the following effect, the question being:—

To find whether a ray of light passing infinitely nearly horizontally through the atmosphere will be bent with a finite curvature, or not bent at all; and whether the curvature approaches to a maximum or to a minimum as the direction of the ray approaches towards horizontality.

Conceive two laminæ, Lamina 1 and Lamina 2, each of the thickness λ. Conceive the density in each as being constant, but that there is a sudden increase of density in passing from the one to the other. Then the ray of light PAO will at A be suddenly bent or deflected from its previous line. This case may be substituted mathematically, when the laminæ are taken infinitely thin, for what actually occurs in the atmosphere.

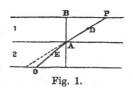

Fig. 1.

Now in the atmosphere the deflection of the ray of light in passing from the middle of one lamina to the middle of the next, as from D to E, is evidently proportional to the thickness assumed for the laminæ, the thickness being small. Hence, if we take δ to represent the angle of deflection at A, we must bear in mind that $\delta \propto \lambda$ for any given angle of incidence, or that δ must be infinitely small when the lamina is infinitely thin. Let the angle of incidence $PAB = i$. Then, by the ordinary law of refraction assumed as applicable to this case,

$$\sin i = \mu \sin (i - \delta),$$

in which μ denotes the index of refraction for passage of a ray from one lamina to the next when the thickness of the laminæ is λ.

Hence $\dfrac{\sin i}{\mu} = \sin i \cos \delta - \cos i \sin \delta$, or by dividing by $\cos i$,

$$\frac{\tan i}{\mu} = \tan i \cos \delta - \sin \delta.$$

But δ must be infinitely small, the laminæ being infinitely thin. Hence for infinitely thin laminæ we have $\sin \delta = \delta$, and $\cos \delta = 1$. Hence the previous equation becomes

$$\frac{\tan i}{\mu} = \tan i - \delta,$$

or

$$\delta = \frac{\mu - 1}{\mu} \tan i.$$

Let DE, or its equal PA, the laminæ being infinitely thin, be denoted by s. Then $s = \lambda \sec i$.

Let the radius of curvature of the ray of light, or the radius of the circle touching the ray in the points D and E, be denoted by R, and then we have

$$\text{Curvature} = \frac{1}{R} = \frac{\delta}{s}.$$

Hence

$$\text{Curvature} = \frac{\mu - 1}{\mu\lambda} \frac{\tan i}{\sec i},$$

or

$$\text{Curvature} = \frac{\mu - 1}{\mu\lambda} \sin i.$$

But since the curvature of the ray of light is independent of the small thickness which we may take for the infinitely thin laminæ, and can only vary with the angle of incidence i, we must have $\mu - 1/\mu\lambda$ in the foregoing equation constant; and so we have

$$\text{Curvature} \propto \sin i,$$

which has its maximum value when i is a right angle; that is, when the ray is passing horizontally, or infinitely nearly so.

This shows that if the ordinarily assumed law of refraction be truly applicable to a ray of light passing extremely nearly horizontally through level laminæ of air of varying density, the curvature of the ray of light must approach to a maximum as the inclination of the ray approaches to horizontality. From this, if true, the step is natural, or inevitable, to the conclusion that, leaving out of account the rotundity of the earth, and conceiving the laminæ of constant density to be level planes, a ray of light directed level so that if it were to traverse a straight path it would pass along an infinitely thin lamina of uniform density, but with less density above and greater below, *would be bent by virtue of the difference of the densities above and below it.*

It must, however, be admitted that there is something perplexing, or not quite satisfactory to the mind, in taking this final step to the perfectly level ray; for as soon as the inclination of the ray becomes zero the whole foundation and framework of the investigation fails, there being then no oblique passage of a ray from one lamina into another, no incident and no refracted ray, and consequently no ratio of sines of angles of incidence and refraction; though all these would be required to be discussed as

if they existed in the case of every ray whose curvature is to be
compared with that of any other.   Still, as both Professor Purser
and the author thought at the time, the investigation made the
physical conclusion as to level rays seem highly probable; since, if
it proves, as it seems to do, that a ray of light descending obliquely
must move along a certain curved path, and that the curvature
must increase as the inclination approaches towards horizontality,
and also that the rate of change of curvature with change of
inclination approaches towards zero as the inclination approaches
towards horizontality, it must follow that a ray of light passing
exactly level will be bent with the same curvature as one infinitely
nearly level.

Several years later (in February 1870) a new investigation
occurred to the author of the present paper. The new one is
much simpler, and it is more general, and its reasoning holds good
alike for level as for inclined rays. In fact the previous investigation, founded on the ratio of the sines of angles of incidence and
refraction, and therefore in principle having no direct applicability
to level rays, comes, when considered in connexion with the new
one, to be a case of this more general one, seeing that under the
undulatory theory of light the proportionality of the sines of the
angles of incidence and refraction is not an ultimate fact or
principle, but a consequence of retardation of the velocity of light
in the denser medium. In the new investigation which will now
be submitted the retardation of the velocity of light in the denser
medium is taken as the basis of the reasoning.

Let MN and OP be two level surfaces in the atmosphere, and
let each of these be supposed to pass
through air of uniform density throughout each of them. They may be conceived to be at a very small distance
apart, and then obviously a ray in descending obliquely from one to the other
will alter its curvature only by a very
slight amount.

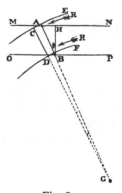

The fundamental assumptions on which
the investigation will be based are the
following three :—

(1) It is assumed that the light at
A has a certain velocity, which may be

Fig. 2.

called v_1, and that the light at B, where the air is denser, has a smaller velocity, which may be called v_2.

(2) It is assumed that these velocities are constant for all inclinations of the ray of light; or, in other words, that the velocity of the ray of light is independent of the inclination of the ray to the horizontal strata of the air.

(3) It is assumed that the direction of the light is perpendicular to the wave front, or that a surface taken crossing every ray in a pencil of rays perpendicularly, and then conceived to advance along the course of each ray with the velocity of that ray, will continue to cross every ray perpendicularly.

Now let AB and CD be two successive positions, indefinitely near to each other, of the advancing front of a ray or pencil of light whose direction of advance is indicated by the lines EA and FB, and by the arrows R in the figure, the direction at all points of AB being normal to the plane represented by AB. Let the inclination of AB to the vertical line BH be denoted by θ, which will then also denote the inclination of the ray to the horizon. Let the thickness of the lamina of air from MN to OP be denoted by λ, or let BH in the figure be denoted by λ.

The lengths AC and BD have to one another the same ratio as the velocities of light at A and B respectively; or

$$AC : BD :: v_1 : v_2.$$

If AB and CD be produced till they meet in G, the length GA is the radius of curvature of the ray at A. Let this radius be denoted by r. Then, since AB is $= \lambda \sec \theta$, we have obviously

$$v_1 - v_2 : v_1 :: \lambda \sec \theta : r.$$

Hence curvature or

$$\frac{1}{r} = \frac{v_1 - v_2}{v_1 \lambda} . \cos \theta,$$

or curvature $\propto \cos \theta$; which shows that the curvature is a maximum when $\theta = 0$, that is, when the ray is level, and that the curvature diminishes to zero as the ray becomes vertical.

The result here brought out,

$$\frac{1}{r} = \frac{v_1 - v_2}{v_1 \lambda} . \cos \theta,$$

is perfectly in agreement with that arrived at in the previous investigation of Prof. Purser, namely $\dfrac{1}{r} = \dfrac{\mu - 1}{\mu\lambda}\sin i$, seeing that $\sin i$ is $= \cos\theta$, and that, according to the undulatory theory of light as confirmed by experimental proofs, it is known that

$$v_1 : v_2 :: \mu : 1,$$

so that $\dfrac{v_1 - v_2}{v_1}$ must be equal to $\dfrac{\mu - 1}{\mu}$. The new method has however, the advantage of quite clearing away the perplexity involved in the other by the collapse of the reasoning when brought to the extreme case of the level ray. In the new method no such collapse occurs; and, in fact, the new method shows clearly how the real fundamental principle (that of retardation of velocity in the denser medium, on which the bending depends, and which holds good quite as much for level rays as for any others) is allowed in the previous investigation gradually to fade out of the reasoning, till, in the case of the level ray, it has absolutely vanished from the conditions which were taken into account. The previous method, like the modes of considering the subject of atmospheric bending of rays which appear to have been most generally entertained hitherto, took a consequence of the important fundamental principle into account instead of the principle itself (that consequence being the proportionality of sines of angles of incidence and refraction in case of oblique transition of light from one lamina to another of different density); but that consequence happens to be not so general as the principle from which it follows, and to be one which becomes nugatory or non-existent in the case of the level ray.

In concluding, the author wishes to state that it seemed to him rather unlikely that so simple a view of the influence of the atmosphere in effecting the bending of rays of light as that which he has now offered could be quite new. He thought that others better acquainted with the science of light than he is must most probably have entertained the same or similar views. He has therefore made inquiries as to the views which have hitherto been put forward regarding the bending of light in the atmosphere and in other mediums of continuously variable index of refraction, or, as they may be better considered in the present investigation, mediums of continuously varying light-velocity*. Much has been

---

* μ^{-1} might be called the index of light-velocity.

written on the subject in general, and on various particular cases of its application; and views very similar in principle with those here offered appear in various ways to have been entertained, or implied more or less explicitly; but he has not learned of anything having been taught which has anticipated the treatment of the subject at present offered so as to deprive it altogether of novelty and interest. The subject, he believes, has been very generally considered under imperfect views; and he will think a good result will have ensued if his drawing the attention of the British Association to it will serve to elicit from others notice of the best views that have hitherto either been fully published, or have been entertained or discussed without complete publication.

POSTSCRIPT.—From Professor Clerk Maxwell I have learned that, in December 1851 or 1852, when on a visit to my brother, Sir William Thomson, he had in his mind the consideration of the path of rays in a medium of continuously variable index of refraction; that he then thought it easiest to calculate the path of the ray by translating the problem into the emission theory, and treating the ray as a moving body acted on by a force depending on the variation of the index of refraction, and so proceeding by an artifice justifiable on the ground that the emission and undulation theories are mutually equivalent in respect to the *course* of rays when the proper alterations of the hypotheses are made; and that my brother showed him, on the other hand, how easy it is to begin with the right hypothesis by making the velocity inversely proportional to μ, and calculating the change of wave-front.

Professor Maxwell, in 1853, sent to the *Cambridge and Dublin Mathematical Journal* a problem about the path of a ray in a medium in which

$$\mu = \frac{\mu_0 a^2}{a^2 + r^2},$$

where μ_0 and a are constant, and r is the distance from a fixed point. Such rays, he points out, move in circles. This problem, he mentions, was intended to illustrate the fact that the principal focal length of the crystalline lens of the eye is very much shorter than anatomists calculate it, from the curvature of its surface and the index of refraction of its substance. The reason, he shows, is the increase of density towards the centre of the lens, so that the

rays pass nearly tangentially through a place where the density is
varying.   Also, in the Cambridge Examinations for 1870, Prof.
Maxwell set a question about the conditions of a horizontal ray of
light having a greater curvature than that of the earth.   A great
deal, he says, has been written about atmospheric refraction by
Bessel, Clairaut, and others; and a question has been set on it in
January of every year at Cambridge for several years back, so that
the subject has been much discussed in various ways; but, he
says, the mode of treatment of the subject in the present paper
does not seem to have been anticipated.—J. THOMSON.

[The following letters from JAMES to WILLIAM THOMSON, and
    from Professor TAIT to JAMES THOMSON, relate to the fore-
    going important paper and are of historical interest.]

6th November 1863.

DEAR WILLIAM,
            The foregoing investigation was worked out a few days
ago by Professor Purser (Prof. of Mathematics in Queen's College,
Belfast) when I proposed the question to him which forms the
heading of the investigation.   As here written out it is stated at
much greater length or with fuller explanations than he gave it.
Perhaps he would think some of the steps needless, but this form
is that in which I have made it clear to my own mind.
    From the result of the investigation:—namely, that the cur-
vature of the ray of light approaches to a maximum as the
inclination of the ray approaches to horizontality, I conclude that
a *level* ray directed so that, if it were to move straight forward, it
would pass along an infinitely thin lamina of uniform density but
with less density above and greater below, *would be bent by the
influence of the difference of densities above and below it.*
    You are aware that I have for several years been turning
attention to the questions here discussed.   It was an idea of mine
that writers on Engineering levelling were not entitled, from the
mere consideration of the refraction of a ray of light descending
obliquely through the atmosphere to run to the conclusion that
a level ray entering a levelling instrument would have any curvature due to refraction, because it is really passing along a lamina

of uniform density along the path of the ray of light, and not passing obliquely from laminæ of less to laminæ of greater density. I thought it probable that the radius of curvature would be infinite for a level ray, as well as for a vertical ray; and that it would have a minimum value for some particular angle of inclination of the ray.

When I mentioned the matter to you, you thought on a slight consideration as I did, that a ray passing horizontally along a lamina of uniform density along the path of the ray would not be bent *.

Still the question appeared to me to be open for investigation, that perhaps the refraction of a ray falling on a lamina at a very small angle of inclination might be so great as to make the ray have a finite curvature when infinitely nearly parallel to the laminæ. Prof. Tait tried to solve the question one evening; observing the importance of the consideration last mentioned, but he did not happen on any easy way of arriving at a solution, and so he let the matter drop, I believe.

Professor Purser's solution as explained or developed more fully in the foregoing is perfectly satisfactory to me. When I pointed out to him the consequence I would draw from it:—that a perfectly level ray would be refracted or bent into a curve of the maximum curvature, he would not agree that such a conclusion would be legitimate, as he said that as soon as the inclination of the ray becomes Zero, the whole foundation and framework of the investigation fails. Still I believe that my inference is perfectly legitimate; for I see that if his investigation proves a ray of light to move along a certain curved path, when infinitely nearly level, as also that the rate of change of curvature is Zero, as the infinitely small inclinations increase or diminish, it must follow that a ray of light passing exactly level will be bent with the same curvature as one infinitely nearly level; as there is no sudden change of any physical conditions that I can imagine at the instant the ray becomes level from being infinitely nearly so.

I think Prof. Purser afterwards came to think that probably my view on this matter is correct. What do you think of this? Of course it is to be recollected that the whole is founded on the supposition (which we have no means at present of *knowing* to be

* [Note:—written August 1872. Dr Lloyd in his *Optics*, 1849, pages 108 and 109, seems deliberately to think so too.]

true) that the ordinary law of refraction; viz. that the sine of angle of refraction is proportional to the sine of the angle of incidence; holds good for rays of light falling with infinitely small inclinations on infinitely thin laminæ of the atmosphere.

If this be true, is it not the case that the conclusion I have deduced is a new principle in optics, namely that a ray of light passing exactly along a lamina with denser matter to one side than to the other will bend itself towards the side where the denser matter is?

I am, Your affte. brother,

JAMES THOMSON.

17 UNIVERSITY SQUARE, BELFAST.
22nd March, 1870.

MY DEAR WILLIAM,

Will you look over the enclosed paper about Atmospheric Refraction of Inclined Rays, and the Path of a Level Ray? You may recollect my showing to you the investigation made out by Prof. Purser on that subject some years ago. The investigation I now give is, I think, greatly preferable; and, so far as I know, is quite new. In fact I am not aware of any investigation on the subject previous to the one which Prof. Purser made out, when I put the matter before him as one of importance. If you care to see the investigation made by Prof. Purser, I can send you a copy of it as written out and explained by me at the time when it was new.

The subject is of much practical importance in respect to putting people concerned in extensive geodesical operations on right modes of considering the possible or probable effects of the atmosphere on long sights taken with the Levelling Instrument....

I am, Your affte. brother,

JAMES THOMSON.

17 DRUMMOND PLACE, EDINBURGH.
25th April 1870.

MY DEAR THOMSON,

Your brother sent me the accompanying paper to read. As I am alluded to in it, I would only say that I *hope* it may be correct, because it gives us if true (by means of total internal reflection) a hint of the nature of transparent bodies

29—2

very close to their surfaces; in fact it reproduces Poisson's assumption in his theory of Capillary Forces.

But I see one very formidable objection—viz. that the direction of propagation of light is only normal to the wave front in *homogeneous* singly refracting media. So that I see at present no reason for believing that if the earth were *plane* and the density of the air depended on the height only, a horizontal beam should not remain horizontal while its wave-front takes all possible inclinations. I have been too busy of late to work at the subject properly, and what I now write is the result of conjecture merely. So it is quite likely that you may disprove it*,—only it will be *necessary to do so* as, if what I hint is possible, your demonstration breaks down.

Yours truly,
P. G. TAIT.

64. ON AN INTEGRATING MACHINE HAVING A NEW KINEMATIC PRINCIPLE.

[From the *Proceedings of the Royal Society*, Vol. XXIV, 1876, p. 262. This paper was followed in *Proc. Roy. Soc.* by three short papers of the same date by Sir William Thomson. The whole series of papers is reprinted in Thomson and Tait, *Natural Philosophy*, Ed. 2, part i, 1879, pp. 488—508 under the Titles :—'An Instrument for calculating the Integral of the Product of two given Functions'; 'Mechanical Integration of Linear Differential Equations of the Second Order with Variable Coefficients'; 'Mechanical Integration of the general Linear Differential Equation of any Order with Variable Coefficients.']

THE kinematic principle for integrating ydx, which is used in the instruments well known as Morin's Dynamometer† and

---

* [Its truth is implied in the law of refraction of the rays, which here constitutes their definition. The argument can moreover be conducted in terms of wave-fronts alone, without mentioning rays.]

† Instruments of this kind, and any others for measuring mechanical work, may better in future be called Ergometers than Dynamometers. The name "dynamometer" has been and continues to be in common use for signifying a spring instrument for measuring *force*; but an instrument for measuring *work*, being distinct in its nature and object, ought to have a different and more suitable designation. The name "dynamometer," besides, appears to be badly formed from the Greek; and for designating an instrument for *measurement of force* I would suggest that the name may with advantage be changed to *dynamimeter*. In respect to the mode of forming words in such cases, reference may be made to Curtius's *Grammar*, Dr Smith's English edition, § 354, p. 220.—J. T., 26th February, 1876.

Sang's Planimeter*, admirable as it is in many respects, involves one element of imperfection which cannot but prevent our contemplating it with full satisfaction. This imperfection consists in the sliding action which the edge wheel or roller is required to take in conjunction with its rolling action, which alone is desirable for exact communication of motion from the disk or cone to the edge roller.

The very ingenious, simple, and practically useful instrument well known as Amsler's Polar Planimeter, although different in its main features of principle and mode of action from the instruments just referred to, ranks along with them in involving the like imperfection of requiring to have a sidewise sliding action of its edge rolling wheel, besides the desirable rolling action on the surface which imparts to it its revolving motion—a surface which in this case is not a disk or cone, but is the surface of the paper, or any other plane face, on which the map or other plane diagram to be evaluated in area is drawn.

Professor J. Clerk Maxwell, having seen Sang's Planimeter in the Great Exhibition of 1851, and having become convinced that the combination of slipping and rolling was a drawback on the perfection of the instrument, began to search for some arrangement by which the motion should be that of perfect rolling in every action of the instrument, corresponding to that of combined slipping and rolling in previous instruments. He succeeded in devising a new form of planimeter or integrating machine with a quite new and very beautiful principle of kinematic action depending on the mutual rolling of two equal spheres, each on the other. He described this in a paper submitted to the Royal Scottish Society of Arts in January 1855, which is published in Vol. IV. of the Transactions of that Society. In that paper he also offered a suggestion, which appears to be both interesting and important, proposing the attainment of the desired conditions of action by the mutual rolling of a cone and cylinder with their axes at right angles.

The idea of using pure rolling instead of combined rolling and slipping was communicated to me by Prof. Maxwell, when I had the pleasure of learning from himself some particulars as to the

---

* Sang's Planimeter is very clearly described and figured in a paper by its inventor, in the *Transactions of the Royal Scottish Society of Arts*, Vol. IV. January 12, 1852.

nature of his contrivance. Afterwards (some time between the years 1861 and 1864), while endeavouring to contrive means for the attainment in meteorological observatories of certain integrations in respect to the motions of the wind, and also in endeavouring to devise a planimeter more satisfactory in principle than either Sang's or Amsler's planimeter (even though, on grounds of practical simplicity and convenience, unlikely to turn out preferable to Amsler's in ordinary cases of taking areas from maps or other diagrams, but something that I hoped might possibly be attainable which, while having the merit of working by pure rolling contact, might be simpler than the instrument of Prof. Maxwell and preferable to it in mechanism), I succeeded in devising for the desired object a new kinematic method, which has ever since appeared to me likely sometime to prove valuable when occasion for its employment might be found. Now, within the last few days, this principle, on being suggested to my brother as perhaps capable of being usefully employed towards the development of tide-calculating machines which he had been devising, has been found by him to be capable of being introduced and combined in several ways to produce important results. On his advice, therefore, I now offer to the Royal Society a brief description of the new principle as devised by me.

The new principle consists primarily in the transmission of motion from a disk or cone to a cylinder by the intervention of a loose ball, which presses by its gravity on the disk and cylinder, or on the cone and cylinder, as the case may be, the pressure being sufficient to give the necessary frictional coherence at each point of rolling contact; and the axis of the disk or cone and that of the cylinder being both held fixed in position by bearings in stationary framework, and the arrangement of these axes being such that when the disk or the cone and the cylinder are kept steady, or, in other words, without rotation on their axes, the ball can roll along them in contact with both, so that the point of rolling contact between the ball and the cylinder shall traverse a straight line on the cylindric surface parallel necessarily to the axis of the cylinder—and so that, in the case of a disk being used, the point of rolling contact of the ball with the disk shall traverse a straight line passing through the centre of the disk—or that, in case of a cone being used, the line of rolling contact of the ball on the cone shall traverse a straight line on the conical surface, directed necessarily towards the vertex of the cone. It will thus readily be seen that, whether the cylinder and the disk or cone be at rest or revolving on their axes, the two lines of rolling contact of the ball, one on the cylindric surface and the other on the disk or cone, when both considered as lines traced out in space fixed relatively to the framing of the whole instrument, will be two parallel straight lines, and that the line of motion of the ball's centre will be straight and parallel to them. For facilitating explanations, the motion of the centre of the ball along its path parallel to the axis of the cylinder may be called the ball's longitudinal motion.

Now for the integration of ydx: the distance of the point of contact of the ball with the disk or cone from the centre of the disk or vertex of the cone in the ball's longitudinal motion is to represent y, while the angular space turned by the disk or cone from any initial position represents x; and then the angular space turned by the cylinder will, when multiplied by a suitable constant numerical coefficient, express the integral in terms of any required unit for its evaluation.

The longitudinal motion may be imparted to the ball by having the framing of the whole instrument so placed that the lines of longitudinal motion of the two points of contact and of the ball's centre, which are three straight lines mutually parallel, shall be inclined to the horizontal sufficiently to make the ball tend decidedly to descend along the line of its longitudinal motion, and then regulating its motion by an abutting controller, which may have at its point of contact, where it presses on the ball, a plane face perpendicular to the line of the ball's motion. Otherwise the longitudinal motion may, for some cases, preferably be imparted to the ball by having the direction of that motion horizontal, and having two controlling flat faces acting in close contact without tightness at opposite extremities of the ball's diameter, which at any moment is in the line of the ball's motion or is parallel to the axis of the cylinder.

It is worthy of notice that, in the case of the disk, ball, and cylinder integrator, no theoretical nor important practical fault in the action of the instrument would be involved in any deficiency of perfect exactitude in the practical accomplishment of the desired condition that the line of motion of the ball's point of contact with the disk should pass through the centre of the disk.

The reason of this will be obvious enough on a little consideration.

The plane of the disk may suitably be placed inclined to the horizontal at some such angle as 45°; and the accompanying sketch, together with the model, which will be submitted to the Society by my brother, will aid towards the clear understanding of the explanations which have been given.

My brother has pointed out to me that an additional operation, important for some purposes, may be effected by arranging that the machine shall give a continuous record of the growth of the integral by introducing additional mechanisms suitable for continually describing a curve such that for each point of it the abscissa shall represent the value of x, and the ordinate shall

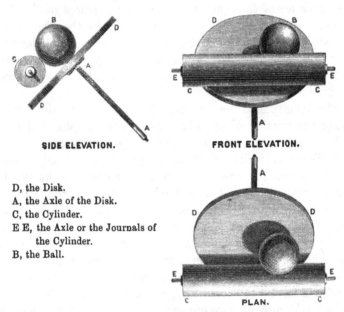

SIDE ELEVATION.   FRONT ELEVATION.

D, the Disk.
A, the Axle of the Disk.
C, the Cylinder.
E E, the Axle or the Journals of
    the Cylinder.
B, the Ball.

PLAN.

represent the integral attained from $x = 0$ forward to that value of x. This, he has pointed out, may be effected in practice by having a cylinder axised on the axis of the disk, a roll of paper covering this cylinder's surface, and a straight bar situated parallel to this cylinder's axis and resting with enough of pressure on the surface of the primary registering or *the indicating* cylinder (the one, namely, which is actuated by its contact with the ball) to make it have sufficient frictional coherence with that surface, and by having this bar made to carry a pencil or other tracing point which will mark the desired curve on the secondary registering or *the recording* cylinder. As, from the nature of the apparatus, the axis of the disk and of the secondary registering or the recording cylinder ought to be steeply inclined to the horizontal, and as, therefore, this bar, carrying the pencil, would have the line of its length, and of its motion alike steeply inclined with that axis, it seems that, to carry out this idea, it may be advisable to have a thread attached to the bar and extending off in the line of the bar to a pulley, passing over the pulley, and having suspended at its other end a weight which will be just sufficient to counteract the tendency of the rod, in virtue of gravity, to glide down along the line of its own slope, so as to leave it perfectly free to be moved up or down by the frictional coherence between itself and the moving surface of the indicating cylinder worked directly by the ball.

The following correspondence with Prof. Clerk Maxwell ensued.

GLENLAIR, DALBEATTIE,
25th June, 1879.

DEAR PROFESSOR THOMSON,

In the description of the disk globe and cylinder integrator as given in the new edition of T and T', the position of the globe is determined by means of a fork with two flat faces, which graze the opposite sides of the sphere without nipping it, or else by one flat face and gravity.

Now it is easy to ensure that the friction of these surfaces on the sphere shall be small in comparison with that at the contacts with the disk and cylinder, but it occurs to me that the motion of the surface of the globe at the points of contact of the fork relative to its centre is perpendicular to the plane containing the two lines of contact of the sphere with the disk and cylinder. Hence if instead of a flat face we place a cylinder having its axis perpendicular to the lines of contact and in a plane through the centre of the sphere parallel to the plane through the lines of contact, and if this cylinder is free to rotate about this axis, there will be rolling without slipping between this (guiding) cylinder and the sphere.

It is evident that when the sphere is changing its "position" by rolling on the disk and cylinder, these being at rest, there will be pure rolling between the sphere and guiding cylinder. When the disk moves, the position of the globe being fixed, the globe spins upon the point of contact with the guiding cylinder.

Hence for all motions of the machine the action between the globe and guiding cylinder (or cylinders if there is another on the other side) is pure rolling.

Yours very truly,

J. CLERK MAXWELL.

OAKFIELD HOUSE, UNIVERSITY AVENUE,
GLASGOW, 16 *July* 1879.

MY DEAR PROFESSOR MAXWELL,

I duly received your letter of the 25th of June, and was much interested by your device of the cylindric roller, for giving a perfect rolling contact at the one or two guide or guides or terminals of guide fork for the ball in the integrator.

What you propose is very nice indeed as a kinematic contrivance, or as a means for accomplishing a condition that might possibly in some way be wanted. The instrument as already devised with sliding contact at the guide face or faces, is not subject, I would say, and I think you mean also, to any practical or theoretical objection from the sliding contact, at the point of contact of the globe with the guide, because all we want is to have *hesion* at the points of contact of the globe with the cylinder and disk, and this is not at all hindered by the frictional sliding motion at the face of contact of the guide with the ball. I suppose we are entitled to say that this frictional resistance at the guide is not a theoretical objection nor a practical objection, unless also we were to treat the frictional resistance of the journals or axle of the cylinder (which receives its motion through the ball) as objectionable. Of course if the point of contact of the ball with the disk and cylinder be considered (in virtue of yielding of the materials) as not being quite mathematical points, we may see that there would be an almost infinitesimally small theoretical imperfection introduced by anything requiring the transmission of mechanical energy from the disk through the globe to the cylinder, or indeed the transmission of energy either way between the disk and globe or between globe and cylinder: and in this sense any expenditure of energy in friction at the guide of the globe would be an almost infinitesimally small theoretical fault. This, however, I think may be left out of account, as being small relative to the ordinary inaccuracies incidental to all kinds of instruments.

Your guide cylindric roller, though nice for geometrical or kinematic consideration, would I think increase the difficulty and cost of construction of the machine, and would introduce new liabilities to slight defects owing to imperfections of construction, in ways from which the plane guide face contact with the ball is quite free.

We are all well. If there is any chance of your being in this region of Scotland this summer, I hope you will recollect that we would be very glad to have a visit from you for as long as you could stay. We are expecting to be at home most of this summer.

Mrs Thomson unites in kind regards,

Yours very truly,

JAMES THOMSON.

65. ON CERTAIN APPEARANCES OF BEAMS OF LIGHT, SEEN AS IF EMANATING FROM CANDLE OR LAMP FLAMES.

[Posthumous paper, from the *Proceedings of the Royal Society*, Vol. LII, 1892, p. 70, communicated by LORD KELVIN, P.R.S., with an Introductory Note.]

ABOUT the end of last January, when my brother was fully occupied in writing his paper on the Trade Winds for the Bakerian Lecture, he called my attention to the well-known beams or ladders of light seen below or above a lamp flame viewed with partially-closed eyelids, and he gave me verbally an explanation of the phenomenon which surprised me very much. By some simple and interesting trials with my own eyes, which he explained to me how to make, I was perfectly convinced that his explanation was correct; and believing it, as I still believe it, to be new, I urged him to write a short paper on the subject for the Royal Society, but not to let it interfere with his work for the Bakerian Lecture; and he undertook to do so as soon as might be after being freed from this work. We hoped, somewhat confidently, that he might be able to give the thus promised paper before the end of the present session of the Royal Society. That hope has not been fulfilled, and I had offered to the Secretaries a communication describing my recollection of what my brother had told me, when his son found a memorandum of date 18th October, 1891, and a little book of notes of date 29th December, which tell the story better than I could have told it, and which, therefore, though not completed in proper form for publication, I now give in the unfinished form in which they have been found, with only a somewhat more clear drawing, and description of drawing, substituted for the rough sketch found in his note of date October 18, 1891.

Proposed probable Paper for the (?) Society, by J. T., "On the Nature and Origin of certain Appearances of Beams of Light as if emanating from Candle or Lamp Flames."

Description of the Drawing.

[The drawing represents a vertical section of the eye, eyelids, and watery prismoids*, through FF', the axis of the eye. The large number of parallel lines outside represent rays of light coming from a flame several feet or yards away in the direction of F', to the eyelids, the prismoids, and the undisturbed outer surface of the cornea between the prismoids. The lines within the eye below FF' represent the convergence to F, the image of the flame, of those of the external rays from the flame which fall on the undisturbed portion of the surface of the cornea. The lines within the eye above FF' represent rays disturbed by the prismoid of the upper eyelid which, incident on the retina at bbb, give the perception as if of light coming from without in the direction of the dotted lines outside the eye. It is this perception that constitutes the appearance of the downward beams or ladders of light, due to the prismoid of the upper eyelid. The rays disturbed by the prismoid of the lower eyelid, in the position represented in the diagram, are all stopped by the lower part of the iris.

Looking now at the diagram, we understand perfectly that if, with the eyeball and flame unchanged, the upper eyelid be gradually raised a little, the uppermost of the rays coming inwards

---

* The refracting watery liquid in the entrant corner between lip of eyelid and cornea may be called the prismoid or liquid prismoid.

from the prismoid will fall on the upper part of the iris and will
be stopped by this screen.   Thus, the length of *bbb* upwards from
F is diminished, until all the beams from the prismoid are stopped
by the iris, and the length of the apparent beams below the flame
correspondingly diminishes to zero.   When the upper eyelid is
wide open the flame is seen without any appearance of the beams
below it.   We also understand readily from the diagram how, if
the lower eyelid is lifted a little without any change in the
position of the upper eyelid, beams both above and below the

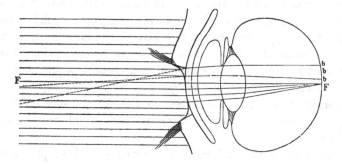

flame are seen.   We also conclude that if, with the eyelids fixed
relatively to the head, the head is moved while the eyeball remains
with its axis in the direction of the flame, we see beams of light
above the flame when the head is turned upwards, and beams of
light below the flame when the head is turned downwards.   Also
that if the eyelids are partially closed, as in the diagram, beams
will be seen both above and below the flame when the head,
carrying the eyelids with it, is turned slightly up from the position
shown in the diagram.   Also that if the eyelids be wide open,
instead of half closed as shown in the diagram, no beams, either
above or below the flame, will be seen when the two eyelids are
equidistant, or nearly equidistant, above and below the middle of
the pupil.   When the head, with the eyelids, is turned downwards,
so as to bring the upper eyelid across the aperture of the pupil
beams of light are seen below the flame; and when the head, with
the eyelids, is turned upwards so as to bring the lower eyelid across
the middle of the pupil, beams of light produced by the prismoid
of the lower eyelid are seen above the flame.]

*Notes on Quasi-Ray Beams of Light from Candles, or
other small Luminous Spots.*

Date of Note, 29th December, 1891.

I have noticed decidedly this morning to the following effect:—

In some cases (the nature of which I intend to note further
on) I found that, when seeing a small gas flame with apparent
descending tail (or quasi-beam of rays), I could, by lowering the
upper eyelid, cut off vision of the flame, while leaving the tail
visible; and, by still further lowering the upper eyelid, I could
cut off the upper part of the tail, leaving the lower part, the part
remote from the flame, quite visible as before. The contrast
between lowering the upper eyelid and lowering a screen (a card,
for instance) in front of the eye was very remarkable. In the
lowering of the card or other screen, the tail vanishes before the
flame is eclipsed; but in lowering the eyelid the flame is
eclipsed first.

In some attitudes I could not bring out these phenomena.
I did find them when awake in bed early in the morning, head
on pillow and light coming down from a gas flame obliquely to
the eye. Point to which eye was directed seemed to do best when
taken at an altitude (angular) somewhat above the gas flame.

Afterwards, this same morning, I found I could see the phenomenon when standing upright and looking at image of gas in
mirror. Ray from image ascending obliquely; eyesight directed
above image in looking-glass.

Again, looking at a gas flame a little above the level of the
eye, I stood erect and elevated my face, directing my eyesight
to above the gas; then lowered the upper eyelid and saw the
downward tail remaining when the gas flame was eclipsed by the
eyelid. The theory of all this is clear to me, and in agreement
with what I have previously devised.—J. T.

Take notice that to get the phenomena above sketched out
to show themselves, the edge of upper eyelid, where roots of
eyelashes are situated, must not shadow the prismoid when the
eyelid is lowered enough to cover the pupil from the direct rays
of the candle or gas flame. After the candle is cut off from the
pupil, the direct rays from the flame must still be reaching the
prismoid. This, I think, tallies with the experimental conditions

under which the tail was seen when the flame was eclipsed by eyelid.—J. T.

P.S.—Same day, 29th December. On a little further consideration I notice that the elevation of the face is of no importance. It is only the elevation of the line of special direction of the eyesight [axis of the eye] relatively to the line from flame to eye that is important.—J. T.

Notes on Quasi-Light Beams.

(For paper.)

Often I fail to see the apparently ascending beam above the candle or gas flame. But I find that by very nearly shutting the eye I can see the ascending beam going up very high and the descending one at same time. On bringing my open hand down from above as if to cut off the ascending beam I see the beam as if between my eye and my hand, and the flame begins to be eclipsed before the beam is cut off, or even diminished.

Note by the President of date June 16.

I had asked many friends well acquainted with optical subjects whether they knew of this explanation of the luminous beams, and all said "no" until yesterday evening, at the *soirée* of the Royal Society, when Professor Silvanus Thompson immediately answered by giving the explanation himself, and telling me that he had given it to his pupils in his lectures on optics, as an illustration of a concave cylindrical lens. He did not know of the explanation ever having been published otherwise than in his lectures. I have myself also looked in many standard books on optics, and could find no trace of intelligence on the subject. It seems quite probable, therefore, that, of all the millions of millions of men that have seen the phenomenon, none, within our three thousand years of scientific history, had ever thought of the true explanation except Professor Silvanus Thompson and my brother*.

* [Cf. *Life of Lord Kelvin*, by Silvanus P. Thompson, vol. ii. p. 919, footnote.]

APPENDIX

66. ON PUBLIC PARKS IN CONNEXION WITH LARGE TOWNS, WITH A SUGGESTION FOR THE FORMATION OF A PARK IN BELFAST*.

[A Paper read before the Belfast Social Inquiry Society on the 2nd March, 1852.]

THE importance of public parks and other open spaces in large towns is such as scarcely to admit of debate, yet the consideration of some of the advantages resulting from them cannot fail to be a matter of interest. They have a great effect in improving the atmosphere of the districts around them, by allowing of the change of air so much needed where many human habitations are closely packed together. Thus the beneficial influence of the parks of London extends to multitudes even of those who seldom use them as places for recreation and exercise. When, however, one leaves the hot and crowded streets of London on a summer day, and, entering one of the parks, passes along pleasant walks among flowering shrubs, green grass, and ornamental patches of water enlivened with numerous water-fowls of various kinds and colours, he feels practically some of the advantages of open spaces in large towns. When one walks in the Glasgow Green, as being the best place for exercise within the town, or the best outlet from it to the country, and when he sees himself surrounded on all sides with tall chimneys vomiting smoke which is wafted past him in clouds, although he may not feel disposed to congratulate himself on the freshness of the air, yet still his reason convinces him that the state of the town, bad as it is, would be much worse were this

---

* [This paper led directly to the purchase, for the town of Belfast, of the large Ormeau Park.]

extensive space covered densely with mills and other buildings like those with which it is surrounded. When the continental tourist visits on a fine summer evening the promenades which surround the city of Frankfort-on-the-Main, and enjoys the balmy freshness of the air and the perfume of the flowers, while he watches with pleasure the fire-flies as they float before him in the dusk, he is forcibly struck with the contrast between the present condition of the place, and its former state, when it was allocated for the fortifications of the city. He rejoices that the modern system of warfare has rendered all such fortifications unavailable, and admires the good taste which has turned to such a useful and agreeable purpose the space thus left vacant. Of all the thorough-fares of Paris, the Boulevards are among the most agreeable, on account of their spaciousness. These are also due to ancient fortifications, which surrounded the city when it was much less extensive than at present. Now, it is well worthy of remark, that the original conservation of most of the open spaces at present existing in towns has been due rather to accidental circumstances than to a wise foresight in the allocation of the land and the arrangement of the buildings. The open spaces in Frankfort and Paris, to which I have just alluded, are examples of this fact. Most of the parks of London, too, were originally the private property of the Crown; and, besides, they were formerly not open spaces preserved in the city, but they were far away from it in the country. It is only in later times that the town has spread itself round them so as to include them within its precincts. The name of the church at Charing Cross, St Martin's-in-the-Fields, affords a striking indication that at the time when it was built it was outside of the metropolis; and yet, the whole line of parks, comprising St James's Park, Green Park, Hyde Park, and Kensington Gardens, lies still farther away from the original city. The Dublin College Park was not, at the time of the foundation of the University, within the town. The charter of Queen Elizabeth, establishing the college, designates it as the College of the Holy Trinity, *near* Dublin. The Glasgow College Green, also, was originally outside of the town; and, even at the present day, the college, though surrounded by the town, is still a separate corporation, not being subject to the municipal authorities.

Although, then, in former times very little foresight was applied in the disposition of the buildings and open spaces of increasing towns, yet in our own times better methods have been commenced. Great good has already resulted from the labours of the Health of Towns' Commissioners, and I may be allowed to read their report and recommendation, respecting public parks and open spaces, given about seven years ago; especially as considerable improvements have been effected, since that time, in some of the places referred to in the report.

"In the course of our inquiries into the sanatory state of large towns and populous districts, where a high rate of mortality and much disease is prevalent, we have noticed the general want of any public walks, which might enable the middle and poorer classes to have the advantage of fresh air and exercise in their occasional hours of leisure. With regard to all open spaces, especially well ordered squares, ornamented by trees or gardens, which already exist in the metropolis and large towns, we strongly recommend their preservation from any encroachment by public or private buildings. Although not open to the public, they contribute largely to the general salubrity of a town; and it has too commonly happened that, as population has increased, almost every open space has been enclosed; thus, at the same time, excluding the people from their former places of exercise and recreation, and preventing that ventilation which would otherwise have been preserved.

"We have found this state of things very generally lamented by the inhabitants of large towns, and a very prevalent desire existing in many of them, and shared by benevolent persons of the more opulent classes elsewhere, to repair this deficiency.

"The great towns of Liverpool, Manchester, Birmingham, Leeds, and very many others, have at present no public walks. Shrewsbury, Newcastle-under-Lyme, Derby, and a few more, possess them.

"The metropolis, except at the west and north-west, where the different parks minister so much to the comfort and health of the people, has no public walks, though the Victoria Park, now in progress, will supply this want towards the east.

"The large population of Southwark and Lambeth, to the south of the Thames, are yet without such a source of enjoyment and salubrity.

"This subject was considered by a Select Committee of the House of Commons in 1833, who strongly recommended steps should be taken to supply the want. In 1840 the sum of £10,000 was voted by Parliament, to assist local efforts for this purpose in provincial towns; and a few places have had grants from that sum for this purpose.

"In any attempt to carry out these objects, we do not anticipate so much difficulty as has by many been apprehended. It sometimes happens that there is a common, or waste lands, in the vicinity, which, by an alteration of the law, and proper compensation given, might be made available for this purpose. The formation of a public walk would, in such case, at the same time minister to the comfort and improve the health of the inhabitants by a proper drainage of the lands in their vicinity. In many cases local exertion and munificence would accomplish the object, if some moderate assistance were given.

"*We therefore recommend that, for the purpose of aiding the establishment of public walks, in addition to the legal facilities adverted to, the local administrative body be empowered to raise the necessary funds for the management and care of the walks when established.*"

Manchester, I am glad to say, has now three public parks, though at the time of the report it had none. Victoria Park, in the east of London, is also now opened, and it confers a great benefit on that locality. The case of Manchester may serve as a very important lesson to other towns of smaller size. While, on the one hand, its three parks, obtained in so short a time, may afford great encouragement, on the other, their situation, so far distant from the centre of the town, ought to be taken as a very serious warning. The town was allowed to extend itself uninterruptedly outwards from a centre, no open space being preserved; and then, when parks came to be imperatively demanded, sites for them were not to be obtained except in the suburbs; and the central parts of the town are but little benefited. Thus, although their formation constitutes a tardy acknowledgment of past error, it provides no adequate remedy for the evil consequences.

With the manufacturing system, which has sprung up so wonderfully in the last fifty or sixty years, and which has caused the rapid extension of so many towns, there has also arisen a greatly increased want of space for ventilation of the towns, and for healthful recreation of the inhabitants. While, on the one hand, it has greatly augmented the wealth of many classes of society, and has supplied the wants of almost all more abundantly than would otherwise have been possible; it has still, on the other hand, entailed on the inhabitants some discomforts or inconveniences to which they were formerly not subject. Of these, the principal are, smoky air, and an increased distance of their residences from the country. Why then, I would ask, when manufactures have done so very much to increase riches, why should the possessors object to spend some small part of those riches in obviating the concomitant evils of the manufacturing system? They do already frequently spend large sums for the sake of enabling themselves and their families to escape from town for greater or less portions of their time. Many of them, however, still find themselves obliged to live entirely in town, or, at least, to spend the greater part of their time in it. Would it not, then, be well that they should do their best to render town a pleasant place of residence? To effect this desirable object, the establishment of public parks is certainly one of the most important steps.

Now, where does the difficulty lie? It must be attributed, in a great degree, to the fact that the want of a park in any particular locality is but little felt at the time when the land can be easily procured, and that, when the want is felt, the space, being already covered with valuable property, cannot be obtained unless at a cost which practically amounts to a prohibition. Persons residing at the outskirts of an increasing town can commonly make their way to the country when they want exercise, and in their houses they get a tolerable share of air fresh from the country. The individual proprietors of land at the outskirts wish to make the most of their property, and so they usually build closely, or, at most, leave vacant only such small patches of gardens and shrubberies as will remunerate them by increasing the value of their adjacent houses. The building operations commonly proceed without any consideration for the previously existing interests of the owners of property in the interior of the town, which, in respect to suitableness for dwellings, is deteriorated the farther it becomes swallowed up among dense masses of houses. Of so much importance does this matter appear to me, that I have long been of opinion, that we ought to have a legislative enactment, laying a moderate tax on all land adjoining a town as soon as, in virtue of the increase of the town, it comes, for the first time, to be built on. The object of the tax would be to supply funds for the establishment of permanent open spaces wherever they would be judged most suitable, as a compensation in kind for the sanatory evil to which I have referred, as being inflicted on the interior of the town by buildings erected at the outskirts. Even the proprietors round the outskirts, as a body, need not complain of the system of taxation which I have proposed. Their land is constantly receiving accessions of value as the town spreads; and this is not in consequence of any exertions of their own, but rather of the industry and energy of the people already established in the town. Besides this, they should recollect that, for every portion of land preserved open, nearer than theirs to the town, the buildings will advance outwards the more quickly, communicating to the land an increased value by their approach.

Although I would advocate the propriety of a legislative enactment such as I have briefly described, yet, as long as it is wanting, I consider that we should by no means stand idly waiting for it, but that we should adopt the best modes of procedure within our power for the accomplishment of our ultimate object. Let us not exemplify the fable of "*Rusticus expectat dum transeat amnis*"—(The clown stands waiting for the river to flow past). The recommendation of the Health of Towns' Commissioners, which I have already quoted, contains what, in the present state of the law, appears to me to be the best mode of procedure available—that, namely, of private subscriptions, aided by public grants.

Having said thus much on remedial measures for the smoky atmosphere of towns, I may be excused in adding here a few statements in regard to preventive measures against the smoke itself.

Numberless have been the schemes proposed and tried for procuring a perfect combustion of coals in furnaces. These, however, have, in general, proved so unsatisfactory as to create the idea that any farther attempts are likely to be fruitless; and that the only thing to be done is to let the fireman bring about the combustion as best he can, by carefulness and regularity in supplying coals to the furnaces. There is no doubt that much can be effected by skill and care on the part of the fireman; but still it is exacting too much from him to demand that he must send out no dense black smoke from the chimney. Using very good coals, and having a furnace large in comparison to the quantity of coals consumed, he may succeed tolerably well in avoiding smoke; but, in ordinary cases, he will be unable to attain anything like a perfect combustion. Whatever impression may be made on people's minds by the numerous failures which have occurred, the application of correct scientific principles to this subject has, as is usually the case, led to the most favourable results. One smoke-consuming furnace, at least, has for several years been in successful operation; not only avoiding the generation of smoke, but also, as I am convinced, economising both fuel and labour. I refer to Juckes' patent self-feeding furnace. The perfect combustion is effected simply by a more regular process of firing than is possible on the part of an attendant, this regularity being combined with the principle of making the coals burn from their upper surface downwards, no black coals being thrown on the top of red-hot ones, and so no air passing from the red coals through the black.

I come now to the part of my subject in which I feel by far the deepest interest—the proposal for the formation of a public park in the town of Belfast. The present condition of this town, and its prospects for the future, seem to me to render the establishment of open spaces highly desirable; and I consider that, just at the present time, the town possesses, for such an undertaking, one of those golden opportunities which are seldom met with, and which, if allowed to pass, must be lost for ever.

During the progress of the town, an extensive space of ground, situated between the Linen Hall and Donegall Pass, has, through the pestiferous influence of a most intolerable nuisance, been kept nearly free from buildings. The best dwelling-houses have been effectually repelled from it by the dam on the Blackstaff, and as yet only a few buildings have been erected in its vicinity, while the town has sent out a branch on each side, and is now spreading itself out beyond in the neighbourhood of the Botanic Garden and the Queen's College. On the nearly vacant space under consideration manufacturing establishments have already begun to rise, and it is clear that, if this process be not stayed in time, the best parts of the town for the transaction of business will become almost completely hemmed in with mills and second or third rate houses, all good approaches from the country being cut off. The people of Belfast have thus the future destiny of the town placed in their hands. They cannot help fixing, within a very short period, whether the town is to become a densely built and smoky place, good enough, indeed, for the prosecution of work, but not fit for the enjoyment of life, or whether it is to possess a public ornament which will be a source of admiration in all time to come. One very desirable requisite for a park is possessed by the ground in question. It has on it at present a considerable number of trees already well grown. The importance of this may be judged from the high value set by the people of London on their trees. We all know that rather than condemn a row of lofty elms to the axe, to make room for the Exhibition Building, they preferred to carry a spacious arch of glass over the summits of the trees. With reference to this subject, too, it may be well to recollect that one of the most attractive and most admired features of the Belfast Botanic Garden is due to the selection, at the time of its formation, of a piece of ground on which trees were already growing. While pleading for the conversion of the effects due to the baneful influence of the Blackstaff into a source of increase to the amenities of the town of Belfast, I may direct attention to one case in which a permanent benefit has been derived by this town from the former existence of a nuisance. The wideness of High Street was determined by the presence of a dirty stream which formerly flowed uncovered down the middle of the street. Let us hope that the evils of the Blackstaff may be turned to good in a tenfold degree.

With reference to the extent of the proposed park, I wish here to state that I would not confine my views to the space between the Linen Hall and Donegall Pass. There lies a most tempting piece of additional land between Donegall Pass and the Lagan, in the direction towards Ormeau Bridge and the Queen's College. The exterior limits of the park may, however, be safely left for future consideration*.

The subject I have now brought before you shows the necessity of attending early to projected public improvements, which can be effected only by the combined action of many persons, and which are of such a kind that the advantages to result from them, if

---

* At the meeting of the Society, in a discussion which followed the reading of the present paper, several other desirable sites for parks were mentioned by various members.

carried out, would be shared in various degrees by multitudes of individuals, as much by those who would give no aid in the undertaking as by those who would make the greatest sacrifices, whether of their time or of their money. It may be, that the entire cost would prove almost inappreciable if compared with the benefits which would accrue; and yet, from the difficulty of apportioning the burden of the undertaking among those who are able and who ought to bear it, the improvement too often remains neglected till the time for effecting it is past, and it has become no longer possible. When this has been the case, regret for the past is vain, unless it impel us to employ more foresight, and to apply more vigorous exertions for the future.

All inquiries into the advantages of such improvements, into the difficulties and impediments which obstruct their introduction, and into the best means of removing these, are essentially social questions, and fall peculiarly within the province of a Social Inquiry Society. The discussion of such questions, I may be permitted to say, is not to be looked on as a mere barren debate. The ultimate purpose is a practical one; and when persons meet together in a true spirit of candid inquiry, and with a genuine wish to promote good objects, their efforts are sure to be, in many cases, crowned with success.

67. NATIONALIZATION OF PUBLIC WORKS.

[From *The Northern Whig of* January 21, 1869.]

THE annual meeting of the Belfast Engineering and Architectural Association was held in the Athenæum on Friday evening, the 15th January, 1869—Professor James Thomson, M.A., C.E., presiding. The meeting being opened, the minutes were read by the secretary, Mr Kelly; after which Professor Thomson, who had been elected as President for the year 1869, delivered his opening address.

He stated that he had selected, as the subject to which he would limit his address, a question which is of great and wide-extending interest in connexion with engineering works throughout the world, and which at present is one of special importance

in reference to engineering affairs in Ireland. It was the question
of the advisability of nations or communities, instead of companies,
being the owners and governors of such works for public use, as,
from their nature, must be more or less of the nature of monopolies,
and do not admit of efficient competition ; and, in particular, the
question now rapidly advancing in public consideration, of the purchase by the State of the Irish railways ; so that the people of
Ireland would virtually become the owners of the railways in their
own country. In general, he was in favour of the principle of
having all such works owned and governed by the nations or
communities for whose use they exist. Railways, extending
throughout an entire country, and being for the use of the
public generally, should belong to the State ; while town water
works and gas works, with their respective pipes under the
public streets, ought properly to belong to the local community
for whose special use they are established. He could not maintain, however, that local government arrangements in Ireland are
as yet sufficiently perfect to render our towns in general quite
ripe for organising and managing advantageously, on their own
behalf, without the intervention of companies, such undertakings
as the gas manufacture. He referred to a paper by John Hancock, Esq., J.P., published in the Transactions of the Social Science
Association for its recent meeting at Belfast, as giving a valuable
exposition of the local government question in its application to
Ireland, and pointed to indications of a growing determination
to have local government arrangements made much more perfect
than hitherto, and kept so. He expressed his confidence in the
safety of trusting every local monopoly to local government properly organised and checked. He proceeded to point out that
the working of railways by companies does not accord with the
general interests of the public. The aim of companies must
naturally be to bring about the maximum of profit to themselves, with the minimum of trouble and risk. Their tendency
will be to accommodate their arrangements to the convenience
of the public, and their fares and charges for passenger and
goods traffic to its advantage, just so far as, and no farther
than, in their opinion will conduce to good dividends to themselves on their shares. A Board of Directors of a railway, who,
by raising the fares and other charges, could show that they
brought about an increased dividend, though accompanied by a

diminution of traffic, would be deemed to have performed good service to their company; but this change would have effected a pecuniary loss to every one of the public who travelled on the railway, or sent goods by it, and would have prevented others from getting the benefit of using the railway in many cases, in which, under some other financial arrangements, they might have used it, and might have contributed something towards its expenses. The benefit conferred on the public, in respect to passenger traffic—if we take for brevity the case of passenger traffic alone—is far above the total of the fares. Because the traveller is himself the best judge of the probable value to him of his intended journey. He will not undertake the journey unless he believes it to be worth to him at least as much as the fare; and so we may consider the fare as indicating the value of any such journey as is just at the point of being indifferent as to whether it be worth undertaking or not; but then it is inevitable that scarcely any can be just at that exact point. Any contemplated journey below it will not be undertaken; and nearly all journeys made will be above it. Many a time a person goes a journey by rail which he would value at five, ten, or twenty times the amount of the fare; often at one and a half, two, or three times. That relates only to actual travellers on the rail who really pay fares, and so contribute towards the reimbursement of the first costs and working expenses of the undertaking. But also those who happen not to require to travel on a certain railway near them, do receive a highly beneficial power or right, which must have to them a real value—the power, namely, of travelling by that railway whenever circumstances may lead them to want to go. It may properly be esteemed by any member of the community as a matter of importance and of value to him to have a railway kept constantly in readiness for him if wanted to carry him rapidly and easily over what otherwise would be a long, fatiguing, and perhaps costly journey. Yet he has contributed nothing towards its construction in consideration of the advantage he holds in virtue of its existence. Thus, on the principle of having a country s railway made and worked by companies, the people benefited by the formation and maintenance of each railway are not all duly brought in as contributors. Those who do travel must then be charged for each journey, if the railway is to be successful to the shareholders, more than would be re-

quired if the contributions were levied for the railway from a
larger portion of the public benefited. A conclusion thence is,
that it is sound policy, in the interest of the general public, not
to arrange for levying the whole of the funds for payment of
interest or profit on the original outlay, and of working expenses,
by fares; but inasmuch as the general public hold a benefit in
having the railway kept ready for them, whether they happen
to have occasion to use it or not, they should pay by taxation,
and especially by local taxation levied on the districts specially
benefited, something towards the attainment of that benefit; and
then that every person in travelling should have lower fares to
pay, and have his interest and convenience generally better provided for, than under the present arrangements of the railways as
owned and worked by companies. Another conclusion is that the
construction of a railway is not to be considered as on the whole
a disastrous misapplication of labour and capital, even if the income for the promoters, raised by charges on passenger and goods
traffic, be totally inadequate to pay fair interest on the money
sunk in their shares. In many such cases it is quite possible
that the only matter to be regretted may be that some other
method more in unison with sound policy for the general interests
of the public had not been contrived and adopted for levying the
funds for the expenses of the undertaking. The policy thus ad-
vocated, though it would be a great change from what we are
accustomed to in railway affairs, would really be no innovation
with reference to well-established and successful modes of conducting other undertakings for public use. Common roads in
Ireland are already exempt from tolls, and are paid for by taxation.
The person using them for a journey, or for conveyance of goods,
has not anything to pay, even for the wear or injury of the road
due to his using it. This freedom to use the road goes far beyond
what in any case would be proposed in railways, because in them,
according to any propositions that are likely seriously to be put
forward, local taxation would only be looked for to supply a fund
supplemental to an important income to be derived from passenger
fares and charges on goods traffic. He pointed to other cases of
very different kinds of public undertakings admitted to be of high
public value, in which the idea of having the necessary funds levied
directly as charges on the use of what is made available would be
altogether untenable. One of these was that of the Ordnance

Survey—an undertaking of great and general public value, and yet impossible to be done at all, if allowed only on the condition that the price of the maps sold must be fixed at such a rate as that the costs of the survey shall be compensated for by the income from the sale of the maps.   Whether the price be fixed at a high rate per map, or at a low rate per map, the cost of the survey could not be compensated for in that way; and yet no one will say that surveys of countries should for that reason be necessarily deemed misapplications of money, skill, and work. Bridges repeatedly have been built at great cost, and for long years, owing to the obstruction of a toll-gate, have continued serving but a small fraction of the use to the public that they might serve; but, on the abolition of the toll, and the substitution of funds levied by taxation, instead of the charges collected per drive or per walk across the bridge, the service afforded by the bridge to the public has been vastly augmented, and yet the cost to the public is in no way increased by this, but it is even diminished by the saving of the remuneration to the toll-collector for his services, no longer needed.   High fares on railways operate as a hindrance to the public from receiving so much as they might of the advantage rendered available by the original construction of the bridges, tunnels, cuttings, embankments, and other parts of the whole works of the line; yet, by preventing a person from travelling, no part of the first cost will be saved, but he will suffer the loss of whatever advantage the journey might have afforded him.   If part of the funds were levied by taxation, and another part by fares or charges on passenger and goods traffic, the fares and other charges might be so much reduced as to effect so great an increase of the traffic as would allow of the amount to be levied by taxation being kept very small, and very little burdensome to the community, while the public would gain the triple advantage of being permitted to make more use of the railways already existing, while paying much less for each journey; and of obtaining valuable railway accommodation in many localities to which railways could not be profitably extended on the principle of their being made and worked by companies.   The results of the elaborate investigations of the Royal Commissioners appointed to inspect the accounts and examine the works of the railways in Ireland, as embodied in their two reports, may lead us to look in hope to the prospect of the railways of Ireland being soon rendered public property, and with confidence that, in that event, if the purchase be effected with due prudence, a vast benefit will accrue to the general public in Ireland.

68. WARMING AND VENTILATING.

[Read before the Belfast Natural History and Philosophical Society, 4 Dec. 1867. From the MS. notes supplemented by a newspaper report.]

IF I had offered at the present season, and in the present state of the weather, a paper on warming alone, I might reasonably have anticipated a favourable reception from my audience. I know however that many people shudder at the very thought of ventilation and I am conscious that it requires no small amount of boldness to come forward, in a frosty beginning of December, to advocate the introduction of fresh air into people's houses in any other than in homeopathic doses. Knowing however, as I well do, the great amount of discomfort, generally, and almost universally, suffered in dwellings and other buildings from vitiated and over-heated air, accompanied often by cold draughts along the floor, or at other places in apartments; and knowing how those evils may in general be mitigated, or altogether removed; and by what means pleasant and healthful air may be usually made available; I have determined to brave the opposition with which the advocacy of ventilation will assuredly be met in many of your minds.

Ventilation is not in reality the dreadful thing that it is often supposed to be. It consists in the attainment of a due and proper change of air in any apartment, passage, or other enclosed space to which the external atmosphere has not free and unlimited access; and it is essential to good ventilation of places to be occupied by people or other animals that the fresh air should be so introduced as not to cause cold or discomfort to the occupants, but on the contrary to render the place more healthful and more agreeable for them to live in.

The principal cases for which means of ventilation are requisite, are:—dwelling houses, public buildings, ships, factories, mines; as also stables, cow houses and the like.

It is only to a small portion of this wide range of cases that my remarks on the present occasion must be confined.

There are many varieties of methods for ventilation, and there is no single one that can be regarded as the best for all cases; but in arranging the method to be pursued in any particular case, various circumstances may require to be taken into account, such as

(1) The uses to which the apartments or spaces have to be applied.

(2) The number of individuals commonly or occasionally occupying them.

(3) The requirements and means for warming and lighting them.

(4) The requirements for ventilation at different seasons of the year.

The source of the motive power for producing ventilation is in the vast majority of instances the ascent of heated air, whether naturally or artificially produced. In hot weather, or in warm climates, ventilation is often attained by the simple method of opening windows or other apertures widely, and trusting to the breezes, or yet more gentle motions, of the external atmosphere, for the attainment of the requisite change of air. This may be called natural ventilation; and as the winds and breezes of the atmosphere are due wholly to differences of temperature at different parts of the earth and atmosphere, the change of air even in this case is to be traced ultimately to the ascent of heated masses of air.

In ordinary dwelling houses, fires burned for the primary purpose of attaining warmth afford incidentally likewise a source of motive power which is a most important means for promoting ventilation.

The fires are constantly withdrawing by their chimneys the air from the apartments, and the flow of air which may thence be attained can easily be made by proper arrangements to suffice in ordinary cases for very good ventilation. What is mainly required consists simply in the providing of properly formed and properly placed apertures for the admission of air under the indrawing tendency of the heated chimney. This may usually be advantageously effected by having a large aperture formed in a wall of the room at a convenient place near the ceiling, and adapted for receiving air from the hall, passage, or staircase, and by aid of some guiding woodwork or other means adapted for shooting it obliquely upwards quite to the ceiling, so that it may flow rapidly along the ceiling in a broad but moderately thin sheet. The air, when thus set into motion along the ceiling, although it is colder and heavier than the air below, yet tends to cling to the ceiling, and does not fall down through the warmer air as a current, nor in large detached masses, but gradually mingles and unites with the general air of the room in the space quite above the heads of the occupants, so that no cold, unpleasant draughts are felt; but the air of the room is kept in a condition of agreeable freshness.

In connection with this, a great abatement of the usual cold draughts flowing along the floor, or of the cold air stagnating there, ensues or tends to ensue, and is rendered easily practicable. Thus, if the air is admitted freely by a properly arranged ventilator, it will not rush in so briskly as before by chinks in the floor or under the door. All such chinks, too, may then be carefully closed by improved fittings; while, if closed in the absence of a ventilating inlet, the draught up the chimney would fail to be sufficient, and the room would become smoky. The aperture in the woodwork where the air issues into the room may be made about two or three feet wide, as horizontally measured in a direction parallel to the wall, and the width across may be made variable at pleasure, but may very well be ordinarily in winter kept at about two or three inches wide. The aperture through the wall should be of greater area, so that it may not obstruct the flow. On the outside of the wall—that is to say, the side next the hall or staircase—there ought to be either louvres or some other suitable screen for preventing persons outside of the room from seeing into its upper part through the ventilator, and for aiding in preventing the too free transmission of sound (as of the speaking of persons within) through the ventilator.

Professor Thomson here exhibited and explained to the meeting by aid of large drawings, the detailed arrangements of several varieties of such ventilators which had been found good and suitable in practice. In proof of the necessity of means for attaining a more equable distribution of warmth in ordinary dwelling apartments than is met with when no special provision is made for that purpose, especially in rooms lighted by gas, Professor Thomson cited some observations which he had made on a winter evening in a room in his own house. The room had no ventilator and was lighted by a pair of small gas burners, and a small fire was burning in the grate. The door of the room having been shut for some time, he found the temperature, shown by a thermometer placed a few inches above the floor, to be 56° F.; while the temperature at the top of a bookcase a little below the ceiling was 72°, and so the difference of temperature between the floor and the ceiling was no less than 16° F. The door of the room was then left wide open for an hour and a half, and the temperatures were observed at the same two places, and found to be, near the floor 52°, and near the ceiling 73°, so that the difference now was increased to 21° and the keeping of the door open had been quite ineffective in cooling the air of the apartment near the ceiling. The air near the ceiling had, indeed, risen in temperature by 1° during the hour and a half when the door was left open. This was sufficient proof that no satisfactory ventilation of a sitting-room with gas lights on a winter evening is attainable by widely opening the door. But, besides, it is to be observed that, when the door is wide open, a cold current of air flows in along the floor, while warm air passes outwards by the upper part of the doorway, and thus people's feet are kept cold by the fresh air, of which they get a very insufficient supply to breathe. This can readily be observed by standing in the open doorway of a warmed sitting-room, and holding a burning taper in the hand at different heights successively, when the blowing of the flame will show the motion of the air very clearly and convincingly.

Professor Thomson mentioned the well-known ventilator introduced by Dr Arnott, which is adapted for drawing into the chimney hot and vitiated air from the upper part of the room. He did not, however, much approve of the principle of that kind of ventilator, because he preferred the principle of scouring out the heated air from the upper part of the room by introducing the fresh air there. He recommended strongly, however, the use, whenever practicable, of chimney pipes for leading the products of combustion of the gas flames to the chimney of the fireplace. These chimney pipes, on account of danger of fire and for some other reasons, ought by no means to be placed for concealment from view between the ceiling of the apartment and the floor of the room above, but they ought to be made ornamental, or as little unsightly as possible, and conducted close below the ceiling without touching it, or in any other suitable situation free from danger of setting fire to the house.

He mentioned several reasons for preferring to draw the fresh air for apartments from the hall and other passages of the house rather than directly from the external atmosphere, among which were that the external air in towns often carries so much of sooty particles—too well known as "blacks"—that it does not always suit well for being introduced at once into apartments without having first been allowed to get rid, in a great degree, of these particles by depositing them in the passages, or in some little used apartment through which the air may be drawn. One or more windows or other apertures ought always to be kept open to admit air to the staircase or other passages of the house. The omission to do this is the usual cause of the back smoke coming down the unused chimneys, which is a very common annoyance in dwellings.

He explained some of the methods which he could most recommend for warming the fresh supplies of air introduced into passages of houses. The warming in a slight degree of the air while entering the passages of the house, or while in them, is not usually to be regarded as necessary, but may sometimes be desirable, especially in very cold weather, or in case of persons liable to suffer much from cold. Professor Thomson then, with the aid of drawings, gave explanations of ordinary and frequent circumstances of the warming and ventilating of public buildings. In doing so he spoke in condemnation of methods depending on the introduction of the fresh air, whether warm or cold, through the floor or by any openings near the occupants; and recommended rather that the fresh air be introduced near the ceiling, and well mixed with the general atmosphere of the room before reaching the occupants; and that it should be introduced in quantity sufficient to maintain a satisfactory degree of freshness in the whole space. Professor Thomson had given much attention to the subject of the ventilation of public buildings, and he offered various recommendations and suggestions for improvements from prevailing practices. The limits of the present abstract of the lecture, however, do not allow more than the mention of this part of the subject, as it has appeared that a greater number of the general public may be practically interested in the means taught for improving the comfort and healthfulness of their own dwellings.

An interesting discussion followed the reading of the paper, in which several members took part.

T.

INDEX

Printed in the United States
By Bookmasters